Herbicide Resistance in Plants
Biology and Biochemistry

Edited by
Stephen B. Powles
Joseph A. M. Holtum

LEWIS PUBLISHERS
Boca Raton   Ann Arbor   London   Tokyo

Library of Congress Cataloging-in-Publication Data

Herbicide resistance in plants : biology and biochemistry / edited by
    Stephen Powles and Joseph Holtum.
       p.   cm.
    Includes bibliographical references and index.
    ISBN 0-87371-713-9
    1. Herbicide resistance.   2. Plants, Effect of herbicides on.
  I. Powles, Stephen.   II. Holtum, Joseph.
SB951.H442   1994
632′.954—dc20   93-44552
                                                                                                  CIP

This book contains information obtained from authentic and highly regarded sources. Reprinted material is quoted with permission, and sources are indicated. A wide variety of references are listed. Reasonable efforts have been made to publish reliable data and information, but the author and the publisher cannot assume responsibility for the validity of all materials or for the consequences of their use.

Neither this book nor any part may be reproduced or transmitted in any form or by any means, electronic or mechanical, including photocopying, microfilming, and recording, or by any information storage or retrieval system, without prior permission in writing from the publisher.

All rights reserved. Authorization to photocopy items for internal or personal use, or the personal or internal use of specific clients, may be granted by CRC Press, Inc., provided that $.50 per page photocopied is paid directly to Copyright Clearance Center, 27 Congress Street, Salem, MA 01970 USA. The fee code for users of the Transactional Reporting Service is ISBN 0-87371-713-9/94/$0.00+$.50. The fee is subject to change without notice. For organizations that have been granted a photocopy license by the CCC, a separate system of payment has been arranged.

CRC Press, Inc.'s consent does not extend to copying for general distribution, for promotion, for creating new works, or for resale. Specific permission must be obtained in writing from CRC Press for such copying.

Direct all inquiries to CRC Press, Inc., 2000 Corporate Blvd., N.W., Boca Raton, Florida 33431.

© 1994 by CRC Press, Inc.
Lewis Publishers is an imprint of CRC Press

No claim to original U.S. Government works
International Standard Book Number 0-87371-713-9
Library of Congress Card Number 93-44552
Printed in the United States of America 1 2 3 4 5 6 7 8 9 0
Printed on acid-free paper

PREFACE

In the half-century since World War II, food production and living standards have risen dramatically in most parts of the world. New agricultural chemicals, drugs, and other xenobiotics have played a major role in ensuring a secured food supply and protection against the ravages of pests, diseases, and weeds. Abundant and sustained food production has been achieved, in part, due to the success of modern herbicides in controlling weedy plant species infesting crops and pastures. While not as visual as some other pests, uncontrolled weed infestations decimate yields and cause many other problems. The era of chemical weed control commenced post-war with the spectacular success of the herbicide 2,4-D in controlling weeds without damaging cereal crops. Since then, the international chemical industry has produced a range of increasingly sophisticated herbicides which have been enthusiastically adopted worldwide, especially in industrialized nations. Weed control by herbicides is now an integral part of most modern agronomic systems delivering food and fiber. Additionally, it is acknowledged by practitioners, but not by all sectors of the community, that herbicides contribute to sustainable land use in that weed control can be obtained with minimal destructive soil cultivation in many systems.

For the reasons outlined, judicious herbicide use is widely practiced and there is a great deal of herbicide expertise among farmers and their advisers. An unfortunate result of the continued success of herbicides is that practitioners and many researchers have concentrated on the herbicide technology involved in weed control and have sometimes lost sight of the biology of the plants. Put in simple terms, there has been economic and practical justification to concentrate on chemicals for weed control because this has delivered results since 1945. Many individuals have considerably greater knowledge of herbicide technology than of the biology of the plants being controlled.

Given the success of herbicide technology and the widespread and persistent use of herbicides it was inevitable that there would be biological repercussions of reliance on a single control method. One repercussion has been the selection and enrichment of genes which endow herbicide resistance in weed populations. Commencing from the landmark observation in 1970 of resistance in *Senecio vulgaris* to triazine herbicides,[1] populations of weeds have developed herbicide resistance in agro-ecosystems around the world. Initially, resistance was somewhat of an academic curiosity as the triazine-resistant biotypes were rare, lacking in vigor,[2] could easily be controlled with alternative herbicides, and continued triazine herbicide use still provided control of other weed species. Conversely, for researchers, the triazine-resistant weed biotypes provided superb experimental material. Research on triazine-resistant weeds has resulted in advances in agro-ecology,[2] genetics,[3] and biochemistry,[4] as well as contributing to an understanding of the photosynthetic reaction center.[5]

The advent and continued march of triazine resistance provided the impetus for a meeting and subsequent book *Herbicide Resistance in Plants* published in 1982.[6] This important book reviewed triazine and other resistance cases up to 1982. However, 1982 also saw the introduction to world agriculture of the first of the new-generation acetolactate synthase inhibitor herbicides[7] and the commencement of widespread usage of new acetyl coenzyme A carboxylase inhibitors.[8] These new herbicide chemistries have rapidly been adopted worldwide and have had a major impact on agriculture. In turn, however, there has also been a dramatic escalation in the types and extent of herbicide resistance, especially to these new herbicide chemistries. Throughout the 1980s there has been an explosion in the number and diversity of cases of herbicide resistance, such that resistance is no longer an academic curiosity. Resistant biotypes are not rare, are mostly not detectably lacking in vigor,[2] and some cannot be controlled with alternative herbicides.

The 1980s have also seen revolutionary advances in plant molecular biology and the ability to manipulate genes. Herbicide resistance research has been at the cutting edge of recombinant DNA technology. Using genetic engineering techniques, a gene endowing resistance to glyphosate has been inserted into the genome of a number of crop species where it imparts glyphosate resistance.[9] Similar success has been achieved with genes endowing resistance to some other herbicide chemistries. Recombinant DNA technology has also allowed the molecular characterization of herbicide-resistance genes in weed populations.[3,7,10] Herbicide-resistance genes are being sequenced and manipulated in academic and industry laboratories around the world; they have provided, and will continue to provide, unequivocal evidence of the precise genetic changes that can endow herbicide resistance.

The rapid escalation in cases of herbicide resistance worldwide and the pace of developments in techniques and understanding of resistance over the past decade more than justifies this current book *Herbicide Resistance in Plants: Biology and Biochemistry*. As presented in the following chapters, current research reveals that resistant weeds are utilizing a plethora of resistance mechanisms to overcome herbicides. Evidence is mounting that plants can utilize various mechanisms to provide resistance to a particular herbicide group. Details of the precise gene mutations which endow resistance are being elucidated. Of considerable significance is that plants are starting to accumulate resistance mechanisms, resulting in the expression of multiple herbicide resistance.[11] The cross-pollinated grass weed, *Lolium rigidum,* has accumulated a number of resistance mechanisms which provide resistance across a wide range of herbicide groups.[11] Inevitably, in the next few years, other weed species, especially cross-pollinated species, will acquire such multiple resistance mechanisms.

In this book active, highly regarded researchers have distilled a great body of complex information into lucid reviews and are gratefully thanked for their sterling efforts. The editing and myriad of tasks necessary to produce a coherent text have been done within our herbicide resistance research team at the Waite Agricultural Research Institute of the University of Adelaide. Heartfelt thanks are extended to Dr. Linda Hall for her unstinting assistance and to Dr. Christopher Preston, Dr. Jack Christopher, and Dr. François Tardif for their expert advice and help with the assembly of this book. We believe that this text will be very useful to a range of people involved in the worthy professions which result in the production of food and fiber. We believe that biologists and others in institutions ranging from state-of-the-art research laboratories through to provincial agricultural extension stations and industrial agrichemical marketing offices will find much of interest and importance to them in this book. Finally, it is hoped that readers will be stimulated to take the lessons and messages presented in this text to help develop practical, integrated weed management practices that deliver sustained food and fiber production without major herbicide resistance problems.

REFERENCES

1. Ryan, G. F. "Resistance of Common Groundsel to Simazine and Atrazine," *Weed Sci.* 18:614-616 (1970).
2. Chapter 11 in this volume. Holt, J. S., and D. C. Thill. "Growth and Productivity of Resistant Plants."
3. Chapter 10 in this volume. Darmency, H. "Genetics of Herbicide Resistance in Weeds and Crops."
4. Hirschberg, J. A. Bleeker, D. J., Kyle, L., McIntosh, L., and C. J. Arntzen. "The Molecular Basis of Triazine Resistance in Higher Plant Chloroplasts," *Z. Naturforsch.* 39c:412-419 (1984).
5. Diesenhofer, J. and H. Michel. "The Photosynthetic Reaction Center from the Purple Bacterium Rhodopseudomonas viridis," *Science* 245:1463-1473 (1989).
6. LeBaron, H. M. and J. Gressel., Eds. *Herbicide Resistance in Plants* (Academic Press, New York, 1982).
7. Chapter 4 in this volume. Saari, L. L., Cotterman, J. C., and D. C. Thill. "Resistance to Acetolactate Synthase Inhibiting Herbicides."
8. Chapter 5 in this volume. Devine, M. D. and R. H. Shimabukuro. "Resistance to Acetyl Coenzyme A Carboxylase Inhibiting Herbicides."
9. Chapter 8 in this volume. Dyer, W. E. "Resistance to Glyphosate."
10. Chapter 2 in this volume. Gronwald, J. W. "Resistance to Photosystem II Inhibiting Herbicides."
11. Chapter 9 in this volume. Hall, L. M., Holtum, J. A. M. and S. B. Powles. "Mechanisms Responsible for Cross Resistance and Multiple Resistance."

THE EDITORS

Stephen Powles spent his childhood in a rich agricultural region of coastal New South Wales and then undertook an agricultural science degree at the University of Western Sydney where he developed a strong interest in applied plant science. This interest was pursued in a masters degree in the Department of Crop Science at Michigan State University and a doctorate at the Australian National University specialising in plant physiology and photosynthesis. Postdoctoral research followed as a CSIRO fellow at the Carnegie Institution Department of Plant Biology at Stanford University and a further year as a Carnegie Fellow at C.N.R.S., Gif-sur-Yvette, France. In 1983 he commenced at the Waite Agricultural Research Institute of the University of Adelaide in South Australia and is now a Reader in the Department of Crop Protection and leader of a large research team investigating fundamental and applied aspects of herbicide resistance in plants and their management in agro-ecosystems. Aside from professional interests in the plant sciences and agriculture, he remains active in field hockey and as a (now) spectator in rugby football.

Joseph Holtum was born and educated in Sydney before undertaking a science degree at James Cook University at Townsville, Queensland where he graduated with honors in plant physiology. He then moved to the Australian National University and undertook a Ph.D. in plant biochemistry. Throughout their Ph.D. studies, Powles and Holtum shared the same laboratory and Ph.D. supervisor as well as campus housing and university rugby team. In 1980, Dr. Holtum left Australia for two years as a post-doctoral fellow at the Chemistry Department of the University of Wisconsin and then five years as a research scientist in the Botany Institute of the University at Munster, Germany. In 1988, he returned to Australia to join Powles at the Waite Agricultural Research Institute where they worked together on herbicide resistance for the next five years. In 1993, Dr. Holtum accepted a tenured lectureship in the Botany Department at James Cook University at Townsville. In addition to his interests in plant biology, Dr. Holtum continues as an ageing but still vigorous rugby footballer.

CONTRIBUTORS

Josephine C. Cotterman
DuPont Agricultural Products
Stine-Haskell Research Center
Newark, Delaware

David Coupland
AFRC Institute of Arable Crops
University of Bristol
Long Ashton Research Station
Long Ashton, Bristol, England

H. Darmency
INRA
Laboratoire de Malherbologie
Dijon, France

M.D. Devine
Department of Crop Science and
   Plant Ecology
University of Saskatchewan
Saskatoon, Saskatchewan
Canada

William E. Dyer
Department of Plant, Soil and
   Environmental Science
Montana State University
Bozeman, Montana

John W. Gronwald
Plant Science Research Unit
USDA-ARS and Department of
   Agronomy and Plant Genetics
University of Minnesota
St. Paul, Minnesota

Linda M. Hall
Department of Crop Protection
Waite Agricultural Research
   Institute
University of Adelaide
Glen Osmond, South Australia

J.S. Holt
Department of Botany and Plant
   Sciences
University of California
Riverside, California

Joseph A.M. Holtum
Department of Crop Protection
Waite Agricultural Research
   Institute
University of Adelaide
Glen Osmond, South Australia

J.M. Matthews
Department of Crop Protection
Waite Agricultural Research
   Institute
University of Adelaide
Glen Osmond, South Australia

Bruce D. Maxwell
Department of Plant, Soil and
   Environmental Science
Montana State University
Bozeman, Montana

A. Martin Mortimer
Department of Environmental and
   Evolutionary Biology
University of Liverpool
Liverpool, United Kingdom

Stephen B. Powles
Department of Crop Protection
Waite Agricultural Research
   Institute
University of Adelaide
Glen Osmond, South Australia

Christopher Preston
Department of Crop Protection
Waite Agricultural Research
  Institute
University of Adelaide
Glen Osmond, South Australia

Leonard L. Saari
DuPont Agricultural Products
Stine-Haskell Research Center
Newark, Delaware

R.H. Shimabukuro
USDA-ARS Biosciences Research
  Laboratory
North Dakato State University
Fargo, North Dakota

Reid J. Smeda
U.S. Department of Agriculture
Agricultural Research Service
Southern Weed Science Lab
Stoneville, Mississippi

Donn C. Thill
Department of Plant, Soil and
  Entomological Sciences
University of Idaho
Moscow, Idaho

Kevin C. Vaughn
U.S. Department of Agriculture
Agricultural Research Service
Southern Weed Science Lab
Stoneville, Mississippi

TABLE OF CONTENTS

Chapter 1. Selection for Herbicide Resistance 1
B.D. Maxwell and A.M. Mortimer

Chapter 2. Resistance to Photosystem II Inhibiting Herbicides 27
J.W. Gronwald

Chapter 3. Resistance to Photosystem I Disrupting Herbicides 61
C. Preston

Chapter 4. Resistance to Acetolactate Synthase Inhibiting Herbicides 83
L.L. Saari, J.C. Cotterman and D.C. Thill

Chapter 5. Resistance to Acetyl Coenzyme A Carboxylase Inhibiting Herbicides 141
M.D. Devine and R.H. Shimabukuro

Chapter 6. Resistance to the Auxin Analog Herbicides 171
D. Coupland

Chapter 7. Resistance to Dinitroaniline Herbicides 215
R.J. Smeda and K.C. Vaughn

Chapter 8. Resistance to Glyphosate 229
W.E. Dyer

Chapter 9. Mechanisms Responsible for Cross Resistance and Multiple Resistance 243
L.M. Hall, J.A.M. Holtum and S.B. Powles

Chapter 10. Genetics of Herbicide Resistance in Weeds and Crops 263
H. Darmency

Chapter 11. Growth and Productivity of Resistant Plants 299
J.S. Holt and D.C. Thill

Chapter 12. Management of Herbicide Resistant Populations 317
J.M. Matthews

Index 337

CHAPTER 1

Selection for Herbicide Resistance

Bruce D. Maxwell and A. Martin Mortimer

INTRODUCTION

The major ecological questions associated with herbicide resistance evolution involve an intricate understanding of the interplay between gene frequency, fitness, inheritance, and gene flow. In this chapter we provide a basis for examining the questions that are relevant to understanding herbicide resistance evolution and that may determine appropriate weed management strategies.

Herbicide resistance in weeds is an example of evolution in plant species as a consequence of environmental changes brought about by the hand of man. In concept at least, the phenomenon is no different from the evolutionary changes that have arisen in insects and pathogens in response to the use of insecticides and fungicides, or in plants to industrial pollutants such as heavy metals, like copper in industrial spoilheaps,[1] or to zinc leached from electricity pylons.[2] Herbicide resistance is a result of selection for traits that allow weed species to survive specific management practices which would otherwise cause mortality. By definition, weeds are plants that are a hazard to the activities of man and which are either preadapted (as endemic or alien species) to an agroecosystem or have evolved a suite of characters that ensures species persistence.[3] Herbicide resistance constitutes a "weedy" trait and resistant weeds are those plant species that express the genetic variation required to evolve mechanisms to escape control. Like other crop pests resistant to pesticides, herbicide-resistant weeds are a symptom of intensive and extensive selection pressure. Gould[4] eloquently pointed out that modern agriculture attempts to share less and less of the crop yield with pests and thereby increases the selection pressure for pest adaptation. The short-term triumphs of new pest control technologies have often carried with them the seeds of long-term failure.

Definitions

Herbicide resistance has been defined in several ways, but is generally used to describe a characteristic of species (as intact plants or plant cells in culture) to withstand substantially higher concentrations of a herbicide than the wild type of the same plant species. This definition is problematic because it is reliant upon two concepts. First, "substantially higher concentration" must be defined; and second, the assumption that resistance is associated with alterations to physiological and/or morphological traits in the phenotype that inhibit herbicide uptake, translocation, site of action, or metabolism. An expanded definition could include a more functional view of resistance, including phenology or dormancy changes within a population that would decrease herbicide exposure. A more pragmatic definition simply states that herbicide resistance is the inherited ability to not be controlled by a herbicide.[6] Maximal expansion of the concept of resistance in agro-ecosystems could include weed species shifts (substitutions) in response to herbicides. The expanded definitions of herbicide resistance are conceptually more satisfying, but implicitly pose more complex issues in the study of resistance evolution in weeds. Resistance associated with avoidance as well as resistance mechanisms expands the range of resistance traits to be considered and will inevitably include quantitative traits which increases the likelihood of epigenetic effects and makes assessment of fitness more difficult to interpret.

The difference in herbicide resistance between phenotypes of the same species is qualitatively assessed in most cases by comparing the herbicide rates required to reduce survival (LD_{50}), biomass (GR_{50} or ED_{50}), or specific enzyme activity (I_{50}) by 50% from untreated plants. However, a more rigorous approach is the development of methodology based on the statistical comparison of dose responses of different phenotypes.[7-9] The selection of resistance due to mechanisms that are not physiological or morphological in nature will be discussed, but emphasis will be on the more common resistance that is associated with physiological changes which have most often been attributed to a single gene that modifies the site of herbicide action.

Fitness (considered in more detail later) is defined as the evolutionary advantage of a phenotype, which is based on its survival and reproductive success.[10]

Natural selection is the differential contribution of individuals to future generations through unequal reproduction and survival. Some progeny have a greater likelihood of surviving and reproducing than others when living in a specific environment. The term "natural" does not imply that the process of natural selection does not occur in agricultural (non-natural) ecosystems.

Requirements For Resistance Evolution

There are two precursors for the evolution of a trait like herbicide resistance in plant populations: the occurrence of (1) heritable variation for the trait and (2) natural selection. Given the existence of genetic variation, the rate of evolution will be determined by the mode of inheritance of resistance traits, together with the intensity of selection. The evolution of resistance under persistent applications of herbicide may be considered as an example of recurrent selection in which there is a progressive, and sometimes rapid, shift in average fitness of populations of weeds exposed to herbicide. At least for single gene traits, the shift in fitness is directly related to the increase in the frequency of the resistance trait (phenotype) in the population.

Three components contribute to selection pressure for herbicide resistance; the efficiency of the herbicide, the frequency of use and the duration of effect. Selection intensity in response to herbicide application is a measure of the relative mortality in target weed populations and/or the relative reduction in the seed production of survivors and will be proportional, in some manner, to herbicide dose. Selection duration is a measure of the period of time over which phytotoxicity is imposed by a herbicide. Both intensity and duration interact to give seasonal variation in the selection pressure imposed on weed species according to their phenology and growth. Selection pressure is not a preferred term because it implies an active force behind selection. Natural selection is passive and therefore the term selection intensity will be used in place of selection pressure, and it will encompass the elements of efficacy, frequency and duration of selection agent (herbicide) use.

BASIC CONCEPTS

Genetic Variation

The common view that genetic variation is almost always present in plant populations rests on the wide-ranging demonstration of very considerable amounts of variation at the level of isoenzymes.[11] This leads to the assertion that given sufficient duration and intensity of selection then evolutionary responses to selection are inevitable. However, the key to evolutionary responses is the existence of the appropriate variation on which selection can act and there is no *a priori* reason why such variation should be present in every population.

There are two ways in which resistance traits may arise within a weed population. A major gene, or genes, may be present at low frequency, or mutate, so that selection acts to change a population which is initially susceptible. Alternatively, recurrent selection may act on continuous (quantitative) variation and achieve a progressive increase in average resistance from generation to generation, with changes in gene frequency at many loci conferring resistance (see Chapter 10).

Genetic variation can arise *de novo* by mutation (or recombination) or be preexisting. We can thus distinguish two situations with regard to genetic variation for herbicide resistance in nonselected populations: (1) factors affecting the acquisition of resistance by novel mutation and (2) factors affecting the probability of preexisting variation for resistance.

Resistance from Mutation

For a population that does not possess alleles for resistance before selection is applied, the probability of a population acquiring resistance through mutation depends on the relationship between mutation frequency and the size of the population. Figure 1 indicates the relationship between mutation frequency (μ) and the size of the population (N) necessary to detect a resistant phenotype in a diploid plant, assuming an equilibrium forward and backward mutation rate at a locus. As mutation frequency diminishes, the number of individuals that need to be screened to expose a single resistant phenotype increases broadly in a logarithmic fashion.

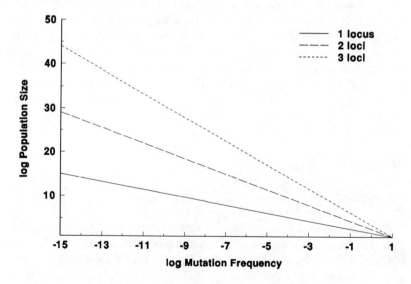

Figure 1. The expected population size required to expose one resistant phenotype (mutant) containing at least one dominant allele conferring resistance. Mutation frequency is assumed to be the same for all loci and forward and backward mutation rates to be equal.

Resistance mutation frequency (μ) increases with the number of loci that can result in resistance. Taking into account the fact that even an advantageous mutant will by no means always spread, it can be argued that only once $N > 1/\mu$ individuals have been exposed to selection is resistance ever likely to emerge in a population.

Mutation frequency may be environmentally influenced, although the concept of directed mutation is controversial.[12-16] Bettini et al.[17] reported that pesticides (including triazine herbicides) applied at sublethal doses to certain susceptible genotypes of *Chenopodium album* resulted in progeny with triazine-resistant characteristics similar to highly resistant plants. These results imply that triazine resistance could be induced by the low doses of pesticide in certain genotypes (discussed fully in Chapter 10). Lenski and Mittler[18] suggested that variable mutation rates may be an evolved response that promotes increased genetic variation under stress, but further stated that there is only weak evidence for directed mutation. Rainey and Moxon[12] suggested several mechanisms that facilitate alterations in the frequency of gene expression that are compatible with neo-Darwinian theory. These mechanisms allow an intrinsically stochastic component in the regulation of gene expression which results in polymorphism and ultimate population heterogeneity. Further research is needed to determine if mechanisms exist that cause increased mutation as a result of stress induced by herbicides or that somehow predispose weeds for rapid selection of resistant genotypes (see Chapter 10).

Resistance from Preexisting Genes

The evolution of resistance will be much more rapid if a population carries resistant alleles at low frequency before selection is imposed. The probability of resistance occurring in a weed population prior to selection with the herbicide depends on the mutation frequency, the selective disadvantage (d) of the genes for resistance in the normal, nonselecting, environment and the size of the breeding population. Theoretically, the phenotypic frequency for deleterious characters in nonselecting environments is approximately μ/d, though in populations of small effective size the average frequency will be reduced.[19] One must also realize that resistance can be associated with more than one trait which effectively increases the mutation rate. There also can be a range in selective disadvantage among resistant genotypes which further complicates estimates of the resistant phenotype frequency in a weed population.

Most herbicides are extensively screened for efficacy in the field before release. Herbicides have rarely been accepted for market without at least 70% control of target weeds averaged over many locations and times. Therefore, it is very likely that resistance traits are present, but undetectable, in weed populations before large-scale selection with herbicides. Matthews and Powles[20] measured a frequency of one resistant plant in every 50 from previously unsprayed populations of *Lolium rigidum* present on farms from across southern Australia. This frequency of resistance is several orders of magnitude greater than the expected mutation rate. Plants of the same species that were collected from nonagricultural regions had a much lower frequency (one resistant plant in every 500) of resistance using the same test. They suggested that the higher frequency of resistance from farm collections may be due to gene flow from resistant plants in nearby fields (subjected to herbicide selection) and possibly to linkages between resistance genes and other traits that would be selected by farm practices. For many other cases, resistant individuals are expected to be present at gene frequencies considered to be at the mutation rate. For example, if the mutation rate for resistance to acetolactate synthase (ALS) inhibitor herbicides in *Kochia scoparia* is 10^{-7} and there are 1 million seeds/ha, then one would expect only 10 ALS-resistant mutants in a 100-ha field. Thus, the probability of detecting resistance in screening trial plots is very low, even in dense weed populations. If the mutation rate is constant over generations then in the absence of herbicide selection, it would take 1000 generations to increase the frequency of ALS-resistant mutants to 1 plant m^{-2} if there was no selective disadvantage associated with the resistant phenotype (d <1.0) However, under constant herbicide selection the proportion of resistant phenotype would increase rapidly to a noticeable level within a few generations. Where few alleles have been suggested in the inheritance of herbicide resistance, it has evolved in 3 to 10 generations under continuous selection (Table 1). These estimates, compared with predictions from the models developed to predict herbicide resistance evolution,[34-36] suggest that before selection, resistance was present at low levels in the weed populations, but the low levels were often near the expected mutation rates ($\mu = 10^{-5}$ to 10^{-12}).

Price et al.[37,38] examined the variability of loci governing enzyme/morphological traits as well as herbicide response in *Avena barbata* and *A. fatua* populations not previously exposed to herbicides. They concluded that the amount of genetic variance for response to barban and bromoxynil was higher than could be expected on the basis of mutation alone. These findings suggest that natural selection maintained genes conferring herbicide resistance at a low frequency in the weed populations.

Table 1. Estimated number of years for natural selection of herbicide resistance in weed populations

Weed species	Chemical selection agent	Years for resistance to be recognized	Ref.
Kochia scoparia	Sulfonylureas	3–5	21,22
Avena fatua	Diclofop-methyl	4–6	23,24
Lolium multiflorum	Diclofop-methyl	7	25
Lolium rigidum	Diclofop-methyl	4	26
Lolium rigidum	Amitrole + atrazine	10	27
Lolium rigidum	Sethoxydim	3	87
Senecio vulgaris	Simazine	10	28
Alopecurus myosuroides	Chlortoluron	10	29
Setaria viridis	Trifluralin	15	30
Avena fatua	Triallate	18–20	31
Carduus nutans	2,4-D or MCPA	20	32
Hordeum leporinum	Paraquat/diquat	25	33

Selection

As discussed, the rate of herbicide resistance development in weeds is initially dependent upon the origin of resistance. The interaction between genetic diversity for the trait and the intensity of selection will largely influence the onset of resistance. Intense selection with herbicides coupled with genetic diversity, provides the elements required for rapid development of herbicide-resistant populations.

A major determinant in the selection of herbicide-resistant biotypes is the effective selection intensity that differentiates resistant individuals (more fit) from susceptible ones (less fit) in the face of selection (the application of herbicide). Measurement of the selection intensity exerted by herbicides has not been undertaken in any significant way. Population geneticists theoretically measure selection as the differential survival of alleles or change in gene frequency after the action of the selection agent. However, weed scientists observe selection at the phenotypic level, and selection coefficients may be variously defined at the gametic (gene) or zygotic (genotypic) level. Thus, the coefficient of selection, s, may be defined most simply as the proportionate contribution of a particular genotype to future generations compared with a standard genotype (usually the most favored or fit), whose contribution is usually taken to be unity. This definition makes assumptions about the genetic architecture of the population and the mode of inheritance of alleles undergoing selection.[39]

The conceptual (theoretical) approach in calculating selection is as follows. A diploid bisexual plant goes through a cycle from production of a zygote through maturation to adult, production of gametes (pollen and ovules) and "mating behavior" to ensure that gametes form new zygotes (plants). Given two alleles (A and A′) at a locus determining resistance, the gene frequency q of A′ is defined as a fraction of the total number of alleles at that locus. For a diploid autosomal locus, the frequency of AA homozygotes is a, of AA′ heterozygotes is b, and that of A′A′ is c; then in zygotes, $q = c + b/2$. The frequency of the allele A (p), is then $1 - q = a + b/2$. If the population is infinite in size, mating is at random and, with no mutation, migration, or selection, one generation later three genotypes will be present in the frequencies p^2, $2pq$, and q^2. This is known as the Hardy-Weinberg equilibrium and a similar relationship will exist for multiple alleles.

If discrete generations of plant growth occur then allelic changes can be represented as $\Delta q_n = q_n + 1 - q_n$. If there is differential survival of zygotes then allelic frequencies will change in relation to the relative fitness of homozygotes and heterozygotes after a generation of growth n $\Rightarrow$ n+1. This constitutes selection and

$$\Delta q = -pq(w_1 p - 1 + 2q - w_3 q)/(w_1 p^2 + 2pq + w_3 q^2) \qquad [1]$$

where w_1 and w_3 are the fitness (relative "performance") of AA and A′A′ compared to the survival of the heterozygote. This equation is often arranged and represented in other ways and does not take into account the absolute size of the population and simply relates proportional changes in gene frequency over a generation.

THE RATE OF RESISTANCE EVOLUTION

May and Dobson[40] have presented an elegant expression of the evolution of pesticide resistance. Following their reasoning the original susceptible allele is denoted as S, and the resistant allele R, each occurring at respective frequencies s_t and r_t in generation t ($s_t + r_t = 1$). The genotype RR is resistant, heterozygotes RS intermediate, and SS susceptible. Herbicide resistance in weeds has been most often associated with the dominant allele indicating a gain in function. Under a given dosage of pesticide the fitness of the three genotypes are denoted w_{RR}, w_{RS}, and w_{SS} and we assume $w_{RR} \geq w_{RS} \geq w_{SS}$.

Re-expression of Equation [1] gives the gene frequency of R in successive generations as

$$r_{t+1} = \frac{W_{RR}r_t^2 + W_{RS}r_t s_t}{W_{RR}r_t^2 + 2W_{RS}r_t s_t + W_{SS}s_t^2} \quad [2]$$

Before the application of recurrent selection by herbicide, the frequency of the resistant allele r_t will be very low (e.g., 10^{-10}). Therefore, the ratio r_t/s_t will be much smaller than either w_{RS}/w_{RR} or w_{SS}/w_{RS} and Equation [2] reduces, in a good approximation, to

$$r_{t+1} = \frac{W_{RS}}{W_{SS}} r_t \quad [3]$$

If the allele R is present in an unselected population at an initial frequency of r_0 then compounding Equation 3 gives the number of generations, n, that must occur before a significant degree of resistance appears (i.e., r achieves the value $r_f \approx 0.5$, for example) as approximately

$$\frac{r_f}{r_0} = \left(\frac{W_{RS}}{W_{SS}}\right)^n \quad [4]$$

If we define T_R to be the absolute time taken for a significant degree of resistance to appear and T_g is the time taken for a generation of population growth, then $n = T_R/T_g$ and, substituting into Equation 4], we have an equation that provides the *approximate* time by which a specified resistance level r_f is reached. Thus,

$$T_R = T_g \ln\left(\frac{r_f}{r_0}\right) \bigg/ \ln\left(\frac{W_{RS}}{W_{SS}}\right) \qquad [5]$$

Equation [5] shows that the rate of evolution depends on the species generation time and logarithmically on (1) the initial frequency of the resistance allele (r_0), (2) the choice of threshold at which significant resistance is recognized (r_f), and (3) the strength of selection (w_{RS}/w_{SS}) which is determined by dosage level and the degree of dominance of R.

Even if r_0 varies in the range 10^{-5} to 10^{-16}, w_{RS}/w_{SS} in the range 10^{-1} to 10^{-4} and T_g is 1 year, this form of analysis leads to the conclusion that T_R lies in a relatively narrow range of around 10 to 100 years assuming recurrent selection. Gressel and Segel[5,88] came to the same conclusion in an earlier and independently derived model. Such a theoretical conclusion indicates that the appearance of herbicide resistance may be relatively rapid. In fact, for weed species such as *L. rigidum* (Chapter 9) and *K. scoparia* (Chapter 4), resistance can evolve over a 3- to 4-year period (Table 1). This probably occurs because of a higher than anticipated gene frequency for resistance[20] and near equal fitness of resistant homozygotes and heterozygotes at the rate of herbicide used in the field.

FACTORS THAT MODIFY THE RATE OF EVOLUTION

Equation [5] gives an approximate time for the appearance of herbicide resistance; however, it takes no account of (1) the mode of inheritance of traits determining resistance, (2) the number of genes that can confer resistance, (3) mating behavior, and (4) the influence of gene flow.

Modes of Inheritance

The genetic basis of resistance will affect the rate at which the majority of the population will come to resistance. In broad overview, the following conclusions can be made assuming that the resistant phenotype is of equal fitness to the susceptible. With the assumption of diploidy and random mating, at low gene frequencies a dominant allele will spread faster than a recessive. A maternally inherited character will spread only marginally faster than a dominant one. The difference in rate is most important in the early stages of selection where traits controlled by dominant alleles and cytoplasmic genes reach appreciable frequencies (1%) much faster than recessive ones. Thereafter, small differences in rate result from the differing modes of inheritance. The much longer time taken by recessive alleles to reach appreciable frequencies means that resistance governed by these genes is much less likely to evolve. Simulation models suggest that conditions exist such that resistance may never evolve if resistance is governed by rare recessive alleles.

Number of Resistance Traits

The number of traits that confer herbicide resistance is related to the amount of genetic variation available for mechanisms associated with phenology (avoiding the herbicide), morphology (limiting herbicide uptake), or physiology (altering herbicide uptake, translocation, activity, or metabolism). Upon the onset of selection with a herbicide, if a number of different resistance traits exist in a population, their respective frequency will depend on (1) their initial frequency, (2) the selective advantage of each genotype, and (3) the genetic variation associated with each trait that will allow it to increase its selective advantage. One may assume that resistant genotypes associated with the least amount of change from the wild type would have the greatest selective advantage.

Powles and Matthews[41] have argued that in very large, genetically diverse populations, such as *L. rigidum* in Australia, a number of alternative resistance mechanisms can be simultaneously selected by regular herbicide use (see Chapter 9). It is believed that multiple herbicide resistance in *L. rigidum* is the result of resistance mechanisms being aggregated in the progeny as a result of cross pollination of surviving plants.[41] The frequency of the different resistance mechanisms increases under selection with the herbicide so that the probability of crosses between plants with different mechanisms also increases. If this scenario is true then more cases of cross and multiple resistance should be observed in the near future in other species.

Quantitative Inheritance

Where many genes confer resistance in a quantitatively inherited manner, the response to selection will be strongly related to the heritability of resistance (see Chapter 10). Heritability is a measure of the amount of genetic control of variation in resistance as opposed to environmental sources.[42] If quantitative inheritance is involved, a progressive response to herbicide application is to be expected so long as populations retain additive genetic variation. Thus, alleles accumulate, each causing a small increase in fitness under herbicide selection, and may systematically promote increased fitness in genotypes as genetic recombination occurs over successive generations. Implicitly the rate of evolution is likely to be slower than for single nuclear encoded genes.

Soil-applied herbicides may be more likely to select for quantitative inheritance because the relatively slow dissipation of the herbicide over time tends to expose late emerging weeds in a population to lower doses that allow accumulation of resistance alleles. If emergence time is under genetic control, resistance could increase rapidly in late emerging plants, but only if they can reproduce. High rates of herbicide application may be lethal to genotypes with a low number of resistance alleles (early selection), and therefore could reset the evolutionary progression toward quantitative inheritance of resistance. In any case the rate of resistance evolution by quantitative inheritance could be influenced quite differently by the same factors that effect single gene-inherited resistance.

Mating Behavior

The extent to which a species is obligate cross-pollinated (as opposed to being self-compatible and self-pollinated) can strongly determine the genotypic structure of a population. Theoretically at least, in a random mating diploid population, one generation of self-fertilization leads to an average reduction in heterozygosity of 50%. The spread and subsequent recognition of herbicide resistance will occur more rapidly in cross-pollinated populations assuming resistance is associated with a single dominant allele and cross pollination is as efficient as self-pollination for viable seed production.

Gene Flow

Gene flow can have an impact on the rate of evolution of resistance. If a single plant or set of spatially noncontiguous plants survive a herbicide application due to a rare mutation that makes them resistant then they will probably go unnoticed in a field unless the allele for resistance is passed to other plants in the field. Gene movement among plants can occur through two primary modes: (1) pollen dispersal and resultant fertilization, and (2) seed dispersal. Gene flow is not strictly an intraspecific phenomenon. Introgression (gene exchange between species) is well documented; however, most species have mechanisms that decrease the likelihood of successful pollination by foreign pollen. The possibility of interorganism gene transfer is discussed below.

The most common application of herbicides is in cropped fields which are adjacent to unsprayed areas. If evolution toward resistance has occurred in sprayed areas then there may be a marked discontinuity in the fitness of plants along a transect across the crop margin. Susceptible plants will occur at the field edge and resistant plants within the crop. In plant populations, a discontinuity in phenotypes may be very sharp, as in the case of zinc-tolerant plants under zinc galvanized fences. The distribution of a trait in individuals away from a local source of selection has been measured on a scale of meters as in *Liatris glindracea*[44] or even centimeters as in *Anthoxanthum odoratum*.[45]

Gene flow may, however, occur between adjacent (parapatric) populations by seed and pollen flow and this may result in a cline in resistance, a gradual change in resistance along a transect. The steepness of the cline (the change in fitness with distance) may be sharp and will depend on two factors, gene flow and the differential in fitness of the various genotypes along the gradient. In addition, the shape of the cline will depend upon whether a steady state in genotype replacement has been reached. The effect of gene flow should be to flatten the cline but will very much depend on ecological factors affecting seed movement and pollen dispersal. In the long term, gene flow may also exert considerable selection pressure for reproductive isolation. This may come about by adaptations favoring assortative mating whereby selection results in changes in flowering time, or pollen incompatibility between populations. The effect of gene flow in promoting assortative mating is likely to be greater in

Factors Influencing Pollen Dispersal

Gene flow by wind-mediated pollen dispersal is influenced by a number of factors that could determine the fitness of the resistant genotype. The number of gametes (pollen grains) that contain the resistant allele(s) determines the probability of successful crosses if all other factors are favorable for sexual reproduction. Release of the gamete so that it corresponds in time with receptive stigmas could be important. Ghersa et al.[46] observed that a diclofop-resistant phenotype of *Lolium multiflorum* grown in Oregon exerted anthers 3 to 5 days after the susceptible biotype. They further suggested that if ryegrass stigmas were receptive throughout this period and exerted anthers correspond with pollen release then stigmas could have been inundated with susceptible pollen before resistant pollen was released, which would encourage gene flow from susceptible to resistant plants, but not the reverse.

Survival of the wind-dispersed gamete between the time of release and arrival at a stigma involves the aerodynamics of the pollen grain as well as atmospheric environmental factors. Successful fertilization entails the relative attributes of the gamete which allows it to land on a stigma, germinate and fertilize the female gametophyte. The success of pollen from one biotype versus another biotype may depend on the relative receptivity to each type of pollen by the stigma.[47]

Pollen dispersal by insects and animals is influenced by many of the same factors as wind-mediated pollen dispersal, but can be much more complicated by factors that influence the insect or animal behavior.[48-50] Even with all the complicating factors, insect- and wind-mediated gene flow by pollen dispersal has been characterized as a diffusion gradient that can be described as a probability distribution where the probability of successful crosses decreases exponentially with distance from the pollen source.[49-51] The probability distribution can be characterized with a probability density equation in the exponential family of functions.[52] One function found to be useful is the Weibull function:[53]

$$p(x) = \text{EXP}\left[-(d+a)/b)^c\right]$$

where $p(x)$ is the proportion of total effective crosses over a range of distances from the pollen (gene) source, d is the distance from the source, and a, b and c are constants. This equation can then be used to calculate the probability of a gene moving a specific distance.

Gene flow by pollen has been measured directly by the use of marker genes in spaced-plant "garden" populations and observations have shown that gene flow may fall to about 5% at 10 m.[54] Copeland and Hardin[55] examining genetic contamination in *Lolium perenne* swards found that gene flow was negligible 6 m from the source and undetectable 12 m away. Many studies which sought to establish isolation distances for crop breeding purposes used similar methodology which often disregards the true potential (probability) of gene flow at distances from a source.

Probability distributions were used to estimate the potential gene flow between field-grown, wind-pollinated diclofop-resistant and -susceptible *L. multiflorum* populations. Of the successful crosses that occurred, 50% were predicted to occur within 1 m, 90% would occur within 3.6 m, and 99% would occur within 7 m of a resistant source population in one generation.[53] The distance for effective crosses to occur was not influenced by direction even though there was a prevailing wind. The ability to fit a negative exponential function to the probability of gene flow was influenced by the boundary layer of air which is effected by the uniformity in the surrounding canopy. The results from these experiments indicate that gene flow from a source population can only generally be described as a diffusion gradient and must consider site-specific characteristics to accurately predict the gene flow potential.

The size of the source population and distribution of individuals within the population can have an influence on gene flow through pollen dispersion.[56,57] However, this has not been verified for herbicide-resistant weeds. Maxwell et al.[58] integrated a pollen dispersion model adapted from Fitt et al.[59] and Mundt and Brophy[60] into a simulation model to predict the evolution of herbicide resistance. Simulations with the model show the interaction of the size and distance from a herbicide-resistant source population and show that a large gene source may overcome a low probability of gene exchange due to distance between source and receiving (treated) populations.[58]

Gene Flow through Seed Dispersal

Seed movement in agro-ecosystems due to agricultural implements can be substantial and dramatic in effect[61,62] but, within fields, gene flow through natural seed movement may be relatively small.[63]

Gene Flow through Associated Organisms

A potential mechanism for increasing genetic variation and fluidity is through the horizontal (nonsexual) transfer of genetic elements or sequence among microorganisms and host plants.[64,65] Transfer and integration of DNA from *Agrobacterium* into infected or colonized host plant genomes has been documented.[66] Bryngelsson et al.[67] reported examples of detectable horizontal genetic transfer between *Brassica napus* and the parasitic fungus, *Plasmodiophora brassicae*. Cytoplasmic genetic elements migrate between fungi when their hyphae fuse,[68] and filamentous fungi can link plants with hyphal connections between their roots.[69] Whether transfer of RNA or DNA elements between different organisms occurs frequently in nature is not known. Many plant-colonizing fungi enter cortical and even vascular tissue of plants by enzymatic digestion or mechanical breaching of cell walls, and form stable, intimate relationships with their host. Although it has been a topic of much speculation,[65] it is not known whether plant hosts exchange cytoplasmic genetic elements with colonizing fungi. These recent discoveries make it conceivable that herbicide-resistant genes could be intra- or interspecifically transferred through fungi or other organisms.

FITNESS

In population genetics, a fitness value is a numerical quantity which describes the survival and subsequent reproductive success of one kind of individual relative to another as a consequence of natural selection.[70] It should always be borne in mind, however, that it is individuals with particular phenotypes that survive and reproduce and that phenotypic response is the product of the interaction of a particular genotype and the environment. An individual plant is a unique genotype with variation at many loci affecting fitness. In practice, however, the fitness of a group of plants having a certain genotype is estimated relative to the fitness of other genotypes lacking key traits of interest. In a single interbreeding population of plants in a homogeneous environment the genotypic response due to allelic changes at a single locus are considered to occur against a constant environmental and genetic background and the expression "genotypic fitness" is used. With this approach, then, for given genotypes (homozygotes, heterozygotes) it is possible to measure and calculate "genotypic fitness" in the field and laboratory.[71]

In plants reproducing by seeds in discrete generations, measures of fitness can be derived from the number of seeds produced by a particular genotype over a generation of growth. For vegetatively reproducing species, measuring fitness is intrinsically more difficult and may require measurement of biomass or plant parts over a time period appropriate to the species in question.

Fitness is expressed in relative terms whereby genotypes are compared among themselves relative to the most successful one, but it is important to recognize that: (1) absolute fitness contributes to the rate of evolution in its own right. Thus, evolutionary rate will be proportional to per capita rates of increase, all other factors being equal. (2) Fitness is also a measure of genotypic performance in a particular environment. As fully discussed in Chapter 11, early studies on triazine-resistant weed biotypes identified a "cost" to resistance reflected in traits such as growth and competitiveness.[72] Triazine-resistant weed biotypes were found to yield 10 to 50% of susceptible biotypes under competitive and noncompetitive conditions in a nonselective environment.[73-75] Triazine resistance may be a special case that amplifies the fitness differences because resistance is due to an altered photosystem II which renders the resistant biotype less photoefficient (see Chapters 2 and 10). As also discussed in Chapter 11, certain new genes introduced through mutation to a population can result in a less well-adapted (fit) phenotype than is associated with the original gene if there is no change in the environment.[76] The new genotype would then be displaced by the wild type, or increase in frequency to some equilibrium in the population. The rate of decrease or increase and the new equilibrium frequency would depend on the relative fitness advantage or disadvantage of the genotype and the rate that the genotype can acquire compensatory traits by crossing with the wild type. If changes occur in the environment or the wild-type genotype, then the new genotype may become more frequent in the population within a few generations. In the absence of herbicide, a new herbicide-resistant genotype should be to some degree less fit than the susceptible genotype but its success will be determined by the rate that it can acquire compensatory traits. If an herbicide is introduced into the system the resistant genotype becomes immediately more fit than the susceptible and rapidly becomes a greater proportion of the population.

As discussed in Chapter 11, many biological processes interact to determine the fitness of a plant. Fitness must be measured over the whole life cycle of a plant to encompass the effects of selection on mortality and seed production (fecundity) of survivors. Thus, for an annual species, the seed produced by a genotype per generation constitutes a fitness estimate for a given environment only at one point in evolutionary time. When determining the fitness of a weed with a persistent seed bank, the rate of loss of seed from the soil needs to be measured. Seed carryover from previous generations plus the seed produced in the current generation contribute to total seed production. Work on *Senecio vulgaris* has shown that resistant and susceptible biotypes differ in their persistence in the soil according to depth of burial and the type of crop husbandry practiced.[77] Measurements of fitness will not be valid for understanding the rate of resistance evolution or management of resistance unless they are conducted under field conditions with the crop and with and without herbicide application.[78] In addition, fitness must be viewed as a dynamic integration of processes that can change with the environment and further natural selection for more fit genotypes within the resistance phenotype.

THE CHALLENGES AND LESSONS FROM HERBICIDE RESISTANCE

Active design of weed management strategies addressing the likelihood and rate of evolution of herbicide resistance is in its infancy at present. This is in part due to the past availability and success of alternative herbicides for populations that evolved resistance. The appearance of cross and multiple resistance in grasses (Chapters 5, 9, 12) increasingly limits successful herbicide options. A major challenge to weed science is to confidently provide viable long-term strategies for the containment of alleles conferring herbicide resistance in weeds of the major cropping regions of the world. These strategies must be underpinned by sound knowledge of the ecology and genetics of weed species undergoing selection. It remains a fact that in many cases this knowledge is lacking for many of the important weeds of agriculture and horticulture. Comparative studies of weeds and the screening of (relatively) unselected populations for resistance alleles is a valuable but arduous task of identifying weed species likely to become resistant.

Equally important is the question, have herbicide-resistant weeds in a single field evolved from concurrent but possibly different mutation events or from gene flow from a single or few sources? The answer to this difficult question is particularly important for the management of resistance.[34] In studies of metal-tolerant populations of plants on metalliferous mine sites, experimental trials suggest that the evolutionary process takes place in three stages. Elimination of the most sensitive genotypes and the observation of weakly resistant individuals is followed by a second stage of selection of the most resistant phenotypes in the population. Finally, interbreeding of survivors may lead to increased fitness in progeny through allelic segregation and recombination. To our knowledge, no detailed studies have been published of a weed population evolving resistance at one particular location. Anecdotal evidence from farmers suggests that the spread of resistance within a population may be extremely rapid as in *L. rigidum* (see Chapter 12), but conversely may be relatively slow (chlortoluron- resistant *Alopecurus myosuroides*). Major gene resistance coupled with substantial gene flow will clearly favor rapid spread in contrast to the restrictions imposed by quantitative inheritance coupled with limited gene flow.

A challenge lies in understanding the genesis of variation in weed populations. In *Avena* spp. there was more genetic diversity in populations found in the habitat of origin than in a population found on agricultural land.[79] This increased diversity may be due to increased niche differentiation driven by higher species diversity in natural plant communities. The question remains, do processes that generally increase genetic variation in populations necessarily increase the probability of herbicide resistance traits in weed populations? The subsequent hypothesis may be; the species that have evolved resistance to herbicides should be those with the most genetic diversity. In support of this hypothesis, one would also expect a number of mechanisms of resistance for a single herbicide assuming no difference in fitness among the different genotypes associated with each resistance mechanism. If there was a fitness difference among biotypes exhibiting different resistance mechanisms, the mechanism that caused the least change in fitness from the wild type (susceptible) would be the most frequent in the weed population. The assumption is that the wild type was selected to be the most fit in the environment without the influence of the herbicide.

Powles and Matthews[41] report that there are a number of different resistance mechanisms that contribute to multiple herbicide resistance in *L. rigidum*, and attribute the rapid evolution of multiple resistance to general genetic diversity within the species. Guttieri et al.[80] analyzed a set of *K. scoparia* samples from different sites and concluded that target site resistance to ALS inhibitors was due to several different amino acid substitutions indicating that genetic variability exists within a single mechanism of resistance. Darmency and Gasquez[81] found as much polymorphism in morphological characters and isozymes in a triazine-resistant biotype as in the susceptible biotype of *Poa annua*, indicating considerable genetic variation, although they did not exclude the possibility of a founder effect producing the polymorphism. However, there have been resistance cases where a very low degree of polymorphism for a number of traits including resistance has been measured.[82,83] Price et al.[37,38] found the level of genetic variation for herbicide response was correlated with the genetic variation for enzymatic and morphological loci. A similar finding was reported by Jana and Naylor.[84] Somody et al.[85] found a range of responses to herbicides in *A. fatua* and *A. sterilis* that also exhibited a range of morphological traits. Until more is known about the degree of genetic diversity in weed populations it will be difficult to test the hypothesis of general genetic diversity being related to diversity in the herbicide resistance trait.

Another important question, beyond the origin of resistance, concerns the probability of multiple herbicide resistance mechanisms (phenotypes) occurring in a weed population. The hypothesis (see Chapter 9) that there can be a number of alternative herbicide-resistant mechanisms present in weed populations before selection with the herbicide, as suggested by Powles and Matthews,[41] was tested with a simulation model.[35] The question can be reduced to determining how long different resistant phenotypes will coexist in a population without herbicide selection and when different levels of fitness are associated with each phenotype. Two resistant phenotypes were considered, R1 and R2, and the susceptible phenotype of the weed was assumed to be equal in fitness to R1, but both R1 and R2 were assumed to be 20% less fit than the crop. Two types of fitness, relative seed production (fecundity), and relative response to a herbicide were used to determine overall relative fitness of R1 and R2. All other fitness parameters in the model were held equal for R1 and R2. Figure 2 shows the simulation of evolution of resistance (R1) from a single resistant mutant under the assumptions that resistance is associated with a single dominant allele, the weed is a self-pollinating annual species, there is no immigration of resistant and susceptible genes, and the herbicide is applied continuously for 10 years at an efficacy of 90%.

Simulations were conducted to assess how rapidly the frequency of a second phenotype (R2) would decrease to half the initial frequency in the population, by independently changing the relative fitness of R1 and R2 for fecundity and herbicide response. The results of the simulations could then provide theoretical evidence for the likelihood of one or many resistant phenotypes (mechanisms) existing in a population that is, or is not, under selection with a herbicide and could provide an understanding of how the different types of fitness can interact to determine phenotype frequencies. In the first simulations, R1 herbicide response fitness was set equal to R2, and fecundity fitness was

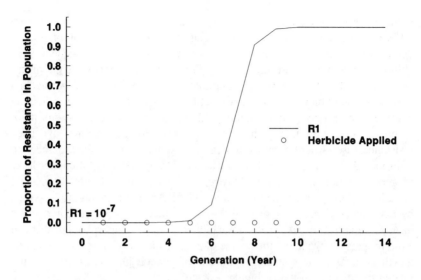

Figure 2. Predicted evolution of a herbicide-resistant phenotype (R1) starting from a mutation proportion (10^{-7}) at generation 0 and exposed to a herbicide with 90% efficacy (removal of susceptible phenotype) for 10 continuous generations. R1 was assumed to be equal in fitness to the susceptible phenotype.

systematically varied from a 1 to a 50% difference favoring the R1 phenotype. The R2 phenotype was predicted to take 66 generations to decrease to half of its original frequency in the population when R2 had 1% less fecundity (1% less fit) than R1 (Figure 3). The number of generations to decrease the frequency of R2 by half in the population decreased sharply and asymptotically when the difference in seed production fitness was increased. These results imply that a fitness difference (exclusive of herbicide response) of 10% or greater would reduce R2 frequency to half its original frequency in less than 10 generations. With this rate of phenotype decrease it is unlikely that R2 would be at a high enough frequency to be identified in a weed population or be able to significantly donate genes through crosses with the other resistant phenotypes.

Since it is unlikely that fecundity differences would be the same over a number of generations due to phenotypic plasticity, another set of simulations were conducted where the fecundity fitness differences between R1 and R2 were stochastically varied between 0 and 20% in favor of R1. Under no herbicide selection (R1 = R2 for herbicide response fitness), it took an average of 8 generations to reduce the frequency of R2 by half. However, in a similar set of simulations, the herbicide response fitness was set to a 10% difference in R1 and R2 in favor of R2 and then fecundity differences were stochastically varied between 0 and 20% in favor of R1. All possible trends (increasing, stable, decreasing) for R2 frequency in the population were represented (Figure 4). When R2 was predicted to decrease, it took an average of 76 generations to decrease the frequency by half. These simulations indicate that a substantial

SELECTION FOR HERBICIDE RESISTANCE

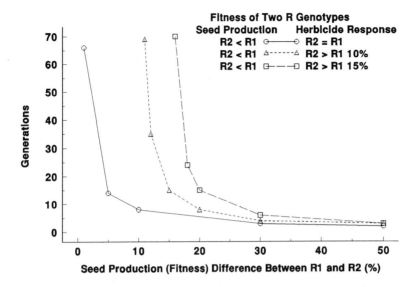

Figure 3. The predicted number of generations to decrease a less fit resistant phenotype (R2) to half its original frequency (10%) in a weed population under selection with a herbicide when fecundity (seed production) and herbicide response fitness were simultaneously varied for R1 and R2.

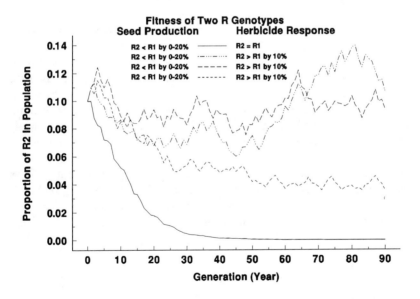

Figure 4. The predicted proportion of a less fit resistant phenotype (R2) in a weed population under selection with a herbicide when fecundity and herbicide response fitness were simultaneously and stochastically varied for R1 and R2.

fecundity fitness difference would be required to decrease the frequency of R2 in the population if R2 had any herbicide response fitness advantage. Indeed, a *L. rigidum* population containing multiple resistance mechanisms when evaluated in field de Wit replacement series competition experiments was found to be equally fit to a relevant susceptible biotype.[86]

In conclusion, the results from the simulations indicate that more than one herbicide resistance phenotype could be expected in a weed population, but only if the different phenotypes are nearly equal in cumulative fitness. The results accentuate the importance of determining the relative fitness of different resistance phenotypes and specifically determining fitness in the presence and absence of the herbicide.

CONCLUSIONS

Several important questions have been raised in this chapter which could have a major impact on how weed populations are managed to slow resistance evolution. Weed scientists must become familiar with the principles and research techniques of population genetics and plant ecology in order to effectively examine the important questions regarding herbicide resistance selection in weeds. Robust population dynamics models can help in this regard.

The evolution of herbicide resistance in weeds has opened windows of realization for plant scientists and others. Resistance is a striking example of an evolutionary response of weeds to control practices. Weeds have historically evolved mechanisms to avoid, tolerate, or resist methods of control. The ancient process of winnowing grains to separate chaff from grain selected for weed seeds that were similar in size to the crop seed, which increased the probability of their being planted in the future crop. Crop mimicry by weeds is a well-documented evolutionary process that allows weeds to escape elimination. However, methods of weed control have never been as physiologically specific or intensively used as herbicides, so that prior to the introduction of herbicides there was more subtle evolution of resistance, tolerance, or avoidance traits. The herbicide resistance phenomenon has taught us that we must design weed management strategies (see Chapter 12) that slow or prohibit the evolution of traits that allow escape from control to maximize the number of tools that can be used to manage weeds.

REFERENCES

1. Bradshaw, A. D. "Evolution of Heavy Metal Resistance — an Analogy for Herbicide Resistance," in *Herbicide Resistance in Plants*, H. M. LeBaron and J. Gressel, Eds. (New York: John Wiley & Sons, 1982), pp. 293-308.

2. Al-Hilayli, S. A. K., T. McNeilly, A. D. Bradshaw, and A. M. Mortimer. "The Effect of Zinc Contamination from Electricity Pylons. Genetic Constraints on Selection for Zinc Tolerance," *Heredity* 70:22-32 (1993).
3. Mortimer, A. M. "The Biology of Weeds," in *Weed Control Handbook: Principles*. 8th Edition, R. J. Hance and K. Holly, Eds. (Oxford: Blackwell Scientific Publications, 1990), pp. 1-42.
4. Gould, F. "The Evolutionary Potential of Crop Pests," *Am. Sci.* 79:496-507 (1991).
5. Gressel, J., and Segel, L. A. "The paucity of plants evolving genetic resistance to herbicides: Possible reasons and implications. *J. Theor. Biol.*, 75:349–371 (1978).
6. Gressel, J. "Need Herbicide Resistance Have Evolved? Generalizations from Around the World," *Proceedings of the 9th Australian Weeds Conference*, J. W. Heap, Ed. (Adelaide: Crop Science Society of South Australia, Inc., 1990), pp. 173-184.
7. Putwain, P. D. "The Resistance of Plants to Herbicides," in *Weed Control Handbook: Principles*. 8th Edition, R. Hance and K. Holly, Eds. (Oxford: Blackwell Scientific Publications, 1990), pp. 217-242.
8. Streibig, J. C., M. Rudemo, and J. E. Jensen. "Dose-response Curves and Statistical Models," in *Herbicide Bioassays*, J. C. Streibig and P. Kudsk, Eds. (London: CRC Press, 1993), pp. 29-56.
9. Maxwell, B. D., K. Puettmann, and B. Murry. Unpublished data (1992).
10. Crow, J. F. *Basic Concepts in Population, Quantitative, and Evolutionary Genetics*, (New York: Freeman and Co., 1986), pp. 29-53.
11. Brown, A. D. H. "Enzyme Polymorphisms in Plant Populations," *Theor. Popul. Biol.* 15:1-42 (1979).
12. Rainey, P., and R. Moxon. "Unusual Mutational Mechanisms and Evolution," *Science* 260:1958 (1993).
13. Dover, G. "Molecular Drive: a Cohesive Mode of Species Evolution," *Nature* 299:111-117 (1982).
14. Dover, G. A., and D. Tautz. "Conservation and Divergence in Multigene Families: Alternatives to Selection and Drift," *Philos. Trans. R. Soc.* 312:275-289 (1986).
15. Durrant, A. "The Environmental Induction of Heritable Change in *Linum*," *Heredity* 17:27-61 (1962).
16. Al Saheal, Y. A., and A. S. Larik. "Genetic Control of Environmentally Induced DNA Variation in Flax Genotrophs," *Genome* 29:643-646 (1987).
17. Bettini, P., S. McNally, M. Savignac, H. Darmency, J. Gasquez, and M. Dron. "Atrazine Resistance in *Chenopodium album*: Low and High Levels of Resistance to the Herbicide are Related to the Same Chloroplast psbA Gene Mutation," *Plant Physiol.* 84:1442-1446 (1987).
18. Lenski, R. E., and J. E. Mittler. "The Directed Mutation Controversy and Neo-Darwinism," *Science* 259:188 (1993).
19. Wright, S. "The Distribution of Gene Frequencies in Populations," *Proc. Natl. Acad. Sci. U.S.A.* 23:307-320 (1937).
20. Matthews, J. M., and S. B. Powles. "Aspects of the Population Dynamics of Selection for Herbicide Resistance in *Lolium rigidum* (Gaud)," Proceedings of the First International Weed Control Congress (1992), Melbourne, Australia, pp. 318-320.
21. Thill, D. C., C. A. Mallory-Smith, M. Alcocer-Ruthling, and W. J. Schumacher. "Sulfonylurea Herbicide Resistant Weeds in North America," *Proceedings of the 9th Australian Weeds Conference*, J. W. Heap, Ed. (Adelaide: Crop Science Society of South Australia, Inc., 1990), pp. 194-195 .

22. McKinley, N. D. "Sulfonylurea Herbicide Resistance in Weeds in Cereals and Non-crop Areas in the U.S. and Canada," *Proceedings of the 9th Australian Weeds Conference*, J. W. Heap, Ed. (Adelaide: Crop Science Society of South Australia, Inc., 1990), pp. 268-269.
23. Piper, T. J. "Field Trials on Diclofop-methyl Tolerant Wild Oats (*Avena fatua*)," *Proceedings of the 9th Australian Weeds Conference*, J. W. Heap, Ed. (Adelaide: Crop Science Society of South Australia, Inc., 1990), pp. 211-215.
24. Mortimer, A. M. "A Review of Graminicide Resistance," Monograph 1 Herbicide Resistance Action Committee — (Monsanto, Brussels, 1992), p. 70.
25. Stanger, C. E., and A. P. Appleby. "Italian ryegrass (*Lolium multiflorum*) Accessions Tolerant to Diclofop," *Weed Sci*. 37:350-352 (1989).
26. Heap, I. M. "Resistance to Herbicides in Annual Ryegrass (*Lolium rigidum*) in Australia," in *Herbicide Resistance in Weeds and Crops*, J. C. Casely, G. W. Cussans, and R. K. Atkin, Eds. (Oxford: Butterworth-Heinemann, 1991), pp. 57-66.
27. Burnet, M. W. M., O. B. Hildebrand, S. B. Powles, and J. A. M. Holtum. "Amitrole, Triazine, Substituted Urea and Metribuzin Resistance in a Biotype of Rigid Ryegrass, (*Lolium rigidum*)," *Weed Sci*. 39:317-323 (1991).
28. Ryan, G. F. "Resistance of Common Groundsel to Simazine and Atrazine," *Weed Sci*. 18:614 (1970).
29. Moss, S. R., and G. W. Cussans. "The Development of Herbicide-resistant Populations of *Alopecurus myosuroides* (Black-grass) in England," in *Herbicide Resistance in Weeds and Crops*, J. C. Casely, G. W. Cussans, and R. K. Atkin, Eds. (Oxford: Butterworth-Heinemann, 1991), pp. 45-55.
30. Morrison, I. N., H. Beckie, and K. Nawolsky. "The Occurrence of Trifluralin Resistant *Setaria viridis* (Green Foxtail) in Western Canada," in *Herbicide Resistance in Weeds and Crops*, J. C. Casely, G. W. Cussans, and R. K. Atkin, Eds. (Oxford: Butterworth-Heinemann, 1991), pp. 67-75.
31. Malchow, W. E., B. D. Maxwell, P. K. Fay, and W. E. Dyer. "Frequency of Triallate Resistance in Montana", Proceedings of the Western Society of Weed Science (Tuscon, Arizona, 1993) 46:75.
32. Harrington, K. C. "Spraying History and Fitness of Nodding Thistle, *Carduus nutans*, Populations Resistant to MCPA and 2,4-D," *Proceedings of the 9th Australian Weeds Conference*, J. W. Heap, Ed. (Adelaide: Crop Science Society of South Australia, Inc., 1990), pp. 201-204.
33. Tucker, E. S., and S. B. Powles. "A Biotype of Hare Barley (*Hordeum leporinum*) Resistant to Paraquat and Diquat," *Weed Sci*. 39:159-162 (1991).
34. Gressel, J., and L. A. Segel. "Modelling the Effectiveness of Herbicide Rotations and Mixtures as Strategies to Delay or Preclude Resistance," *Weed Technol*. 4:186-198 (1990).
35. Maxwell, B. D., M. L. Rouch, and S. R. Radosevich. "Predicting the Evolution and Dynamics of Herbicide Resistance in Weed Populations," *Weed Technol*. 4:2-13 (1990).
36. Putwain, P. D., and A. M. Mortimer. "The Resistance of Weeds to Herbicides: Rational Approaches for Containment of a Growing Problem," *Proceedings of the British Crop Protection Conference — Weeds*. (Farnham, U.K.: The British Crop Protection Council, 1989), pp. 285-294.
37. Price, S., J. Hill, and R. Allard. "Genetic Variability for Herbicide Reaction in Plant Populations," *Weed Sci*. 31:652-657 (1983).

38. Price, S., R. Allard, J. Hill, and J. Naylor. "Associations Between Discrete Genetic Loci and Genetic Variability for Herbicide Reaction in Plant Populations," *Weed Sci.* 33:650-653 (1985).
39. Crow, J. F., and M. Kimura. *An Introduction to Population Genetics Theory*, (New York: Harper and Row, 1970), p. 591.
40. May, R. M., and A. P. Dobson. "Population Dynamics and the Rate of Evolution of Pesticide Resistance," In *Pesticide Resistance: Strategies and Tactics for Management*, (Washington, D.C.: Natl. Acad. Press, 1986), pp. 170-193.
41. Powles, S. B., and J. Matthews. "Multiple Herbicide Resistance in Annual Ryegrass (*Lolium rigidum*): A Driving Force for the Adoption of Integrated Weed Management," In *Resistance '91: Achievements and Developments in Combating Pesticide Resistance*, I. Denholm, A. L. Devonshire, and D. W. Hollomon, Eds. (London: Elsevier Applied Science, 1991), pp. 75-87.
42. Falconer, D. S. *Introduction to Quantitative Genetics 2nd Edition*, (London: Longman, 1981).
43. Maxwell, B. D., M. L. Roush, and S. R. Radosevich. "Prevention and Management of Herbicide Resistant Weeds," *Proceedings of the 9th Australian Weeds Conference*, J. W. Heap, Ed. (Adelaide: Crop Science Society of South Australia, Inc., 1990), pp. 260-267.
44. Schaal, B. A. "Population Structure and Local Differentiation in *Liatris cylindracea*," *Am. Nat.* 109:511-528 (1975).
45. Snaydon, R. W., and M. S. Davies. "Rapid Population Differentiation in a Mosaic Environment. IV. Populations of *Anthoxanthum odoratum* at Sharp Boundaries," *Heredity* 37:9-25 (1976).
46. Ghersa, C. M., M. A. Ghersa, M. L. Roush, and S. R. Radosevich. "Fitness Studies of Italian Ryegrass Resistant to Diclofop-methyl: 2. Pollen Phenology and Gene Flow," Proceedings of the Western Society of Weed Science (Salt Lake City, Utah, 1991) 44:42.
47. Mattsson, O., R. B. Knox, J. Heslop-Harrison, and Y. Heslop-Harrison. "Protein Pellicle of Stigmatic Papillae as a Probable Recognition Site in Incompatibility Reactions," *Nature* 247:298-300 (1974).
48. Levin, D. A., and H. W. Kerster. " Gene Flow in Seed Plants," *Evol. Biol.* 7:139–220 (1974)
49. Manasse, R., and P. Kareiva. "Quantifying the Spread of Recombinant Genes and Organisms," In *Assessing Ecological Risks of Biotechnology*, L. Ginzburg, Ed. (Boston: Butterworth-Heinemann, 1991), pp. 215-231.
50. Morris, W. F. "Predicting the Consequence of Plant Spacing and Biased Movement for Pollen Dispersal by Honey Bees," *Ecology* 74:493-500 (1993).
51. Weinberger, H. F. "Asymptotic Behavior of a Model in Population Genetics," *Lect. Notes Math.* 648:47-96 (1978).
52. Manasse, R. "Ecological Risk of Transgenic Plants: Effects of Spatial Dispersion on Gene Flow," *Ecol. Appl.* 2:431-438 (1992).
53. Maxwell, B. D. "Predicting Gene Flow from Herbicide Resistant Weeds in Annual Agriculture Systems," *Bull. Ecol. Soc. Am. (Abstracts)* 73:264 (1992).
54. Griffiths, D. J. "The Liability of Seed Crops of Perennial Ryegrass (*Lolium perenne*) to Contamination by Wind-borne Pollen," *J. Agric. Sci.* 40:19-38 (1950).
55. Copeland, L. O., and E. E. Hardin. "Outcrossing in Ryegrass (*Lolium* spp.) as Determined by Fluorescence Tests," *Crop Sci.* 10:254-257 (1970).

56. Handel, S. "Pollination Ecology, Plant Population Structure, and Gene Flow," In *Pollination Biology*, L. Real, Ed. (Orlando: Academic Press, 1983), pp.163-211.
57. Cambell, D. R., and N. M. Waser. "Variation in Pollen Flow Within and Among Populations of *Ipomopsis aggregata*," *Evolution* 39:418-431 (1989).
58. Maxwell, B. D., M. L. Roush, and S. R. Radosevich. "Predicting the Evolution and Dynamics of Herbicide Resistance in Weed Populations," *Weed Technol.* 4:2-13 (1990).
59. Fitt, D. L., P. H. Gregory, A. D. Todd, H. A. McCartney, and O. C. Macdonald. "Spore Dispersal and Plant Disease Gradients; A Comparison Between Two Empirical Models," *J. Phytopathol.* 118:227-242 (1987).
60. Mundt, C. C., and L. S. Brophy. "Influence of Number of Host Genotype Units on the Effectiveness of Host Mixtures for Disease Control: A Modeling Approach," *Phytopathology* 78:1087-1094 (1988).
61. McCanny, S. J., and P. B. Cavers. "Spread of Proso Millet (*Panicum miliaceum*) in Ontario, Canada. 2. Dispersal by Combines," *Weed Res.* 28:67-72 (1988).
62. Maxwell, B. D., and C. Ghersa. "The Influence of Weed Seed Dispersion Versus the Effect of Competition on Crop Yield," *Weed Technol.* 6:196-204 (1992).
63. Howard, C. L., A. M. Mortimer, P. Gould, P. D. Putwain, R. Cousens, and G. W. Cussans. "The Dispersal of Weeds: Seed Movement in Arable Agriculture," Brighton Crop Protection Conference — Weeds (1991), pp. 821-828.
64. Cullis, C. A. "DNA Rearrangements in Response to Environmental Stress," *Adv. Genet.* 28:73-97 (1990).
65. Pirozynski, K. A. "Coevolution by Horizontal Gene Transfer: a Speculation on the Role of Fungi." In *Coevolution of Fungi with Plants and Animals*, K. A. Pirozynski and D. L. Hawksworth, Eds. (London: Academic Press, 1988), pp. 247-268.
66. Ream, W. "*Agrobacterium tumefaciens* and Interkingdom Genetic Exchange," *Annu. Rev. Phytopathol.* 27:583-618 (1989).
67. Bryngelsson, T., M. Gustafsson, B. Green, and C. Lind. "Uptake of Host DNA by the Parasitic Fungus *Plasmodiophora brassicae*," *Physiol. Mol. Plant Pathol.* 33:163-171 (1988).
68. Shain, L., and J. B. Miller. "Movement of Cytoplasmic Hypovirulence Agents in Chestnut Blight Cankers," *Can. J. Bot.* 70:557-561 (1992).
69. Newman, E. I. "Mycorrhizal Links Between Plants: their Functioning and Ecological Significance," *Adv. Ecol. Res.* 18:243-270 (1988).
70. Cook, L. M. *Coefficients of Natural Selection* (London: Hutchinson, 1971).
71. White, R. J., and R. M. White. "Some Numerical Methods for the Study of Genetic Changes," In *Genetic Consequences of Man Made Change*, J. A. Bishop and L. M. Cook, Eds. (London: Academic Press, 1981), pp. 295-341.
72. Holt, J. S. "Fitness and Ecological Adaptability of Herbicide-resistant Biotypes," In *Managing Resistance to Agrochemicals — From Fundamental Research to Practical Strategies*, (Washington, D.C.: American Chemical Society Symposium Series 421, ASC Books, 1990), pp. 419-429.
73. Ahrens, W. H., and E. W. Stoller. "Competition, Growth Rate, and CO_2 Fixation in Triazine-susceptible and -resistant Smooth Pigweed (*Amaranthus hybridus*)," *Weed Sci.* 31:438-444 (1983).
74. Conard, S. G., and S. R. Radosevich. "Ecological Fitness of *Senecio vulgaris* and *Amaranthus retroflexus* Biotypes Susceptible or Resistant to Triazine," *J. Appl. Ecol.* 16:171-177 (1979).

75. Holt, J. S., and S. R. Radosevich. "Differential Growth of Two Common Groundsel (*Senecio vulgaris*) Biotypes," *Weed Sci.* 31:112-115 (1983).
76. Haldane, J. B. S. "More Precise Expressions for the Cost of Natural Selection," *J. Genet.* 57:351-360 (1960).
77. Watson, D., A. M. Mortimer, and P. D. Putwain. "The Seed Bank Dynamics of Triazine Resistant and Susceptible Biotypes of *Senecio vulgaris* — Implications for Control Strategies," *Proceedings of the British Crop Protection Conference — Weeds.* (Farnham, U.K.: The British Crop Protection Council, 1987), pp. 917-925.
78. Roush, M. L., S. R. Radesovich, and B. D. Maxwell. "Future Outlook for Herbicide Resistance Research," *Weed Technol.* 4:208-214 (1990).
79. Jain, S. K., and K. N. Rai. "Population Biology of *Avena*. VIII. Colonization as a Test of the Role of Natural Selection in Population Divergence," *Am. J. Bot.* 67:1342-1346 (1980).
80. Guttieri, M. J., C. V. Eberlein, C. A. Mallory-Smith, D. C. Thill, and D. L. Hoffman. "DNA Sequence Variation in Domain A of the Acetolactate Synthase Genes of Herbicide-resistant and -susceptible Biotypes," *Weed Sci.* 40:670-676 (1992).
81. Darmency, H., and J. Gasquez. "Inheritance of Triazine Resistance in *Poa annua*: Consequences for Population Dynamics," *New Phytol.* 89:487-493 (1981).
82. Gasquez, J., and J. P. Compoint. "Enzymatic Variation in Populations of *Chenopodium album* Resistant and Susceptible to Triazine," *Agroecosystem* 7:1-10 (1981).
83. Warwick, S. I., and P. B. Marriage. "Geographical Variation in Populations of *Chenopodium album* Resistant and Susceptible to Atrazine. I. Between and Within-Population Variation in Growth Response to Atrazine," *Can. J. Bot.* 60:483-493 (1982).
84. Jana, S., and J. M. Naylor. "Adaptation for Herbicide Tolerance in Populations of *Avena fatua*," *Can. J. Bot.* 60:1611-1617 (1982).
85. Somody, C. N., J. D. Nalewaja, and S. D. Miller. "Wild Oat (*Avena fatua*) and *Avena sterilis* Morphological Characteristics and Response to Herbicides," *Weed Sci.* 32:353-359 (1984).
86. Matthews, J. M., and S. B. Powles. Unpublished data.
87. Tardif, F. J., J. A. M. Holtum, and S. B. Powles. "Occurrence of a Herbicide-Resistant Acetyl-Coenzyme A Carboxylase Mutant in Annual Ryegrass (*Lolium rigidum*) Selected by Sethoxydim," *Planta* 190:176-181 (1993).
88. Gressel, J., and Segel, L. A. "Herbicide rotations and mixtures: effective strategies to delay resistance. In *Fundamental and Practical Approaches to Combating Resistance,* Green, M. B. et al., Eds., American Chemical Society Series (Washington, D.C.: American Chemical Society, 1991).

CHAPTER 2

Resistance to Photosystem II Inhibiting Herbicides*

John W. Gronwald

INTRODUCTION

As early as the 1950s, there were predictions that herbicide resistance would eventually develop in weeds.[1-3] At that time, pesticide resistance had already appeared in insects and pathogens. However, it was not until 1968 that the first herbicide-resistant weed was discovered; a biotype of *Senecio vulgaris* L. that was no longer controlled by the PS II inhibitor simazine.[4]

In the 25 years since the discovery of the simazine-resistant *S. vulgaris*, there have been numerous reports of weed biotypes exhibiting resistance to PS II inhibiting herbicides.[5-9] Triazine resistance is the most prevalent type of herbicide resistance found in weeds.[7-9] A worldwide survey completed in 1989 indicated that 57 weed species (40 dicots, 17 monocots) had biotypes exhibiting triazine resistance.[9] Worldwide, it is estimated that there are approximately three million hectares infested with triazine-resistant weeds.[7] Most of the resistant biotypes are found in fields in North America, Canada, and Europe that have been in continuous maize monoculture and where triazines have been applied repeatedly for several years.[8-10]

---

*Mention of a trademark, vendor, or proprietary product does not constitute a guarantee of warranty of the product by the U.S. Department of Agriculture or the University of Minnesota and does not imply its approval to the exclusion of other products or vendors that may be suitable. Minnesota Agricultural Experiment Station publication No. 20,751.

With few exceptions, resistance to PS II inhibiting herbicides is due to a modification at the target site, the D1 protein of the PS II complex.[8] In these cases, the weeds have "evolved" a resistance mechanism different from that of the crops in which they are found. The crops are resistant because they can detoxify these herbicides.

In the 1980s, a different type of resistance to PS II herbicides appeared. In *Alopecurus myosuroides* Huds. biotypes found in England[11,12] and Germany,[13] *Lolium rigidum* Gaud. biotypes found in Australia,[14,15] and an *Abutilon theophrasti* Medic. biotype found in Maryland,[16] resistance is due to enhanced herbicide detoxification. In these cases, the resistant weeds "evolved" detoxification mechanisms similar to those found in resistant crops.

This review will discuss both mechanisms of resistance to PS II inhibiting herbicides. The literature regarding resistance due to a modified target site is quite extensive and there have been several general reviews of this topic.[17-26] The author will review this area highlighting selected topics and new developments. A major focus of this review will concern resistance due to enhanced detoxification since this has not been covered previously.

MODE OF ACTION

In the 1950s, the phenylurea monuron and the *s*–chloro–triazine simazine were found to inhibit the Hill reaction in isolated chloroplasts.[27-30] During the following decade, it was discovered that these chemicals block the Hill reaction by inhibiting electron transport on the reducing side of PS II.[31-33] Subsequently, a number of other chemicals were found to act in the same manner. These include the uracils, triazinones, biscarbamates, nitriles, nitrophenols, substituted pyridazinones, phenylcarbamates, anilides, and cyanoacrylates.[21,25,34] Some of these also have other modes of action. For example, the nitrophenols and nitriles act as uncouplers and the substituted pyridazinones inhibit fatty acid desaturation and carotenoid biosynthesis.[25]

Photosystem II

During the past decade, considerable progress has been made in understanding photosynthetic electron transport associated with PS II.[35,36] In higher plants, PS II is a multiprotein complex located primarily in the grana or appressed lamellae and consists of at least 23 different subunits.[36] Current knowledge of PS II in higher plants is in large part due to the elucidation of the three-dimensional structure of the purple photosynthetic bacterium (*Rhodopseudomonas viridis*) reaction center by X-ray crystallography.[37-39] The reaction center of PS II is analogous in structure and function to the reaction center of *R. viridus*. A simplified schematic of the PS II reaction center is illustrated in Figure 1. Two integral membrane proteins, D1 and D2, form the reaction center. Associated with the D1/D2 heterodimer are bound redox components that are involved in

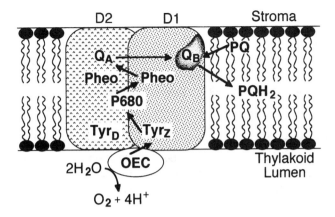

Figure 1. Simplified schematic of the photosystem II reaction center complex. Modified and redrawn from Fuerst and Norman.[25] Arrows indicate the direction of electron flow. Abbreviations: OEC, oxygen-evolving complex; Tyr_z, tyrosine residue 161 that acts as electron donor to P680; P680, reaction center chlorophyll *a* dimer; Pheo, pheophytin; Q_A, plastoquinone tightly bound to the D2 protein; Q_B, exchangeable plastoquinone bound to the D1 protein; PQ, plastoquinone; PQH_2, plastohydroquinone. Other proteins associated with the reaction center complex such as cyt b559 and light-harvesting chlorophyll proteins are not shown.

electon transfer. These include the reaction center chlorophyll P680 (a special chlorophyll a dimer), pheophytin (the primary electron acceptor), nonheme iron, and the bound plastoquinones Q_A and Q_B that serve as secondary electron acceptors. Q_A is tightly bound to the D2 protein while Q_B is an exchangeable plastoquinone located in a binding niche on the D1 protein.

The PS II complex catalyzes the light-dependent oxidation of water with the reduction of plastoquinone.[36] Light energy absorbed by the chlorophyll-protein molecules of the light harvesting complex is transferred to the reaction center chlorophyll, P680, causing a charge separation. An electron is passed to pheophytin and then to Q_A on the D2 protein. The P680$^+$ chlorophyll receives an electron from a tyrosine residue (referred to as Z or Tyr_z) on the D1 protein which in turn receives an electron from water via the oxygen-evolving complex. Two electrons are transferred sequentially from Q_A to Q_B via a nonheme iron. The first electron forms the semiquinone anion of the bound plastoquinone. Upon transfer of the second electron and the acceptance of two protons from the stroma, fully reduced plastoquinone (plastohydroquinone) is formed. This molecule leaves the Q_B site on the D1 protein, diffuses across the thylakoid membrane, and donates two electrons to the cytochrome b_6/f complex with the release of two protons into the thylakoid lumen. Electrons from the cytochrome b_6/f complex are then transferred to photosystem I via plastocyanin.

The D1 polypeptide (also referred to as the 32-kDa protein or the Q_B protein) has been extensively investigated. This highly conserved protein is encoded by the chloroplast *psb*A gene.[40] Based on a hydropathy index plot and the homology in amino acid sequence between the D1 protein and the L subunit of the purple photosynthetic bacteria reaction center, a model describing the folding of the D1 protein in thylakoid membranes has been proposed (Figure 2).[41,42] According to this model, the D1 protein has five transmembrane segments. The Q_B binding niche is formed by the short, parallel alpha helix that connects transmembrane helices IV and V. In the light, the D1 protein undergoes a continuous cycle of degradation and resynthesis with a half-life ranging from 30 min to a few hours depending on conditions.[36,43-45] The continuous turnover of the D1 protein is necessary because of photoinhibition (i.e., photodamage) that occurs during the normal operation of the PS II reaction center.[36,42,46] Under conditions that promote photoinhibition (e.g., strong illumination), the D1 protein incurs greater photodamage and turns over at a faster rate.[42,46-49]

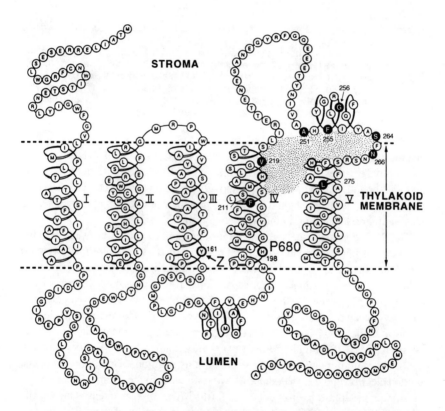

Figure 2. Model of the folding of the D1 polypeptide from spinach. Adapted from Barber and Andersson.[42] The model is based on a hydropathy index plot of the amino acid sequence from the D1 polypeptide of spinach and the homology in amino acid sequence between the D1 protein and the L subunit of purple photosynthetic bacteria. Amino acid residues indicated by single letter code. Amino acid residues modified in herbicide-resistant mutants of algae, cyanobacteria, and higher plants are depicted as darkened circles. General domain for binding of Q_B and PS II inhibiting herbicides indicated by stippled area. His 215 and His 272 form ligands to the nonheme iron and His 190 binds P680. Z (tyrosine residue 161) acts as the primary electron donor to oxidized P680.

Most evidence suggests that photoinhibition is due to the formation of radicals or singlet oxygen by redox components associated with the D1 protein.[36,42] The degradation of the photodamaged D1 protein involves proteolysis.[36,43] Recent evidence suggests that the PS II complex contains a serine-type endoprotease that degrades this protein.[50] There is controversy concerning the specific sites of proteolytic cleavage on the D1 protein, but it appears that one initial cleavage site is located near the Q_B binding niche.[46,51,52]

Target Site

In 1977, Tischer and Strotmann[53] reported that the diverse chemicals that act as PS II inhibitors (triazines, phenylureas, pyridazinones, biscarbamates) compete for a common binding site on thylakoid membranes. Later, photoaffinity labeling studies conducted with [^{14}C]azido derivatives of atrazine,[54] monuron,[55] and a triazinone[56] demonstrated that this common binding site was a 32-kDa protein which we now know to be the D1 protein.

In 1981, Velthuys[57] hypothesized that PS II herbicides act by competing for the plastoquinone binding site thereby blocking electron transport on the reducing side of PS II. Supporting evidence was provided by Vermaas et al.,[58] who showed that an azido-quinone derivative, which competed with plastoquinone for a binding site on thylakoid membranes, prevented atrazine and ioxynil binding. Other research indicated that short-chain plastoquinol analogues competed with diuron for a binding site on chloroplast thylakoids.[59] These and other reports[60,61] have provided strong evidence that PS II inhibiting herbicides displace plastoquinone at the Q_B binding site on the D1 protein and thereby block electron flow from Q_A to Q_B. On the basis of a stereochemical model, Gardner[62] proposed that the triazines and ureas act as nonreducible analogues of plastoquinone while the phenol-type herbicides (nitriles, dinitrophenols) act as nonreducible analogues of the semiquinone anion of plastoquinone.

Various models defining the topology of the Q_B binding niche and the interactions of plastoquinone and PS II herbicides at this site have been proposed.[41,63-66] These models are based on knowledge of: (1) the interaction of quinones and herbicides with the three-dimensional structure of the reaction center of purple photosynthetic bacteria, (2) photoaffinity labeling of residues in the Q_B niche with radiolabeled azido herbicides, and (3) investigations with herbicide-resistant mutants.

According to current models, plastoquinone binding in the Q_B niche involves hydrogen bonding between the carbonyl oxygens of plastoquinone and the amide backbone of His 215 and the hydroxyl of Ser 264 (Figure 3A).[25,41,65] Phe 265 can also hydrogen bond to the same carbonyl as Ser 264. Atrazine binding in the Q_B niche appears to be due to hydrogen bonding with Ser 264 and Phe 265 as well as hydrophobic interactions with Phe 255 (Figure 3B). In the case of the phenol-type herbicides, the nature of the interaction at the Q_B site is not as well characterized, but it is believed that a major component of the binding involves an interaction with His 215. Trebst[41] categorized herbicides that bind

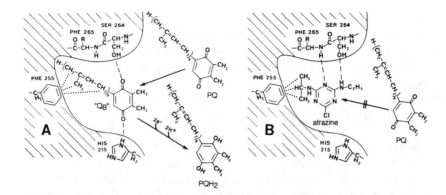

Figure 3. Schematic of the interaction of plastoquinone (A) and atrazine (B) in the Q_B-binding niche of the D1 protein. From Fuerst and Norman.[25] Hydrogen bonds and hydrophobic interactions are represented by dashed and dotted lines, respectively. Atrazine binding in the Q_B niche prevents the binding of plastoquinone. Abbreviations: PQ, plastoquinone; Q_B, bound plastoquinone; PQH_2, plastohydroquinone.

in the Q_B niche into two families based on their interaction with amino acids at this site: the urea/triazine family that exhibits a strong interaction with Ser 264 and the phenol family that interacts strongly with His 215.

Herbicidal Activity

PS II inhibitors block photosynthetic electron transport and hence prevent the reduction of $NADP^+$ required for CO_2 fixation. However, the herbicidal activity of these chemistries is not due to the interruption of photosynthesis, but instead is due to the oxidative stress generated when photosynthetic electron transport is blocked. The net effect of blocking electron transport is the destruction of the PS II reaction center and the photooxidation of lipid and chlorophyll molecules.[67,68] Chloroplasts have multiple mechanisms to scavenge toxic oxygen species, but in this case they are overwhelmed by the level of oxidative stress generated. In addition to generating oxidative stress, it has also been proposed that the herbicidal activity of PS II inhibiting herbicides is partly due to interference with the normal turnover of the D1 protein.[35,69-71] As described above, the D1 protein undergoes rapid, light-dependent turnover to repair oxidative damage that occurs during normal operation of the PS II reaction center. The binding of diuron and atrazine at the Q_B site interferes with the degradation of the D1 protein.[35,69-71] Herbicide binding may block the access of a protease to the cleavage site near the Q_B niche, or it may cause a conformational change that restricts accessibility to this site. According to this hypothesis, the binding of the triazines and ureas at the Q_B niche not only causes photodamage to the D1 protein by preventing electron transfer from Q_A to Q_B, but it compounds herbicidal activity by preventing replacement of the damaged D1 protein.

SELECTIVITY MECHANISMS IN CROP PLANTS

In most cases, crop resistance to certain herbicides is due to the ability of the crop to metabolize the herbicides and thereby prevent injury.[72] Two enzymes that play major roles in conferring resistance to PS II herbicides (as well as numerous other herbicides) are glutathione S-transferase (GST) and cytochrome P450 (Cyt P450, also referred to as mixed function oxidase).[72-76] GSTs are soluble dimeric enzymes that catalyze nucleophilic attack by the thiolate anion of glutathione at electrophilic sites on certain herbicides.[72-74] The formation of the glutathione conjugate detoxifies the herbicide. Cyt P450, an enzyme complex associated with the endoplasmic reticulum, catalyzes the monooxygenation of various substrates, including certain herbicides.[75-76] The monooxygenation or phase I reactions catalyzed by Cyt P450 include N-dealkylation, aryl hydroxylation, alkyl hydroxylation, sulfoxidation, and O-dealkyation. These phase I reactions are usually followed by phase II reactions in which the phase I metabolite is conjugated with water-soluble molecules such as glucose or glutathione. Two examples illustrating the role of GST and Cyt P450 in the metabolism of PS II herbicides (atrazine, chlortoluron) are discussed below.

Maize is highly resistant to s-chloro-triazines such as atrazine and simazine because of multiple mechanisms of detoxification (Figure 4).[72] Rapid detoxification of atrazine via glutathione conjugation is the primary mechanism of atrazine resistance. Maize has three GST isozymes capable of catalyzing atrazine conjugation with glutathione.[77] Atrazine can also be metabolized via N-dealkylation. Mono-N-dealkylation reduces the binding affinity of atrazine for the D1 protein and thereby partially reduces its phytotoxicity.[72] However, removal of both N-alkyl groups is necessary for complete detoxification. As a mechanism for conferring herbicide resistance, N-dealkylation is considerably less efficient than glutathione conjugation. Although Cyt P450 is thought to catalyze N-dealkylation of the s-chloro-triazines in maize, as well as in moderately tolerant species such as pea, this has not been demonstrated *in vitro*. In maize, atrazine can also be detoxified by a nonenzymatic reaction with the cyclic hydroxamate DIMBOA (2,4-dihydroxy-7-methoxy-1,4(2H)-benzoxazin-3(4H)-one) which is found in the roots of this species. This mechanism makes a minor contribution to atrazine resistance and generally only when herbicide uptake occurs via roots.

The substituted phenylurea chlortoluron has been used for over 20 years in Europe for selective weed control in wheat.[78] Wheat can detoxify chlortoluron via two routes of oxidative metabolism; aryl (ring-methyl) hydroxylation and N-dealkylation (Figure 5).[79-83] N-dealkylation makes a relatively minor contribution toward chlortoluron resistance. Complete detoxification of chlortoluron via N-dealkylation requires the removal of both methyl groups. Metabolism via ring-methyl hydroxylation is primarily responsible for chlortoluron resistance in wheat.[81-83] There is strong evidence that ring-methyl hydroxylation of chlortoluron is mediated by Cyt P450.[81-86] N-dealkylation also appears to be mediated by Cyt

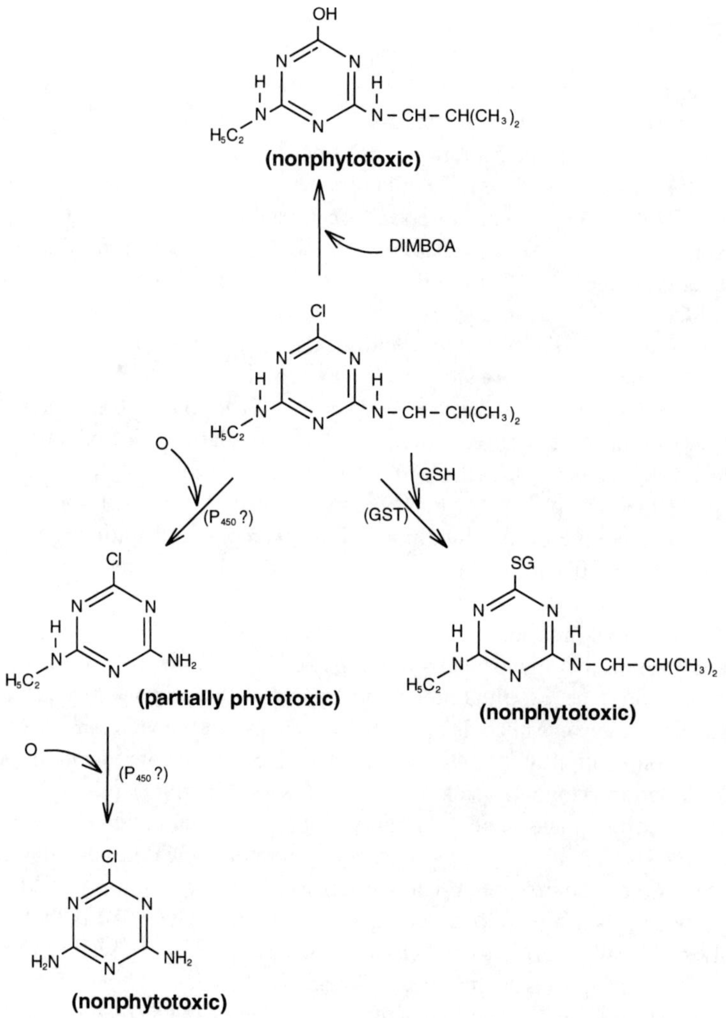

Figure 4. Metabolism of atrazine by maize. From Shimabukuro.[72] Abbreviations: DIMBOA, (2,4-dihydroxy-7-methoxy-1,4(2H)-benzoxazin-3(4H)-one); GST, glutathione S-transferase; GSH, glutathione, P-450, Cyt P450; O, oxygen.

P450, but it is possible that this reaction is mediated by a peroxygenase.[85,86] Evidence for Cyt P450 involvement in chlortoluron metabolism in wheat is based in part on the synergistic effect of the Cyt P450 inhibitor 1-aminobenzotriazole (ABT) on chlortoluron phytotoxicity to wheat.[81-84] Additional evidence comes from *in vitro* studies with isolated microsomal fractions from wheat. Monooxygenation of chlortoluron in these fractions exhibits characteristics typical of Cyt P450-mediated reactions.[84-86]

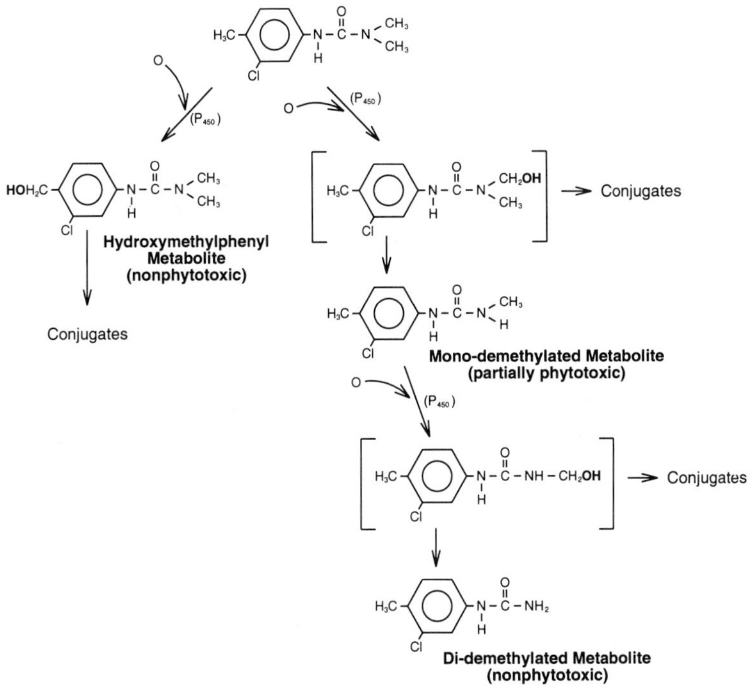

Figure 5. Metabolism of chlortoluron in wheat. From Gonneau et al.[84] Abbreviations: P-450, cytochrome P450; O, oxygen. Unstable compounds are indicated in brackets. The products of oxidative metabolism undergo conjugation with glucose.

MECHANISMS OF RESISTANCE IN WEEDS

Herbicide resistance in weeds can be due to a number of mechanisms: (1) modified target site, (2) enhanced detoxification (metabolism), (3) reduced absorption/translocation, (4) sequestration or compartmentation, and (5) repair of the toxic effects of herbicides.[14] To date, only two mechanisms of resistance to PS II inhibiting herbicides have been identified; modified target site and enhanced detoxification.

Target Site Based

Mechanism

Of the 57 weed species exhibiting biotypes resistant to PS II inhibitors (primarily the *s*-triazines), all but a few cases are resistant due to a modification at the target site.[8] Research conducted with triazine-resistant weeds during the 1980s clearly established that target site-based resistance is due to a modification of amino acid residues in the Q_B-binding niche on the D1 protein.[20,22,25] This modification reduces the affinity of PS II herbicides at this site so that they can no longer effectively compete for the exchangeable plastoquinone Q_B.

In all cases of target site resistance that have occurred in the field, resistance is due to a point mutation of the *psb*A gene resulting in a substitution of Gly for Ser at residue 264.[20,22,25] This modification greatly reduces the affinity of atrazine at the Q_B-binding site, since binding affinity is strongly dependent on hydrogen bonding with the hydroxyl side chain of Ser 264 (Figure 3B). Typically, this modification causes a 1000-fold reduction in atrazine affinity at the Q_B-binding site and greater than a 100-fold increase in atrazine resistance at the whole plant level.[87,88] Although the Ser 264 to Gly mutation confers a high level of resistance to the *s*-triazines and moderate resistance to the triazinones, it confers little or no increase in resistance to the phenylurea diuron.[87,88] Presumably this is because the binding determinants for diuron in the Q_B pocket are not strongly dependent on the interaction with Ser 264.

Associated with the Ser 264 to Gly mutation are a number of pleiotropic effects. Triazine-resistant weeds exhibit a modified galactolipid composition and an increase in the degree of unsaturation of fatty acids.[89,90] In addition, plants with this mutation exhibit altered chloroplast ultrastructure.[91,92] Chloroplasts in triazine-resistant weeds are similar to "shade chloroplasts" which develop under low light intensities. Compared to triazine-susceptible chloroplasts, the chloroplasts of triazine-resistant biotypes exhibit increased grana stacking and a reduced chlorophyll *a*/*b* ratio.

It is also well established that the Ser 264 to Gly mutation reduces the rate of electron transfer between Q_A and Q_B.[93,94] An earlier report indicated that electron transfer between Q_A and Q_B was reduced by approximately 10-fold.[93] However, a more recent study involving several species and the reanalysis of previous data suggests that the Ser 264 to Gly mutation reduces the rate of electron transfer between Q_A and Q_B by about threefold.[94]

The Ser 264 to Gly mutation of *psb*A results in increased sensitivity (i.e., negative cross resistance) to phenol-type herbicides and, in some cases, to other herbicides such as the benzothiadiazinone bentazon.[95-101] Most research indicating negative cross resistance involved *in vitro* studies with isolated thylakoids. However, in a few instances, this phenomenon has also been observed at the whole plant level.[88,100,101] The mechanism underlying negative cross resistance is not fully understood. In the case of the phenol-type herbicides, the altered lipid composition of the thylakoids of atrazine-resistant biotypes may influence accessibility to the Q_B niche[96] or the substitution of Gly for Ser at residue 264 may increase binding affinity at the Q_B niche.[101] A recent report indicated that the hydroxyl group of Ser 264 has a destabilizing effect on the binding of ioxynil, a phenol-type herbicide.[102]

In the field, only the Ser 264 to Gly mutation has been detected.[25] However, in tobacco and potato, plants with a Ser 264 to Thr mutation of the *psb*A gene have been selected in tissue culture.[103,104] In contrast to the Ser 264 to Gly mutation, the Ser to Thr mutation confers resistance to both atrazine and diuron. Other mutations conferring resistance to PS II herbicides have been selected in algae (*Chlamydomonas reinhardtii*, *Euglena gracilis*) and cyanobacterium.[22] Eight different amino acid substitutions that confer resistance have been identified in or near the Q_B-binding niche (between residues 211–275) (Table 1, Figure 2) in algae, cyanobacterium, and higher plants. Depending on the residue altered, different levels and patterns of resistance are observed.

Ecological Fitness

As fully discussed in Chapter 11, most studies indicate that the Ser 264 to Gly mutation of the *psbA* gene causes a significant reduction in relative ecological fitness of plants carrying this mutation. Triazine-resistant biotypes with this mutation exhibit a reduction in CO_2 fixation, quantum yield, and seed and biomass production.[93,105-114]

There are, however, a few studies where differences in fitness were not observed or where the resistant biotype was found to be more fit.[115,116] Many of the earlier studies evaluating the effect of the *psbA* (Ser 264 to Gly) mutation on fitness were conducted with biotypes that contained different nuclear genomes.

Table 1. Amino acid changes in the D1 protein of various herbicide–resistant mutants and the type of resistance conferred

Amino acid change	Resistant to:	Organism
Single Mutations:		
Phe 211 —> Ser	Atrazine/DCMU	*Synechococcus*
Val 219 —> Ile	Metribuzin/DCMU/ioxynil	*Chlamydomonas, Synechococcus*
Ala 251 —> Val	Metribuzin	*Chlamydomonas*
Phe 255 —> Tyr	Atrazine, cyanoacrylate	*Chlamydomonas, Synechococcus*
Gly 256 —> Asp	Atrazine/DCMU/bromacil	*Chlamydomonas*
Ser 264 —> Gly	Atrazine	*Amaranthus, Synechococcus*
Ser 264 —> Ala	Metribuzin/atrazine DCMU/bromacil Atrazine/DCMU	*Anacystis, Chlamydomonas, Euglena, Synechocystis, Synechococcus*
Ser 264 —> Thr	Triazine	*Nicotiana, Euglena*
Ser 264 —> Asn	Triazine	*Nicotiana*
Asn 266 —> Thr	Ioxynil	*Synechocystis*
Leu 275 —> Phe	Metribuzin/bromacil/DCMU	*Chlamydomonas*
Double mutations:		
Phe 255 —> Tyr and Ser 264 —> Ala	Urea/triazine	*Synechococcus*
Phe 255 —> Leu and Ser 264 —> Ala	DCMU plus reversal of atrazine tolerance	*Synechocystis*
Phe 211 —> Ser and Ala 251 —> Val	Atrazine	*Synechocystis*

From Trebst[22]

In some cases the R and S biotypes were collected from different geographical locations. Such studies do not offer unequivocal proof that the *psb*A mutation reduces fitness since differences in the nuclear genome have not been eliminated. However, recent investigations conducted with near isonuclear lines of *Brassica napus, S. vulgaris,* and *Solanum nigrum* clearly showed that the *psb*A gene mutation reduced fitness.[117-121] In addition, studies conducted with nearly isonuclear and nonisonuclear biotypes of *S. vulgaris* demonstrated that differences in the nuclear genome can compensate to some extent for reduced fitness conferred by the *psb*A mutation.[118,119] Therefore, results of fitness studies which do not use near-isogenic lines should be interpreted with caution (see Chapter 11).

It is also evident that the effect of the *psb*A (Ser 264 to Gly) mutation on relative fitness is influenced by the environment, especially temperature. Several reports have indicated that triazine resistance is associated with increased sensitivity to high temperatures. With increasing temperature, biomass production and photosynthetic efficiency were reduced to a greater extent in triazine R compared to S biotypes.[122-129] This differential effect of temperature has been attributed to the *psb*A (Ser 264 to Gly) mutation which reduces the rate of electron transfer from Q_A to Q_B in PS II.[125] At high temperatures, net carbon assimilation was limited by electron transport in a resistant *B. napus* biotype, but not in the S biotype.[128] The altered lipid composition of the thylakoids of triazine R biotypes may also play a role in this differential temperature response.[127]

Although there are a number of studies which conclude that the *psb*A (Ser 264 to Gly) mutation increases heat sensitivity of R biotypes, two recent reports do not support this conclusion.[119,120] In research conducted with nearly isonuclear R and S biotypes of *S. vulgaris*,[119] and *S. nigrum*,[120] no differential effect of temperature on biomass production and photosynthetic efficiency of S versus R biotypes was observed. Furthermore, the study conducted with nearly isonuclear atrazine S and R *S. vulgaris* biotypes showed that the nuclear genome influenced temperature sensitivity.[119] These authors questioned earlier reports concerning the effect of the *psb*A mutation on temperature sensitivity since they were conducted with nonisonuclear triazine R and S biotypes.[119]

A recent report by Dekker and Burmester[130] indicated that relative fitness of R and S biotypes is dependent on the time of day and plant age. Net CO_2 assimilation was measured in nearly isonuclear triazine R and S biotypes of *B. napus* during a diurnal cycle. The experiments were conducted under controlled environmental conditions where photosynthetic photon flux density was increased and decreased on either side of a midday maximum. Young R plants (3- to 4-leaf stage) exhibited slightly greater photosynthetic rates during the early and late portions of the diurnal light period, whereas S plants exhibited slightly greater rates at midday and during the photoperiod as a whole (Figure 6A). However, with the onset of the reproductive phase (8 1/2- to 9 1/2-leaf stage), R plants assimilated more carbon than S plants during all periods of the diurnal cycle with the exception of the late part of the day (Figure 6B). These authors suggested that atrazine R *B. napus* may have

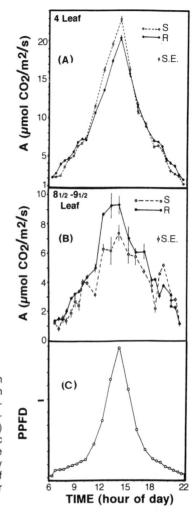

Figure 6. Change in photosynthetic carbon assimilation (μmol CO_2 m²/s) with time of day in resistant (—) and susceptible (- - -) *Brassica napus* in the 4-leaf (A) and 8 1/2- to 9 1/2-leaf (B) stages. The changing photosynthetic photon flux density (PPFD) during the photoperiod is shown in (C). Midday maximum PPFD was in the range of 1150 to 1300 μE m^{-2} s^{-1}. From Dekker and Burmester.[130]

an adaptive advantage over its S counterpart during the latter stages of plant ontogeny and under cool, low-light environments.

Although there is considerable evidence that the *psbA* (Ser 264 to Gly) mutation reduces plant fitness, there is a lack of agreement concerning the specific mechanism. As discussed above, the Ser 264 to Gly mutation reduces the rate of electron transfer between Q_A and Q_B.[93,94] However, it is not certain how this relates to the reduction in photosynthetic capacity and relative fitness. It has been argued that compared to other steps in the pathway of whole chain photosynthetic electron transport, the rate of electron transport from Q_A to Q_B is very rapid even in R chloroplasts and should not limit photosynthesis.[94] There are conflicting reports as to whether the Ser 264 to Gly mutation reduces whole chain electron transport rate in isolated chloroplasts.[106,108,113,116] However, an investigation conducted with whole leaves and thylakoids of nearly isonuclear atrazine R and S *B. napus* biotypes clearly showed that the slower rate of Q_A to Q_B electron transfer in the R biotype was responsible for lower quantum yield and maximum rate of photosynthesis.[110] The lower quantum yield in the R biotype was attributed to the inefficient use of separated charge in the PS II reaction center. Because of the altered kinetics of Q_A to Q_B electron transfer in atrazine R chloroplasts, over 30% of the charge separations recombined rather than being used in linear electron transport.

Recent evidence suggests that the altered kinetics of Q_A to Q_B electron transfer associated with the Ser 264 to Gly mutation increases susceptiblility of PS II to photoinhibition.[131,132] In a study with nearly isonuclear atrazine R and S lines of *B. napus*, Hart and Stemler[131,132] found that biomass production was similar in R and S biotypes under low photon flux densities. Furthermore, there were no differences in quantum yield between R and S biotypes under these conditions. However, at moderate to high light intensities, biomass production and quantum yield were reduced in the R biotype. Based on these and other results, the authors concluded that the reduced fitness of the R biotype under moderate to high light intensity reflected increased susceptibility to photoinhibition. The slower rate of Q_A to Q_B electron transfer in atrazine R biotypes may increase the likelihood of radical formation in the Q_B-binding niche or the PS II reaction center.[46] Increased susceptibility to photoinhibition would reduce fitness, particularly under high light environments, because of the high energy cost associated with the turnover of the D1 protein.[42]

Inheritance and Frequency of Resistance Genes

Chapter 10 fully considers the inheritance of triazine resistance. As expected for a mutation involving a chloroplast gene (*psbA*), triazine resistance is maternally inherited.[133-135] The expected incidence for a random recessive mutation of the *psbA* gene causing triazine resistance is very low (10^{-10} to 10^{-20}).[99,136] However, it appears that in some weed populations the incidence of atrazine resistance is considerably higher than this. It is possible that this mutation confers an adaptative advantage under certain environmental conditions and therefore is maintained in the gene pool at a higher than expected frequency. Dekker and Burmester[130] proposed that the greater than expected incidence of this mutation may reflect the advantage that the mutation confers under less favorable (cool, low light) environments.

Arntzen and Duesing[136] hypothesized that the higher than expected mutation rate of the *psbA* gene in a resistant *S. nigrum* biotype is due to the presence of a plastome mutator, a nuclear gene that increases the frequency of chloroplast DNA mutation. Plastome mutators have been hypothesized to exist in plants.[137] The mechanism whereby this putative gene increases chloroplast DNA mutation rate is not known, but it has been suggested that it codes for an error-prone protein involved in chloroplast DNA replication or repair.[136,137] According to the hypothesis of Arntzen and Duesing,[136] triazine resistance develops in those subpopulations of weeds that have the plastome mutator. One prediction based on this hypothesis is that triazine R biotypes would rapidly develop resistance to other PS II inhibitors since the presence of the mutator gene would increase the likelihood of other mutations in the Q_B-binding pocket. Solymosi and Lehoczki[138] proposed that the rapid appearance of resistance to pyridate and pyrazon in triazine R populations of *Chenopodium album* L. in Hungary was due to the presence of a plastome mutator.

Gasquez, Darmency, and colleagues[139-141] have observed that the appearance of atrazine resistance in certain populations of *C. album* involves a two-step process which depends on the presence of "R precursor" genotypes that exhibit a high random mutation rate (10^{-4} to 3×10^{-3}) of the *psb*A gene (see Chapter 10). It is proposed that this precursor genotype mutates in the absence of herbicides to form an "intermediate" biotype which has the Ser 264 to Gly mutation in the *psb*A gene. This "intermediate" biotype exhibits a high level of atrazine resistance (1000-fold) at the chloroplast level. However, for reasons still unclear, it exhibits only an intermediate level of atrazine resistance (10-fold) at the whole plant level. When treated with xenobiotics (organophosphates, carbamates, sublethal doses of atrazine), the "intermediate" biotype produces progeny that are highly resistant to atrazine both at the whole plant and target site levels. The mechanism underlying this two-step selection for resistance is unclear since the "intermediate" biotype exhibits the *psb*A (Ser 264 to Gly) mutation but does not express high levels of atrazine resistance at the whole plant level. Bettini et al.[139] suggested that the induced expression of a high level of whole plant resistance in the "intermediate" biotype involves an extrachloroplastidic genetic mechanism. Whether this phenomenon is expressed in other species is not known.

Enhanced Metabolism

During the 1980s, resistance to PS II inhibiting herbicides due to enhanced metabolism appeared in weeds.[11-16] To date, biotypes of three weed species exhibiting this type of resistance have been identified and partially characterized.

Abutilon theophrasti

In 1984, an *A. theophrasti* biotype resistant to recommended field rates of simazine and atrazine was found in Maryland.[142] This biotype was located in a field that had been in continuous no-till maize production for 6 years and that had been treated annually with simazine and/or atrazine. The approximately 200-ha farm on which the triazine-resistant *A. theophrasti* was found is now heavily infested with this weed and other herbicide classes are used to control it.[143]

The first clue that resistance to *s*-chloro-triazines in the Maryland *A. theophrasti* biotype was not due to a modification at the target site was the level of resistance. Compared to a susceptible *A. theophrasti* biotype, the Maryland biotype was only 10-fold more resistant to atrazine.[16] Typically, resistance due to a modified target site confers at least a 100-fold increase in atrazine resistance at the whole plant level.[25] The differential response of the R and S biotypes to hydroponic treatment with atrazine is shown in Figure 7. Although resistant to simazine and atrazine, the Maryland biotype is not cross resistant to other PS II inhibitors such as bentazon, cyanazine, linuron, and metribuzin.[144] Another indication that the Maryland biotype exhibits a novel form of atrazine resistance is the inheritance of this trait. Atrazine resistance was not maternally inherited, but instead is controlled by a single nuclear gene exhibiting partial dominance.[145]

Studies with isolated thylakoids indicated that atrazine was equally inhibitory to PS II electron transport in both R and S biotypes.[16] Further investigation revealed that resistance is due to an enhanced capacity to detoxify atrazine via glutathione conjugation. As compared to an atrazine S (wild type) biotype from Minnesota, the R Maryland biotype exhibited a sixfold greater rate of glutathione conjugation of atrazine. Hence, the resistant *A. theophrasti* biotype mimics the crop in which it appeared (maize) in its capacity to detoxify atrazine via glutathione conjugation (Figure 4). The enhanced rate of glutathione conjugation is not due to elevated glutathione content, but instead is due to the overexpression of two GST isozymes that exhibit activity with atrazine.[146]

Compared to an atrazine S *A. theophrasti* biotype from Minnesota, the atrazine R biotype from Maryland is somewhat less vigorous. It germinates later and shoot fresh weight is approximately 30% less during the early stages of vegetative growth.[16] However, because the R (Maryland) and S (Minnesota) biotypes are from different geographical locations, no conclusions can be drawn regarding the effect of overexpression of GST isozymes on fitness. In order to properly evaluate fitness, studies with S and R *A. theophrasti* biotypes from the same field in Maryland will be required.

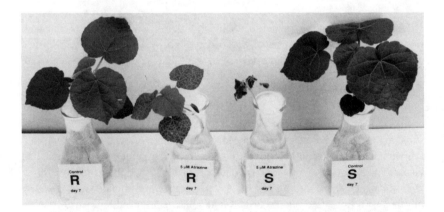

Figure 7. Comparison of the response of atrazine-resistant and -susceptible *Abutilon theophrasti* biotypes to 7 days of exposure to 5 μM atrazine via hydroponic solution. The I_{50} values for the inhibition of shoot growth in the atrazine-resistant and -susceptible biotypes were 3.0 and 0.3 μM, respectively.[16]

Lolium rigidum

L. rigidum is an ubiquitous weed found in the cropping regions of southern Australia.[14,147] Multiple resistance to several herbicide classes has been found in numerous *L. rigidum* biotypes throughout this region.[14,147] Two of these biotypes, WLR 2 and VLR 69, exhibit resistance to PS II inhibiting herbicides and have been partially characterized (see Chapter 9).

WLR 2 was discovered in western Australia along a railroad that had been treated annually for 10 years with 2.56 kg/ha atrazine plus 2.56 kg/ha amitrole.[15,148] Not only is WLR 2 resistant to these herbicides, but it is also cross resistant to several other PS II inhibitors: chloro-*s*-triazines (simazine, propazine, cyanazine), methylthio-*s*-triazines (prometryn, ametryn), phenylureas (chlortoluron, isoproturon, diruon), and the triazinone metribuzin.[15,148-150] The level of resistance varies depending on the particular herbicide, but is less than 10-fold. For example, as compared to a S biotype, WLR 2 is approximately 9-, 3-, and 7-fold more resistant to simazine, atrazine, and chlortoluron, respectively.[148-150]

VLR 69 was discovered in a perennial ryegrass (*Lolium perenne*) field that had been subjected to selection pressure by five herbicide classes over a period of 21 years.[148-150] The selection pressure included 17 years of diuron treatment and 5 years of atrazine treatment. VLR 69 exhibits resistance to PS II inhibitors (*s*-triazines, phenylureas, metribuzin), as well as aryloxyphenoxypropionates (diclofop, fluazifop, haloxyfop), cyclohexanediones (tralkoxydim), imidazolinones (imazaquin), and sulfonylureas (chlorsulfuron, triasulfuron). Similar to the WLR 2 biotype, VLR 69 exhibits about a 6-fold increase in resistance to simazine and chlortoluron.[148-150] Resistance to PS II inhibitors in biotypes WLR 2 and VLR 69 is due to enhanced detoxification via oxidative metabolism.[148-150] Both biotypes exhibit an enhanced capacity to metabolize simazine and chlortoluron via N-dealkylation. Both herbicides are metabolized (N-demethylated) approximately twice as fast in the R biotypes compared to S biotypes.

Indirect evidence suggests that enhanced N-dealkylation in the R biotypes (WLR 2, VLR 69) is mediated by Cyt P450(s).[148-150] In both biotypes, the enhanced rate of N-dealkylation of simazine and chlortoluron can be inhibited by the Cyt P450 inhibitor ABT. Furthermore, the R biotypes become susceptible if ABT or piperonyl butoxide (PBO) are applied with simazine or chlortoluron. The synergistic effect of ABT on the phytotoxicity of chlortoluron in biotype WLR 2 is shown in Figure 8. So far, however, Cyt P450 activity capable of metabolizing PS II herbicides *in vitro* cannot be demonstrated.[151]

Assuming the involvement of Cyt P450(s) in the enhanced detoxification of simazine and chlortoluron in WLR 2 and VLR 69, it is not known whether this is due to the expression of a mutant Cyt P450 that exhibits broad substrate specificity or the overexpression of one or several Cyt P450 isozymes that are able to catalyze N-dealkylation of various herbicides.

A preliminary study of the inheritance of enhanced simazine metabolism in the WLR 2 biotype indicated that resistance is a polygenic trait under nuclear

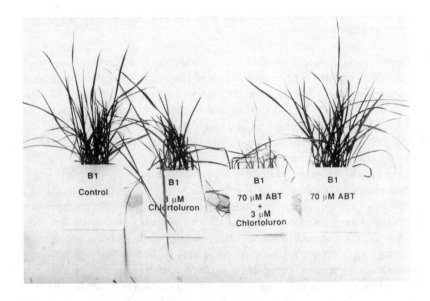

Figure 8. Effect of chlortoluron, 1-aminobenzotriazole (ABT), and the combination (chlortoluron plus ABT) on growth of the chlortoluron-resistant *Lolium rigidum* WLR 2 biotype. The WLR 2 biotype was referred to as B1 at the time the photograph was taken. From Burnet.[148]

control.[148] In addition, there were no obvious indications of reduced fitness associated with this type of resistance.[148] However, more definitive answers regarding inheritance and fitness await further research.

Alopecurus myosuroides

The substituted phenylurea chlortoluron has been used in Europe for over 20 years to control *A. myosuroides* in cereal crops.[78] During the 1980s, there was confirmation of *A. myosuroides* biotypes resistant to chlortoluron in several wheat fields in England[12] and Germany.[13] More recently, *A. myosuroides* biotypes resistant to the phenylureas have been reported in wheat fields in Spain[152] and the Netherlands.[153]

The nature of the problem has been best characterized for the resistant *A. myosuroides* biotypes found in England.[11,12,154-158] Poor control of *A. myosuroides* by chlortoluron was first described in 1982.[155] Currently, there are over 46 fields with *A. myosuroides* biotypes resistant to chlortoluron.[158] All fields have histories of repeated applications of phenylureas (chlortoluron, isoproturon) and most evidence suggests that the appearance of this trait has been gradual.[78]

The best-characterized chlortoluron-resistant biotype in England is "Peldon" which exhibits a moderate level of resistance to the phenylureas (see Chapter 9). Compared to a susceptible biotype, "Rothamsted", "Peldon" is 10-, 3-, and 5-fold more resistant to chlortoluron, isoproturon, and linuron, respectively.[155]

"Peldon" also exhibits cross resistance to dinitroanilines (pendimethalin), sulfonylureas (chlorsulfuron), aryloxyphenoxypropionates (diclofop), imidazolinones (imazamethabenz), carbamothioates (triallate), carbamates (barban), triazines (simazine, cyanazine), chloroacetanilides (metazachlor), and cyclohexanediones (tralkoxydim).[155]

Chlortoluron resistance in "Peldon" appears to be due to enhanced oxidative metabolism via N-demethylation and ring-methyl hydroxylation.[155,156] Hence, the chlortoluron resistance mechanisms in "Peldon" mimic those found in the crop (wheat) in which it is found (Figure 5). Circumstantial evidence suggests that the enhanced oxidative capacity found in "Peldon" involves Cyt P450(s). Consistent with this hypothesis is the greater level of Cyt P450 found in the microsomes of the R biotype compared to the S biotype.[156] Furthermore, the Cyt P450 inhibitor ABT inhibits the enhanced chlortoluron metabolism in "Peldon", and when applied with chlortoluron causes this biotype to become susceptible.[155]

Preliminary investigations of the inheritance of chlortoluron resistance suggest that the trait is polygenic and under nuclear control.[78] This is consistent with the gradual appearance of resistance and the variable levels and patterns of cross resistance observed among biotypes collected at different locations.

Although the effect of resistance on fitness has not been fully evaluated, preliminary studies indicate no obvious decrease in fitness.[78] No differences in growth were detected between R "Peldon" and S "Rothamsted" in the absence of competition, nor were there any indications of differences in competitive ability.[78]

HERBICIDE-RESISTANT CROPS

Shortly after the discovery that atrazine resistance in weeds was due to a modified target site, efforts were initiated to develop atrazine-resistant crops.[159-162] It was envisioned that the incorporation of this trait into normally susceptible crops would allow atrazine, a relatively inexpensive herbicide, to be used for weed control in additional crops and also prevent carry-over injury.

The first success in this area involved the use of traditional breeding techniques to transfer atrazine resistance (Ser 264 to Gly) from *Brassica campestris* L. to the oilseed crop canola (*B. napus*).[163,164] This was accomplished through a combination of multiple backcrosses and cytoplasmic selection for atrazine resistance. The result of this effort was the commercial release of the triazine-resistant canola variety "OAC Triton" in Canada in the mid-1980s.[164] However, because this variety contained the Ser 264 to Gly mutation, it exhibited a 10 to 20% decrease in yield compared to triazine-susceptible varieties.[164,165] It is estimated that less than 2% of the *B. napus* hectarage in Canada is planted with triazine-resistant varieties because of the yield penalty associated with the trait.[166] The use of the resistant varieties is cost effective only in situations where the weed pressure from closely related cruciferous weeds such as *Sinapis arvensis* L. is high. Four triazine-resistant cultivars of *B. napus* are currently registered in Canada.[166]

The use of classical breeding techniques to transfer triazine resistance from wild relatives to crops is only possible in a limited number of species. As a result, other approaches have been employed. One of these involves somatic hybridization of protoplasts from triazine R and S biotypes followed by selection for atrazine-resistant cytoplasm.[167-169] This has been used successfully to incorporate atrazine resistance into *Brassica* species, broccoli (*B. oleracea* var. italica) and cauliflower (*B. oleracea* subsp. botrytis). As expected, the triazine-resistant *Brassica* species exhibited reduced yield.

Recombinant DNA technology has also been employed to transfer atrazine resistance to crops. In theory, a triazine-resistant *psbA* gene, obtained either from a resistant species or generated by site-directed mutagenesis, could be used to transform a crop. However, this approach has not been employed in crops because of the difficulty in transforming the chloroplast genome of plants with high frequency.[159] Nevertheless, it may be possible to overcome this limitation by transforming the nucleus with the *psbA* gene. A moderate level of atrazine resistance was imparted to tobacco by nuclear transformation with the *psbA* gene.[170] This was accomplished by transformation with a chimeric gene consisting of the atrazine-resistant structural *psbA* gene attached to a nuclear promoter and a chloroplast transit peptide-encoding sequence. However, the tobacco plants transformed in this manner did not exhibit a high level of atrazine resistance because the chloroplasts still contained the susceptible *psbA* gene.

Alternatively, resistance to PS II inhibitors could be conferred by the introduction of nuclear genes encoding herbicide detoxifying enzymes. In one case, this approach has been quite successful. A gene encoding a nitralase that detoxifies bromoxynil (Figure 9) was isolated and cloned from a soil-borne microorganism (*Klebsiella ozaenae*).[171-174] The cloned nitralase gene was placed under the control of a plant promoter and transferred to tobacco.[174] The transformed plants exhibited a high level of resistance to bromoxynil. Cotton plants have also been transformed with this gene and field testing of bromoxynil-resistant cotton was conducted during the summers of 1992 and 1993. It is expected the bromoxynil-resistant cotton will be marketed by Calgene Inc. in 1994.[175]

Although there was initially a large interest in both academia and industry in the development of atrazine-resistant crops, this is no longer the case. In large part, this is because the triazines are applied at relatively high rates and have been reported to contaminate ground and surface water.[162] Currently, most efforts directed toward the development of herbicide-resistant crops are focused on chemicals considered to be more environmentally and toxicologically benign, such as glyphosate, or chemicals that are applied at low rates, such as the sulfonylureas and imidazolinones (see Chapters 4 and 8).

CONCLUSIONS AND FUTURE RESEARCH DIRECTIONS

Resistance to PS II inhibiting herbicides due to a modification at the target site is the most prevalent type of herbicide resistance in terms of numbers of species,

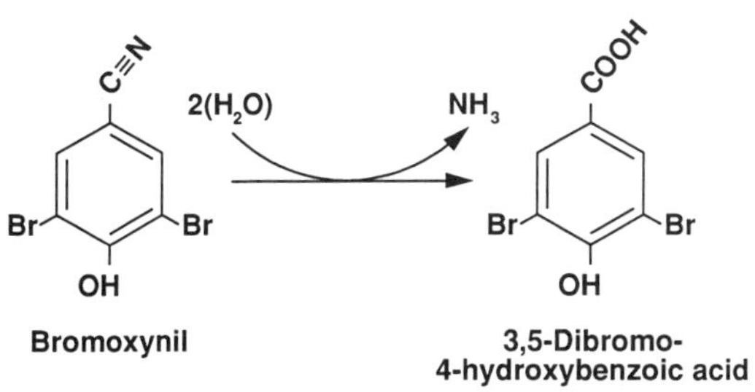

Figure 9. The metabolism of bromoxynil by a bromoxynil-specific nitralase isolated from *Klebsiella ozaenae*. From Stalker et al.[174] Copyright 1988 by AAAS.

biotypes, and geographical distribution.[9] Although this type of resistance is very common, for the most part it has not presented serious problems to crop production because there are usually alternative herbicides available to control these weeds.[8,10] In many cases where triazine-resistant weeds are a problem, the triazines are still used but herbicides with different modes of action are tank-mixed with the triazines or applied as separate treatments. Although this approach will usually solve the problem, it increases the cost of weed control. For example, in order to control triazine-resistant weeds in maize in France, pyridate is applied postemergence after the initial pre-plant incorporated atrazine treatment. This additional treatment increases the cost of weed control fourfold.[141]

In addition to the availability of other chemical treatments to control triazine resistance due to a modification at the target site, other factors have contributed to the limited impact of this type of resistance in agricultural production systems. Triazine-resistant biotypes exhibit reduced fitness, which means that their numbers will decrease in the absence of herbicide selection pressure.[112] The maternal inheritance of triazine resistance prevents the spread of this trait by pollen.[133-135] Furthermore, triazine-resistant weeds do not usually exhibit cross resistance to other herbicide classes that act at a different target site.[8]

Research during the past 15 years concerning triazine resistance has greatly advanced understanding of the biochemistry, genetics, and molecular biology of PS II. Current efforts in this area are focused on defining the topology of the Q_B-binding niche on the D1 protein. Much of this work involves the use of *psb*A mutants to determine the specific interactions of various PS II herbicides with amino acid residues in the Q_B niche. This approach has been taken because it is not yet possible to define the three-dimensional structure of the PS II reaction center by X-ray crystallography as was done with the reaction center of purple photosynthetic bacteria.[37] Knowledge regarding the topology of the Q_B-binding niche will be useful for the biorational design of new PS II inhibiting herbicides.

While the evidence that the *psb*A mutation (Ser 264 to Gly) reduces ecological fitness is quite convincing, there are still some unanswered questions. Further research using isonuclear or nearly isonuclear R and S biotypes is needed to better define the influence of various environments on relative fitness. The mechanisms whereby the nuclear genome compensates for the reduced fitness conferred by the *psb*A (Ser 264 to Gly) mutation need to be investigated. The recently proposed hypothesis that reduced fitness in triazine-resistant biotypes is due to increased susceptibility to photoinhibition deserves further attention. Is this a general response observed in triazine-resistant biotypes and, if so, to what extent does it account for reduced fitness?

The appearance of weed biotypes exhibiting resistance to PS II inhibiting herbicides due to enhanced herbicide metabolism has been a relatively recent development. This resistance mechanism is not as widespread and has been found in only a limited number of species. The level of resistance associated with enhanced detoxification (10-fold or less) is considerably lower than that associated with a modification at the target site. However, this level of resistance ensures the failure of weed control at recommended herbicide rates.

Although currently limited in scope, resistance to PS II inhibiting herbicides due to enhanced detoxification has the potential of becoming a more serious problem than resistance due to modified target site. Preliminary indications are that this type of resistance is not associated with reduced ecological fitness.[16,148,158] Furthermore, resistance is under nuclear control and hence can be spread by pollen.[145,148,158] The impact of enhanced detoxification appears to be particularly serious where the mechanism involves oxidative metabolism as observed in biotypes of *L. rigidum* and *A. myosuroides*.[148,158] Resistance to PS II inhibitors in these biotypes can be associated with broad-spectrum cross resistance to herbicides exhibiting other modes of action. For example, the "Peldon" biotype of *A. myosuroides* in England which was selected due to repeated chlortoluron and isoproturon applications not only exhibits resistance to these herbicides, but it is also resistant to several other herbicide classes that are registered for its control in wheat.[158] Current evidence suggests that chlortoluron resistance in this biotype is due to enhanced detoxification. Whether the cross resistance to other herbicides in "Peldon" is due to a common mechanism such as enhanced detoxification is not known. The repeated applications of chlortoluron and isoproturon may have selected for a complement of herbicide-detoxifying enzymes, perhaps coordinately regulated by a single gene. Alternatively, the application of these herbicides over several growing seasons may have selected for multiple mechanisms of resistance such as modified target sites, restricted uptake/translocation, and compartmentation (see Chapter 9). In this case, mechanisms for resistance would vary among herbicide classes.

Control of weeds that exhibit cross resistance to PS II herbicides due to enhanced detoxification may present a serious challenge. The strategy that has been widely employed for controlling resistance to PS II herbicides due to a modified target site (i.e., using herbicides with different modes of action) may not solve the problem. One possible approach to solve this problem would involve the use of herbicide synergists, compounds that inhibit the enhanced detoxification capacity in the resistant weeds. Insecticide synergists, such as PBO, have been used successfully in a number of cases to control insects that have developed insecticide resistance due to enhanced detoxification.[176] The development of herbicide synergists has been given only limited attention. Although no longer commercially available, tridiphane was used in maize to synergize atrazine in certain grass weeds by blocking GST activity.[74] A number of investigations have indicated that the Cyt P450 inhibitor ABT synergizes the activity of several herbicides that are detoxified via oxidative metabolism.[81-83] However, ABT is not suitable for field application. Relatively high concentrations of this toxic inhibitor are required and it is generally nonselective, i.e., it inhibits Cyt P450 in both weeds and crops. The development of a selective inhibitor of enhanced oxidative metabolism in resistant weeds will be difficult because, in many cases, it appears that the weeds have "evolved" the same detoxification mechanisms found in the crop.[16,148,158] In the long term, solutions to the problem of herbicide resistance due to enhanced detoxification (or other mechanisms) cannot rely solely on strategies involving the use of chemicals. As recognized by weed scientists, the implementation of integrated weed management practices will be needed to deal with this problem.[177-179]

The two enzymes implicated in the development of resistance to PS II herbicides due to enhanced detoxification are Cyt P450 and GST. Our knowledge of these enzymes is limited. Both Cyt P450 and GST appear to be part of multigene families with constitutive and induced isozymes.[73-75] Factors that regulate the expression of these enzymes are poorly understood. Further research is needed to investigate the roles that Cyt P450 and GST play in the development of resistance to PS II inhibitors as well as other herbicides. Not only will this research aid in the development of strategies to control resistance caused by enhanced metabolism, but it will increase understanding of the biochemistry and molecular biology of these enzymes. One positive outcome of the development of triazine resistance due to a modified target site was the advancement of knowledge of PS II and the D1 protein. Likewise, a focused effort on Cyt P450 and GST will not only greatly increase understanding of the role that these enzymes play in the development of herbicide resistance, but will also identify other important endogenous functions of these enzymes in plant metabolism.

REFERENCES

1. Blackman, G. E. "Selective Toxicity and the Development of Selective Weedkillers," *J. R. Soc. Arts* 98:499-517 (1950).
2. Abel, A. L. "The Rotation of Weedkillers," in *Proceedings of the British Weed Control Conference.* (Farnham, U.K.: The British Crop Protection Council, 1954), pp. 249-255.
3. Harper, J. L. "The Evolution of Weeds in Relation to Resistance to Herbicides," *Proceedings of the British Weed Control Conference.* (Farnham, U.K.: The British Crop Protection Council, 1956), pp. 179-188.
4. Ryan, G. F. "Resistance of Common Groundsel to Simazine and Atrazine," *Weed Sci.* 18:614-616 (1970).

5. Bandeen, J. D., G. R. Stephenson, and E. R. Cowett. "Discovery and Distribution of Herbicide-resistant Weeds in North America," in *Herbicide Resistance in Plants*, H. M. LeBaron and J. Gressel, Eds. (New York: John Wiley & Sons, 1982), pp. 9-30.
6. Gressel, J., H. U. Ammon, H. Fogelfors, J. Gasquez, Q. O. N. Kay, and H. Kees. "Discovery and Distribution of Herbicide-resistant Weeds Outside North America," in *Herbicide Resistance in Plants*, H. M. LeBaron and J. Gressel, Eds. (New York: John Wiley & Sons, 1982), pp. 31-55.
7. Holt, J. S., and H. M. LeBaron. "Significance and Distribution of Herbicide Resistance," *Weed Technol.* 4:141-149 (1990).
8. LeBaron, H. M., and J. McFarland. "Herbicide Resistance in Weeds and Crops: An Overview and Prognosis," in *Managing Resistance to Agrochemicals: From Fundamental Research to Practical Strategies*, M. B. Green, H. M. LeBaron, and W. K. Moberg, Eds. (Washington, D.C.: American Chemical Society Symposium Series No. 421, 1990), pp. 336-352.
9. LeBaron, H. M. "Distribution and Seriousness of Herbicide-resistant Weed Infestations Worldwide," in *Herbicide Resistance in Weeds and Crops*, J. C. Caseley, G. W. Cussans, and R. K. Atkin (Oxford: Butterworth-Heinemann, 1991), pp. 27-43.
10. Stephenson, G. R., M. D. Dykstra, R. D. McLaren, and A. S. Hamill. "Agronomic Practices Influencing Triazine-resistant Weed Distribution in Ontario," *Weed Technol.* 4:199-207 (1990).
11. Moss, S. R., and G. W. Cussans. "Variability in the Susceptibility of *Alopecurus myosuroides* (Black-grass) to Chlortoluron and Isoproturon," *Aspects of Applied Biology 9, The Biology and Control of Weeds in Cereals*, pp. 91-98. (The Association of Applied Biologists, National Vegetable Research Station, Wellesbourne, Warwick, U.K., 1985).
12. Kemp, M. S., and J. C. Caseley. "Synergistic Effects of 1-Aminobenzotriazole on the Phytotoxicity of Chlorotoluron and Isoproturon in a Resistant Population of Black-grass (*Alopecurus myosuroides*)," *Proceedings of the British Crop Protection Conference — Weeds*. (Farnham, U.K.: The British Crop Protection Council, 1987), pp. 895-899.
13. Niemann, P., and W. Pestemer. "Resistenz Verschiedener Herkunfte von Ackerfuchsschwanz (*Alopecurus myosuroides*) Gegenuber Herbizidbehandlungen," *Nachr. Dtsch. Pflanzenschutzdienst (Berlin)* 36:113-118 (1984).
14. Powles, S. B., J. A. M. Holtum, J. M. Matthews, and D. R. Liljegren. "Herbicide Cross-resistance in Annual Ryegrass (*Lolium rigidum* Gaud): The Search for a Mechanism," in *Managing Resistance to Agrochemicals: From Fundamental Research to Practical Strategies,* M. B. Green, H. M. LeBaron, and W. K. Moberg, Eds. (Washington, D.C.: American Chemical Society Symposium Series No. 421, 1990), pp. 394-406.
15. Burnet, M. W. M., O. B. Hildebrand, J. A. M. Holtum, and S. B. Powles. "Amitrole, Triazine, Substituted Urea, and Metribuzin Resistance in a Biotype of Rigid Ryegrass (*Lolium rigidum*)," *Weed Sci.* 39:317-323 (1991).
16. Gronwald, J. W., R. N. Andersen, and C. Yee. "Atrazine Resistance in Velvetleaf (*Abutilon theophrasti*) due to Enhanced Atrazine Detoxification," *Pestic. Biochem. Physiol.* 34:149-163 (1989).
17. Arntzen, C. J., K. Pfister, and K. E. Steinback. "The Mechanism of Chloroplast Triazine Resistance: Alterations in the Site of Herbicide Action," in *Herbicide Resistance in Plants,* H. M. LeBaron and J. Gressel, Eds. (New York: John Wiley & Sons, 1982), pp. 185-214.

18. Gressel, J. "Herbicide Tolerance and Resistance: Alteration of Site of Activity," in *Weed Physiology*, Vol. II, *Herbicide Physiology*, S. O. Duke, Ed., (Boca Raton, FL: CRC Press, 1985), pp. 159-189.
19. van Rensen, J. J. S. "Herbicides Interacting with Photosystem II," in *Herbicides and Plant Metabolism*, A. D. Dodge, Ed. (Cambridge: Cambridge University Press, 1989), pp. 21-36.
20. Mets, L., and A. Thiel. "Biochemistry and Genetic Control of the Photosystem II Herbicide Target Site," in *Target Sites of Herbicide Action*, P. Böger and G. Sandmann, Eds. (Boca Raton, FL: CRC Press, 1989), pp 1-24.
21. Dodge, A. D. "Photosynthesis," in *Target Sites for Herbicide Action*, R. C. Kirkwood, Ed. (New York: Plenum Press, 1991), pp. 1-27.
22. Trebst, A. "The Molecular Basis of Resistance of Photosystem II Herbicides," in *Herbicide Resistance in Weeds and Crops*, J. C. Caseley, G. W. Cussans, and R. K. Atkin, Eds. (Oxford: Butterworth-Heinemann, 1991), pp. 145-164.
23. Warwick, S. I. "Herbicide Resistance in Weedy Plants: Physiology and Population Biology," *Annu. Rev. Ecol. Syst.* 22:95-114 (1991).
24. Draber, W., K. Tietjen, J. F. Kluth, and A. Trebst. "Herbicides in Photosynthesis Research," *Angew. Chem. Int. Ed. Engl.* 30:1621-1633 (1991).
25. Fuerst, E. P., and M. A. Norman. "Interactions of Herbicides with Photosynthetic Electron Transport," *Weed Sci.* 39:458-464 (1991).
26. Vaughn, K. C., and S. O. Duke. "Biochemical Basis of Herbicide Resistance," in *Chemistry of Plant Protection 7*, (Heidelberg: Springer-Verlag, 1991), pp. 141-169.
27. Wessels, J. S. C., and R. van der Veen. "The Action of Some Derivatives of Phenylurethan and of 3-phenyl-1,1-dimethylurea on the Hill Reaction," *Biochim. Biophys. Acta* 19:548-549 (1956).
28. Cooke, A. R. "A Possible Mechanism of Action of the Urea Type Herbicides," *Weeds* 4:397-398 (1956).
29. Exer, B. "Über Pflanzenwachstumsregulatoren: Der Einfluss von Simazin auf der Pflanzenstoffwechsel," *Experientia* 14: 136-137 (1958).
30. Moreland, D. E., W. A. Gentner, J. L. Hilton, and K. L. Hill. "Studies on the Mechanism of Herbicidal Action of 2-chloro-4,6-Bis(Ethylamino)-s-triazine," *Plant Physiol.* 34:432-435 (1959).
31. Duysens, L. N. M., and H. E. Sweers. "Mechanism of Two Photochemical Reactions in Algae as Studied by Means of Fluorescence," in *Studies on Microalgae and Photosynthetic Bacteria. Jpn. Soc. Plant Physiol.* (Tokyo: University of Tokyo Press, 1963), pp. 353-372.
32. Zweig, G., I. Tamas, and E. Greenberg. "The Effect of Photosynthesis Inhibitors on Oxygen Evolution and Fluorescence of Illuminated *Chlorella*," *Biochim. Biophys. Acta* 66:196-205 (1963)
33. Murata, N., M. Nishimura, and A. Takamiya. "Fluorescence of Chlorophyll in Photosynthetic Systems. II. Induction of Fluorescence in Isolated Spinach Chloroplasts," *Biochim. Biophys. Acta* 120: 23-33 (1966).
34. Phillips, J. N., and J. L. Huppatz. "Cyanoacrylate Inhibitors of the Hill Reaction. I. Nature of the Inhibitor/Receptor Site Interaction," *Agric. Biol. Chem.* 48:51-54 (1984).
35. Mattoo, A. K., J. B. Marder, and M. Edelman. "Dynamics of the Photosystem II Reaction Center," *Cell* 56:241-246 (1989).
36. Andersson, B., and S. Styring. "Photosystem II: Molecular Organization, Function, and Acclimation, " *Curr. Topics Bioenergetics* 16:1-81 (1991).

37. Deisenhofer, J., O. Epp, K. Miki, R. Huber, and H. Michel. "Structure of the Protein Subunits in the Photosynthetic Reaction Centre of *Rhodopseudomonas viridis* at 3Å Resolution," *Nature* 318:618-624 (1985).
38. Michel, H., and J. Deisenhofer. "Relevance of the Photosynthetic Reaction Center from Purple Bacteria to the Structure of Photosystem II," *Biochemistry* 27:1-7 (1988).
39. Deisenhofer, J., and H. Michel. "The Photosynthetic Reaction Center from the Purple Bacterium *Rhodopseudomonas viridis*," *Science* 245:1463-1473 (1989).
40. Morden, C. W., and S. S. Golden. "*psb*A Genes Indicate Common Ancestry of Prochlorophytes and Chloroplasts," *Nature* 337:382-385 (1989).
41. Trebst, A. "The Three-Dimensional Structure of the Herbicide Binding Niche on the Reaction Center Polypeptides of Photosystem II," *Z. Naturforsch.* 42c:742-750 (1987).
42. Barber, J., and B. Andersson. "Too Much of a Good Thing: Light can be Bad for Photosynthesis", *TIBS* 17:61-66 (1992).
43. Mattoo, A. K., H. Hoffmann-Falk, J. B. Marder, and M. Edelman, "Regulation of Protein Metabolism: Coupling of Photosynthetic Electron Transport to *in vivo* Degradation of the Rapidly Metabolized 32-Kilodalton Protein of the Chloroplast Membranes," *Proc. Natl. Acad. Sci. U.S.A.* 81:1380-1384 (1984).
44. Kyle, D. J., I. Ohad, and C. J. Arntzen. "Membrane Protein Damage and Repair: Selective Loss of a Quinone-Protein Function in Chloroplast Membranes," *Proc. Natl. Acad. Sci. U.S.A.* 81:4070-4074 (1984).
45. Godde, D., H. Schmitz, and M. Weidner. "Turnover of the D-1 Reaction Center Polypeptide from Photosystem II in Intact Spruce Needles and Spinach Leaves," *Z. Naturforsch.* 46c:245-251 (1991).
46. Kyle, D. J. "The Biochemical Basis for Photoinhibition of Photosystem II" in *Topics in Photosynthesis, Vol. 9, Photoinhibition*, D. J. Kyle, C. B. Osmond, and C. J. Arntzen, Eds. (Amsterdam: Elsevier, 1987), pp. 197-226.
47. Schuster, G., R. Timberg, and I. Ohad. "Turnover of Thylakoid Photosystem II Proteins During Photoinhibition of *Chlamydomonas reinhardtii*," *Eur. J. Biochem.* 177:403-410 (1988).
48. Ohad, I., N. Adir, H. Koike, D. J. Kyle, and Y. Inoue. "Mechanism of Photoinhibition *in Vivo*: A Reversible Light-induced Conformational Change of Reaction Center II is Related to an Irreversible Modification of the D1 Protein," *J. Biol. Chem.* 265:1972-1979 (1990).
49. Richter, M., W. Rühle, and A. Wild. "Studies on the Mechanism of Photosystem II Photoinhibition. II. The Involvement of Toxic Oxygen Species," *Photosynth. Res.* 24:237-243 (1990).
50. Virgin, I., A. H. Salter, D. F. Ghanotakis, and B. Andersson. "Light-induced D1 Protein Degradation is Catalyzed by a Serine-type Protease," *FEBS Lett.* 287:125-128 (1991).
51. Barbato, R., C. A. Shipton, G. M. Giacometti, and J. Barber. "New Evidence Suggests that the Initial Photoinduced Cleavage of the D1-Protein may not Occur near the PEST Sequence," *FEBS Lett.* 290:162-166 (1991).
52. Salter, A. H., I. Virgin, Å. Hagman, and B. Andersson. "On the Molecular Mechanism of Light-induced D1 Protein Degradation in Photosystem II Core Particles," *Biochemistry* 31:3990-3998 (1992).
53. Tischer, W., and H. Strotmann. "Relationship Between Inhibitor Binding by Chloroplasts and Inhibition of Photosynthetic Electron Transport," *Biochim. Biophys. Acta* 460:113-125 (1977).

54. Pfister, K., K. E. Steinback, G. Gardner, and C. J. Arntzen., "Photoaffinity Labeling of an Herbicide Receptor Protein in Chloroplast Membranes," *Proc. Natl. Acad. Sci. U.S.A.* 78:981-985 (1981).
55. Boschetti, A., M. Tellenbach, and A. Gerber. "Covalent binding of 3-azidomonuron to thylakoids of DCMU-sensitive and -resistant Strains of *Chlamydomonas reinhardtii*," *Biochim. Biophys. Acta* 810:12-19 (1985).
56. Oettmeier, W., K. Masson, H.-J. Soll, and W. Draber. "Herbicide Binding at Photosystem II: A New Azido-triazinone Photoaffinity Label," *Biochim. Biophys. Acta* 767:590-595 (1984).
57. Velthuys, B. R. "Electron-dependent Competition Between Plastoquinone and Inhibitors for Binding to Photosystem II," *FEBS Lett.* 126:277-281 (1981).
58. Vermaas, W. F. J., C. J. Arntzen, L.-Q. Gu, and C.-A. Yu. "Interactions of Herbicides and Azidoquinones at a Photosystem II Binding Site in the Thylakoid Membrane," *Biochim. Biophys. Acta* 723:266-275 (1983).
59. Oettmeier, W., and H.-J. Soll. "Competition Between Plastoquinone and 3-(3,4-dichlorophenyl)-1,1-Dimethylurea at the Acceptor Side of Photosystem II," *Biochim. Biophys. Acta* 724:287-290 (1983).
60. Vermaas, W. F. J., G. Renger, and C. J. Arntzen. "Herbicide/Quinone Binding Interactions in Photosystem II," *Z. Naturforsch.* 39c:368-373 (1984).
61. Vermaas, W. F. J., and C. J. Arntzen. "Synthetic Quinones Influencing Herbicide Binding and Photosystem II Electron Transport: The Effects of Triazine-Resistance on Quinone Binding Properties in Thylakoid Membranes," *Biochim. Biophys. Acta* 725:483-491 (1983).
62. Gardner, G. "A Stereochemical Model for the Active Site of Photosystem II Herbicides," *Photochem. Photobiol.* 49:331-336 (1989).
63. Shigematsu, Y., F. Sato, and Y. Yamada. "A Binding Model for Phenylurea Herbicides Based on Analysis of a Thr264 Mutation in the D-1 Protein of Tobacco," *Pestic. Biochem. Physiol.* 35:33-41 (1989).
64. Oettmeier, W., U. Hilp, W. Draber, C. Fedtke, and R. R. Schmidt. "Structure-Activity Relationships of Triazinone Herbicides on Resistant Weeds and Resistant *Chlamydomonas reinhardtii*," *Pestic. Sci.* 33:399-409 (1991).
65. Tietjen, K. G., J. F. Kluth, R. Andree, M. Haug, M. Lindig, K. H. Müller, H. J. Wroblowsky, and A. Trebst. "The Herbicide Binding Niche of Photosystem II — A Model," *Pestic. Sci.* 31:65-72 (1991).
66. Erickson, J. M., K. Pfister, M. Rahire, R. K. Togasaki, L. Mets, and J.-D. Rochaix. "Molecular and Biophysical Analysis of Herbicide-resistant Mutants of *Chlamydomonas reinhardtii*: Structure-function Relationship of the Photosystem II D1 Polypeptide," *Plant Cell* 1:361-371 (1989).
67. Pallett, K. E., and A. D. Dodge. "Studies into the Action of Some Photosynthetic Inhibitor Herbicides," *J. Exp. Bot.* 31:1051-1066 (1980).
68. Barry, P., A. J. Young, and G. Britton. "Photodestruction of Pigments in Higher Plants by Herbicide Action. I. The Effect of DCMU (diuron) on Isolated Chloroplasts," *J. Exp. Bot.* 41:123-129 (1990).
69. Gaba, V., J. B. Marder, B. M. Greenberg, A. K. Mattoo, and M. Edelman. "Degradation of the 32 kD Herbicide Binding Protein in Far Red Light," *Plant Physiol.* 84:348-352 (1987).
70. Kuhn M., and P. Böger. "Studies on the Light-induced Loss of the D1 Protein in Photosystem-II Membrane Fragments," *Photosynth. Res.* 23:291-296 (1990).

71. Gong, H., and I. Ohad. "The PQ/PQH$_2$ Ratio and Occupancy of Photosystem II-Q$_B$ Site by Plastoquinone Control the Degradation of D1 Protein during Photoinhibition *in Vivo*," *J. Biol. Chem.* 266:21293-21299 (1991).
72. Shimabukuro, R. H. "Detoxication of Herbicides" in *Weed Physiology, Vol. II. Herbicide Physiology*, S. O. Duke, Ed. (Boca Raton, FL: CRC Press, 1985), pp. 215-240.
73. Timmerman, K. P. "Molecular Characterization of Corn Glutathione S-Transferase Isozymes Involved in Herbicide Detoxication,"*Physiol. Plant.* 77:465-471 (1989).
74. Lamoureux, G. L., R. H. Shimabukuro, and D. S. Frear. "Glutathione and Glucoside Conjugation in Herbicide Selectivity," in *Herbicide Resistance in Weeds and Crops*, J.C. Caseley, G. W. Cussans, and R. K. Atkin, Eds. (Oxford: Butterworth-Heinemann, 1991), pp. 227-261.
75. O'Keefe, D. P., J. A. Romesser, and K. J. Leto. "Plant and Bacterial Cytochromes P-450: Involvement in Herbicide Metabolism" in *Phytochemical Effects of Environmental Compounds*, J. A. Saunders, L. Kosak-Channing, and E. E. Conn, Eds. (New York: Plenum Press, 1987), pp. 151-173.
76. Jones, O. T. G. "Cytochrome P$_{450}$ and Herbicide Resistance," in *Herbicide Resistance in Weeds and Crops*, J. C. Caseley, G. W. Cussans, and R. K. Atkin, Eds. (Oxford: Butterworth-Heinemann, 1991), pp. 213-226.
77. Dean, J. V., J. W. Gronwald, and M. P. Anderson. "Glutathione S-transferase Activity in Nontreated and CGA-154281-treated Maize Shoots," *Z. Naturforsch.* 46c:850-855 (1991).
78. Moss, S. R., and G. W. Cussans. "The Development of Herbicide-resistant Populations of *Alopecurus myosuroides* (Black-grass) in England," in *Herbicide Resistance in Weeds and Crops*, J. C. Caseley, G. W. Cussans, and R. K. Atkin, Eds. (Oxford: Butterworth-Heinemann, 1991), pp. 45-55.
79. Gross, D., T. Laanio, G. Dupuis, and H. O. Esser. "The Metabolic Behavior of Chlortoluron in Wheat and Soil," *Pestic. Biochem. Physiol.* 10:49-59 (1979).
80. Cabanne, F., P. Gaillardon, and R. Scalla. "Phytotoxicity and Metabolism of Chlortoluron in Two Wheat Varieties," *Pestic. Biochem. Physiol.* 23:212-220 (1985).
81. Cabanne, F., D. Huby, P. Gaillardon, R. Scalla, and F. Durst. "Effect of the Cytochrome P-450 Inactivator 1-aminobenzotriazole on the Metabolism of Chlortoluron and Isoproturon in Wheat," *Pestic. Biochem. Physiol.* 28:371-380 (1987).
82. Gaillardon, P., F. Cabanne, R. Scalla, and F. Durst. "Effect of Mixed Function Oxidase Inhibitors on the Toxicity of Chlortoluron and Isoproturon in Wheat," *Weed Res.* 25:397-402 (1985).
83. Cole, D. J., and W. J. Owen. "Influence of Monooxygenase Inhibitors on the Metabolism of the Herbicides Chlortoluron and Metolachlor in Cell Suspension Cultures," *Plant Sci.* 50:13-20 (1987).
84. Gonneau, M., B. Pasquette, F. Cabanne, and R. Scalla. "Metabolism of Chlortoluron in Tolerant Species: Possible Role of Cytochrome P-450 Mono-oxygenases," *Weed Res.* 28:19-25 (1988).
85. Mougin, C., F. Cabanne, M.-C. Canivenc, and R. Scalla. "Hydroxylation and N-demethylation of Chlortoluron by Wheat Microsomal Enzymes," *Plant Sci.* 66:195-203 (1990).
86. Mougin, C., N. Polge, R. Scalla, and F. Cabanne. "Interactions of Various Agrochemicals with Cytochrome P-450-dependent Monooxygenases of Wheat Cells," *Pestic. Biochem. Physiol.* 40:1-11 (1991).

87. Pfister, K., and C. J. Arntzen. "The Mode of Action of Photosystem II-specific Inhibitors in Herbicide-resistant Weed Biotypes," *Z. Naturforsch.* 34c:996-1009 (1979).
88. Fuerst, E. P., C. J. Arntzen, K. Pfister, and D. Penner. "Herbicide Cross-resistance in Triazine-resistant Biotypes of Four Species," *Weed Sci.* 34:344-353 (1986).
89. Pillai, P., and J. B. St. John. "Lipid Composition of Chloroplast Membranes from Weed Biotypes Differentially Sensitive to Triazine Herbicides," *Plant Physiol.* 68:585-587 (1981).
90. Lehoczki, E., E. Polos, G. Laskay, and T. Farkas, "Chemical Compositions and Physical States of Chloroplast Lipids Related to Atrazine Resistance in *Conyza canadensis* L." *Plant Sci.* 42:19-24 (1985).
91. Vaughn, K. C., and S. O. Duke. "Ultrastructural Alterations to Chloroplasts in Triazine-resistant Weed Biotypes," *Physiol. Plant.* 62:510-520 (1984).
92. Burke, J. J., R. F. Wilson, and J. R. Swafford. "Characterization of Chloroplasts Isolated from Triazine-susceptible and Triazine-resistant Biotypes of *Brassica campestris* L.," *Plant Physiol.* 70:24-29 (1982).
93. Bowes, J., A. R. Crofts, and C. J. Arntzen. "Redox Reactions on the Reducing Side of Photosystem II in Chloroplasts with Altered Herbicide Binding Properties," *Arch. Biochem. Biophys.* 200:303-308 (1980).
94. Jansen, M. A. K., and K. Pfister. "Conserved Kinetics at the Reducing Side of Reaction-center II in Photosynthetic Organisms; Changed Kinetics in Triazine-resistant Weeds," *Z. Naturforsch.* 45c:441-445 (1990).
95. Oettmeier, W., K. Masson, C. Fedtke, J. Konze, and R. R. Schmidt. "Effect of Different Photosystem II Inhibitors on Chloroplasts Isolated from Species Either Susceptible or Resistant Toward s-Triazine Herbicides," *Pestic. Biochem. Physiol.* 18: 357-367 (1982).
96. Durner, J., A. Thiel, and P. Böger. "Phenolic Herbicides: Correlation Between Lipophilicity and Increased Inhibitory Sensitivity of Thylakoids from Higher Plant Mutants," *Z. Naturforsch.* 41c:881-884 (1986).
97. van Oorschot, J. L. P., and P. H. van Leeuwen. " Inhibition of Photosynthesis in Intact Plants of Biotypes Resistant or Susceptible to Atrazine and Cross-resistant to Other Herbicides," *Weed Res.* 28:223-230 (1988).
98. Gressel, J., and L. A. Segel. "Negative Cross Resistance, a Possible Key to Atrazine Resistance Management: A Call for Whole Plant Data," *Z. Naturforsch.* 45c:470-473 (1990).
99. Gressel, J. "Why Get Resistance? It Can be Prevented or Delayed," in *Herbicide Resistance in Weeds and Crops*, J. C. Caseley, G. W. Cussans, and R. K. Atkins, Eds. (Oxford, U.K.: Butterworth-Heinemann, 1991), pp. 1-25.
100. De Prado, R., M. Sanchez, J. Jorrin, and C. Dominguez. "Negative Cross-resistance to Bentazone and Pyridate in Atrazine-resistant *Amaranthus cruentus* and *Amaranthus hybridus* Biotypes," *Pestic. Sci.* 35:131-136 (1992).
101. Arlt, K., and B. Jüttersonke. "The Negative Cross Resistance of Weed Species Particularly *Chenopodium album* L.", *Z. Pflanzenkr. (Pflanzenpathol.) Pflanzenschutz Sonderh.* 13:483-486 (1992).
102. Naber, J. D., and J. J. S. van Rensen. "Activity of Photosystem II Herbicides is Related with their Residence Times at the D1 Protein," *Z. Naturforsch.* 46c:575-578 (1991).
103. Sigematsu, Y., F. Sato, and Y. Yamada. "The Mechanism of Herbicide Resistance in Tobacco Cells with a New Mutation in the Q_B Protein." *Plant Physiol.* 89:986-992 (1989).

104. Smeda, R. J. "The Physiological and Molecular Characteristics of Atrazine Resistance in Photoautotrophic Potato Cells," Ph.D. Thesis, Purdue University, West Lafayette, IN (1990).
105. Conard, S. G., and S. R. Radosevich. "Ecological Fitness of *Senecio vulgaris* and *Amaranthus retroflexus* Biotypes Susceptible or Resistant to Atrazine," *J. Appl. Ecol.* 16:171-177 (1979).
106. Holt, J. S., A. J. Stemler, and S. R. Radosevich. "Differential Light Responses of Photosynthesis by Triazine-resistant and Triazine-susceptible *Senecio vulgaris* Biotypes," *Plant Physiol.* 67:744-748 (1981).
107. Ahrens, W. H., and E. W. Stoller. "Competition, Growth Rate, and CO_2 Fixation in Triazine-susceptible and -resistant Smooth Pigweed (*Amaranthus hybridus*)," *Weed Sci.* 31: 438-444 (1983).
108. Ort, D. R., W. H. Ahrens, B. Martin, and E. W. Stoller. "Comparison of Photosynthetic Performance in Triazine-resistant and Susceptible Biotypes of *Amaranthus hybridus*," *Plant Physiol.* 72:925-930 (1983).
109. Holt, J. S., S. R. Radosevich, and A. J. Stemler. "Differential Efficiency of Photosynthetic Oxygen Evolution in Flashing Light in Triazine-resistant and Triazine-susceptible Biotypes of *Senecio vulgaris* L.," *Biochim. Biophy. Acta* 722:245-255 (1983).
110. Jursinic, P. A., and R. W. Pearcy. "Determination of the Rate Limiting Step for Photosynthesis in a Nearly Isonuclear Rapeseed (*Brassica napus* L.) Biotype Resistant to Atrazine," *Plant Physiol.* 88:1195-1200 (1988).
111. Holt, J. S. "Reduced Growth, Competitiveness, and Photosynthetic Efficiency of Triazine-resistant *Senecio vulgaris* from California," *J. Appl. Ecol.* 25:307-318 (1988).
112. Holt, J. S. "Fitness and Ecological Adaptability of Herbicide-resistant Biotypes," in *Managing Resistance to Agrochemicals: From Fundamental Research to Practical Strategies,* M. B. Green, H. M. LeBaron, and M. K. Moberg, Eds. (Washington, D.C.: American Chemical Society, 1990), pp. 419-429.
113. Stowe, A. E., and J. S. Holt. "Comparison of Triazine-resistant and -susceptible Biotypes of *Senecio vulgaris* and Their F1 Hybrids," *Plant Physiol.* 87:183-189 (1988).
114. van Oorschot, J. L. P., and P. H. Van Leeuwen. "Comparison of Photosynthetic Capacity Between Intact Leaves of Triazine-resistant and Susceptible Biotypes of Six Weed Species," *Z. Naturforsch.* 39c:440-442 (1984).
115. Schönfeld, M., T. Yaacoby, O. Michael, and B. Rubin. "Triazine Resistance without Reduced Vigor in *Phalaris paradoxa*," *Plant Physiol.* 83:329-333 (1987).
116. Jansen, M. A. K., J. H. Hobé, J. C. Wesselius, and J. J. S. van Rensen. "Comparison of Photosynthetic Activity and Growth Performance in Triazine-resistant and Susceptible Biotypes of *Chenopodium album*," *Physiol. Vég.* 24:475-484 (1986).
117. Gressel, J., and G. Ben-Sinai. "Low Intraspecific Competitive Fitness in a Triazine-resistant, Nearly Nuclear-Isogenic Line of *Brassica napus*," *Plant Sci.* 38: 29-32 (1985).
118. McCloskey, W. B., and J. S. Holt. "Triazine Resistance in *Senecio vulgaris* Parental and Nearly Isonuclear Backcrossed Biotypes is Correlated with Reduced Productivity," *Plant Physiol.* 92:954-962 (1990).
119. McCloskey, W. B., and J. S. Holt. "Effect of Growth Temperature on Biomass Production of Nearly Isonuclear Triazine-resistant and -susceptible Common Groundsel (*Senecio vulgaris* L.)," *Plant Cell Environ.* 14:699-705 (1991).

120. Jacobs, B. F., J. H. Duesing, J. Antonovics, and D. T. Patterson. "Growth Performance of Triazine-resistant and -susceptible Biotypes of *Solanum nigrum* over a range of temperatures," *Can. J. Bot.* 66:847-850 (1988).
121. Beversdorf, W. D., D. J. Hume, and M. J. Donnelly-Vanderloo. "Agronomic Performance of Triazine-resistant and Susceptible Reciprocal Spring Canola Hybrids," *Crop Sci.* 28:932-934 (1988).
122. Ducruet, J. M., and Y. Lemoine. "Increased Heat Sensitivity of the Photosynthetic Apparatus in Triazine-resistant Biotypes from Different Plant Species," *Plant Cell Physiol.* 26:419-429 (1985).
123. Vencill, W. K., C. L. Foy, and D. M. Orcutt. "Effects of Temperature on Triazine-resistant Weed Biotypes," *Environ. Exp. Bot.* 27:473-480 (1987).
124. Ricroch, A., M. Mousseau, H. Darmency, and J. Pernes. "Comparison of Triazine-resistant and -susceptible Cultivated *Setaria italica*: Growth and Photosynthetic Capacity," *Plant Physiol. Biochem.* 25:29-34 (1987).
125. Ducruet, J. M., and D. R. Ort. "Enhanced Susceptibility of Photosynthesis to High Leaf Temperature in Triazine-resistant *Solanum nigrum* L.: Evidence for Photosystem II D1 Protein Site of Action," *Plant Sci.* 56:39-48 (1988).
126. Darmency, H., and J. Pernes. "Agronomic Performance of a Triazine Resistant Foxtail Millet (*Setaria italica* (L.) Beauv.)," *Weed Res.* 29:147-150 (1989).
127. Havaux, M. "Comparison of Atrazine-resistant and -susceptible Biotypes of *Senecio vulgaris* L.: Effect of High and Low Temperatures on the *in vivo* Photosynthetic Electron Transfer in Intact Leaves," *J. Exp. Bot.* 40:849-854 (1989).
128. Dekker, J. H., and T. D. Sharkey. "Regulation of Photosynthesis in Triazine-resistant and -susceptible *Brassica napus*." *Plant Physiol.* 98:1069-1073 (1992).
129. Fuks, B., F. van Eycken, and R. Lannoye. "Tolerance of Triazine-resistant and Susceptible Biotypes of Three Weeds to Heat Stress: A Fluorescence Study," *Weed Res.* 32:9-17 (1992).
130. Dekker, J. H., and R. G. Burmester. "Pleiotrophy in Triazine-resistant *Brassica napus:* Ontogenetic and Diurnal Influences on Photosynthesis." *Plant Physiol.* 100:2052-2058 (1992).
131. Hart, J. J., and A. Stemler. "High Light-induced Reduction and Low Light-enhanced Recovery of Photon Yield in Triazine-resistant *Brassica napus* L.," *Plant Physiol.* 94:1301-1307 (1990).
132. Hart, J. J., and A. Stemler. "Similar Photosynthetic Performance in Low Light-grown Isonuclear Triazine-resistant and -susceptible *Brassica napus* L.," *Plant Physiol.* 94:1295-1300 (1990).
133. Darr, S., V. S. Machado, and C. J. Arntzen. "Uniparental Inheritance of a Chloroplast Photosystem II Polypeptide Controlling Herbicide Binding," *Biochim. Biophys. Acta* 634:219-228 (1981).
134. Souza Machado, V., and J. D. Bandeen. "Genetic Analysis of Chloroplast Atrazine Resistance in *Brassica campestris* — Cytoplasmic Inheritance", *Weed Sci.* 30:281-285 (1982).
135. Hirschberg, J., A. Bleecker, D. J. Kyle, L. McIntosh, and C. J. Arntzen. "The Molecular Basis of Triazine-herbicide Resistance in Higher-Plant Chloroplasts," *Z. Naturforsch.* 39c:412-420 (1984).
136. Arntzen, C. J., and J. H. Duesing. "Chloroplast-encoded Herbicide Resistance," in *Advances in Gene Technology: Molecular Genetics of Plants and Animals,* K. Downey, R. W. Voellmy, F. Ahmad, and J. Schultz, Eds. (New York: Academic Press, 1983), pp. 273-294.

137. Rédei, G. P., and S. B. Plurad. "Hereditary Structural Alterations of Plastids Induced by a Nuclear Mutator Gene in *Arabidopsis*," *Protoplasma* 77:361-380 (1973).
138. Solymosi, P., and E. Lehoczki. "Characterization of a Triple (Atrazine-pyrazon-pyridate) Resistant Biotype of Common Lambsquarters (*Chenopodium album* L.)," *J. Plant Physiol.* 134:685-690 (1989).
139. Bettini, P., S. McNally, M. Sevignac, H. Darmency, J. Gasquez, and M. Dron. "Atrazine Resistance in *Chenopodium album*: Low and High Levels of Resistance to the Herbicide are Related to the Same Chloroplast *psb*A Gene Mutation," *Plant Physiol.* 84:1442-1446 (1987).
140. Darmency, H., and J. Gasquez. "Appearance and Spread of Triazine Resistance in Common Lambsquarters (*Chenopodium album*)," *Weed Technol.* 4:173-177 (1990).
141. Darmency H., and J. Gasquez. "Fate of Herbicide Resistance Genes in Weeds," in *Managing Resistance to Agrochemicals: From Fundamental Research to Practical Strategies*, M. B. Green, H. M. LeBaron, and W. K. Moberg, Eds. (Washington, D.C.: American Chemical Society, 1990), pp. 353-363.
142. Ritter, R. L. "Triazine Resistant Velvetleaf and Giant Foxtail Control in No-tillage Corn," *Proc. Northeast Weed Sci. Soc.* 40:50-52 (1986).
143. Ritter, R.L., University of Maryland, personal communication.
144. Gronwald, J. W. Unpublished data.
145. Andersen, R. N., and J. W. Gronwald. "Noncytoplasmic Inheritance of Atrazine Tolerance in Velvetleaf (*Abutilon theophrasti*)," *Weed Sci.* 35:496-498 (1987).
146. Anderson, M. P., and J. W. Gronwald. "Atrazine Resistance in a Velvetleaf (*Abutilon theophrasti*) Biotype Due to Enhanced Glutathione *S*-transferase Activity," *Plant Physiol.* 96:104-109 (1991).
147. Powles, S. B., and P. D. Howat. "Herbicide-resistant Weeds in Australia," *Weed Technol.* 4:178-185 (1990).
148. Burnet, M. W. M. "Mechanisms of Herbicide Resistance in *Lolium rigidum*", Ph.D. Thesis, University of Adelaide, Adelaide, Australia (1992).
149. Burnet, M. W. M., B. R. Loveys, J. A. M. Holtum, and S. B. Powles. "Increased Detoxification is a Mechanism of Simazine Resistance in *Lolium rigidum*," *Pestic. Biochem. Physiol.* 46:207-218 (1993).
150. Burnet, M. W. M., B. R. Loveys, J. A. M. Holtum, and S. B. Powles. "A Mechanism of Chlorotoluron Resistance in *Lolium rigidum*," *Planta* 190:182-189 (1993).
151. C. Preston and S. B. Powles, unpublished.
152. De Prado, R., J. Menendez, M. Tena, J. Caseley, and A. Taberner. "Response to Substituted Ureas, Triazines and Chloroacetanilides in a Biotype of *Alopecurus myosuroides* Resistant to Chlorotoluron," in *Brighton Crop Protection Conference — Weeds* (Farnham, U.K.: The British Crop Protection Council, 1991), pp. 1065-1070.
153. Oorschot, J. L. P., and P. H. van Leeuwen. "Use of Fluorescence Induction to Diagnose Resistance of *Alopecurus myosuroides* Huds. (Black-Grass) to Chlorotoluron," *Weed Res.* 32:473-482 (1992).
154. Clarke, J. H., and S. R. Moss. "The Distribution and Control of Herbicide Resistant *Alopecurus myosuroides* (Black-grass) in Central and Eastern England," in *Brighton Crop Protection Conference — Weeds* (Farnham, U.K.: The British Crop Protection Council, 1989), pp. 301-308.

155. Kemp, M. S., S. R. Moss, and T. H. Thomas. "Herbicide Resistance in *Alopecurus myosuroides*" in *Managing Resistance to Agrochemicals: From Fundamental Research to Practical Strategies*, M. B. Green, H. M. LeBaron, and W. K. Moberg, Eds. (Washington, D.C.: American Chemical Society Symposium Series 421, 1990), pp. 376-393.
156. Kemp, M. S. and J. C. Caseley. "Synergists to Combat Herbicide Resistance" in *Herbicide Resistance in Weeds and Crops*. J. C. Caseley, G. W. Cussans, and R. K. Atkin, Eds. (Oxford: Butterworth-Heinemann, 1991), pp. 279-292.
157. Moss, S. R. "Herbicide Cross-resistance in Slender Foxtail (*Alopecurus myosuroides*)," Weed Sci. 38:492-496 (1990).
158. Clarke, J. H., and S. R. Moss. "The Occurrence of Herbicide Resistant *Alopecurus myosuroides* (Black-grass) in the United Kingdom and Strategies for its Control," *Brighton Crop Protection Conference — Weeds*, (Farnham, U.K.: The British Crop Protection Council, 1991), pp. 1041-1048.
159. Mazur, B. J., and S. C. Falco. "The Development of Herbicide Resistant Crops," *Annu. Rev. Plant Physiol. Plant Mol. Biol.* 40:441-470 (1989).
160. Schulz, A., F. Wengenmayer, and H. M. Goodman. "Genetic Engineering of Herbicide Resistance in Higher Plants," *Crit. Rev. Plant Sci.* 9:1-15 (1990).
161. Duke, S. O., A. L. Christy, F. D. Hess, and J. S. Holt. "Herbicide-resistant Crops" (Council for Agricultural Science and Technology; Ames, Iowa, 1991).
162. Gressel, J. "The Needs for New Herbicide-resistant Crops," in *Achievements and Developments in Combating Pesticide Resistance*, J. Denholm, A. L. Devonshire, and D.W. Holloman, Eds. (London: Elsevier, 1992), pp. 283-294.
163. Beversdorf, W. D., J. Weiss-Lerman, L. R. Erickson, and V. Souza Machado. "Transfer of Cytoplasmically-inherited Triazine Resistance From Bird's-rape to Cultivated Oilseed Rape (*Brassica campestris* and *B. napus*)," *Can. J. Genet. Cytol.* 22:167-172 (1980).
164. Grant, I. and W. D. Beversdorf. "Agronomic Performance of Triazine-resistant Single-cross Hybrid Oilseed Rape (*Brassica napus* L.)," *Can. J. Plant Sci.* 65:889-892 (1985).
165. Forcella, F. "Herbicide-resistant Crops: Yield Penalties and Weed Thresholds for Oilseed Rape (*Brassica napus* L.)," *Weed Res.* 27:31-34 (1987).
166. Hume, D. J., Crop Science Dept., University of Guelph, Guelph, Ontario, personal communication.
167. Jourdan, P. S., E. D. Earle, and M. A. Mutschler. "Synthesis of Male Sterile, Triazine-resistant *Brassica napus* by Somatic Hybridization Between Cytoplasmic Male Sterile *B. oleracea* and Atrazine-Resistant *B. campestris*," *Theor. Appl. Genet.* 78:445-455 (1989).
168. Jourdan, P. S., E. D. Earle, and M. A. Mutschler. "Atrazine-resistant Cauliflower Obtained by Somatic Hybridization Between *Brassica oleracea* and ATB-*B. napus*," *Theor. Appl. Genet.* 78:271-279 (1989).
169. Christey, M. C., C. A. Makaroff, and E. D. Earle. "Atrazine-resistant Cytoplasmic Male-sterile-*nigra* Broccoli Obtained by Protoplast Fusion Between Cytoplasmic Male-sterile *Brassica oleracea* and Atrazine-resistant *Brassica campestris*," *Theor. Appl. Genet.* 83:201-208 (1991).
170. Cheung, A. Y., L. Bogorad, M. Van Montagu, and J. Schell. "Relocating a Gene for Herbicide Tolerance: A Chloroplast Gene is Converted into a Nuclear Gene," *Proc. Natl. Acad. Sci. U.S.A.* 85:391-395 (1988).

171. McBride, K. E., J. W. Kenny, and D. M. Stalker. "Metabolism of the Herbicide Bromoxynil by *Klebsiella pneumoniae* Subsp. *ozaenae*," *Appl. Environ. Microbiol.* 52:325-330 (1986).
172. Stalker, D. M., L. D. Malyj, and K. E. McBride. "Purification and Properties of a Nitrilase Specific for the Herbicide Bromoxynil and Corresponding Nucleotide Sequence Analysis of the *bxn* Gene," *J. Biol. Chem.* 263:6310-6314 (1988).
173. Stalker, D. M., and K. E. McBride. "Cloning and Expression in *Escherichia coli* of a *Klebsiella ozaenae* Plasmid-Borne Gene Encoding a Nitrilase Specific for the Herbicide Bromoxynil," *J. Bacteriol.* 169:955-960 (1987).
174. Stalker, D. M., K. E. McBride, and L. D. Malyj. "Herbicide Resistance in Transgenic Plants Expressing a Bacterial Detoxification Gene," *Science* 242:419-423 (1988).
175. Stalker, D.M., Calgene Inc., personal communication.
176. Plapp, F.W., Jr. "Genetics and Biochemistry of Insecticide Resistance in Arthropods: Prospects for the Future," in *Pesticide Resistance: Strategies and Tactics for Management*, (Washington, D.C.: National Academy Press, 1986), pp. 74-86.
177. Caseley, J. C. "Improving Herbicide Performance with Synergists that Modulate Metabolism," in *IWCC Proceedings, First International Weed Control Congress, Vol. 2*, pp. 113-115 (Weed Science Society of Victoria, Inc. Melbourne, Australia, 1992).
178. Powles, S. B., and J. M. Matthews. "Multiple Herbicide Resistance in Annual Ryegrass (*Lolium rigidum*). A Driving Force for Adoption of Integrated Weed Management," in *Achievements and Developments in Combating Pest Resistance*, I. Denholm, A. Devonshire, and D. Holloman, Eds. (London: Elsevier Press, 1992), pp. 75-87.
179. Putwain, P. D., and A. M. Mortimer. "The Resistance of Weeds to Herbicides: Rational Approaches for Containment of a Growing Problem," *Brighton Crop Protection Conference — Weeds*, (Farnham, U.K.: The British Crop Protection Council, 1989), pp. 285-294.

CHAPTER 3

Resistance to Photosystem I Disrupting Herbicides

Christopher Preston

INTRODUCTION

Photosystem I

Photosystem I (PS I) is a membrane-bound protein complex which catalyzes the light-driven oxidation of plastocyanin (PC) and reduction of ferredoxin (Fd). PS I is made up of 10 or 11 proteins not including the chlorophyll a/b binding proteins, of which there are at least three in several copies each.[1] PS I transfers an electron from PC to Fd requiring a free-energy change of about 0.7 eV. A photon of light captured by the chlorophyll bed excites an electron in the pigment. This excitation energy is passed to P700, the trap chlorophyll of PS I, which has a slightly lower energy than the other chlorophylls. P700 passes the excited electron to an intermediate electron acceptor called A_0, which is probably a specialized chlorophyll a molecule,[2] and then P700 recovers an electron from plastocyanin to return to the ground state. The electron on A_0 rapidly traverses several carriers of increasing (less negative) redox potential prior to being transferred to Fd. The first carrier, described as A_1, is a phylloquinone, or vitamin K molecule,[3] which is followed by three iron-sulfur centers known as F_X, F_A, and F_B. The energetics of the different PS I electron carriers is illustrated in Figure 1.

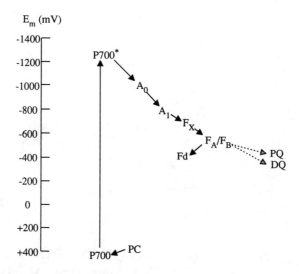

Figure 1. The energetics of electron carriers in Photosystem I including paraquat (PQ) and diquat (DQ) as electron acceptors. Modified from Reference 1.

Due to the low redox potentials of the reducing side of PS I, there are relatively few chemicals able to interact with this side of PS I. Some chemicals able to accept electrons are the viologens (paraquat, diquat, triquat, benzyl viologen, and others), Saranine T, some anthroquinone sulfonates, and methyl purple.[4-7] The sites where these molecules accept electrons have not all been determined; however, several accept from the iron-sulfur centers and some probably from Fd.[7,8] There are no chemicals analogous to the Photosystem II inhibiting herbicides (Chapter 2) which block electron transport in PS I. A larger number of chemicals such as reduced phenazine methosulfonate, reduced dichlorophenolindophenol, diaminodurene, and p-phenylenediamines are able to donate, either directly or indirectly, to PS I,[6,7] but these have been of little interest for the development of herbicides.

Only one class of chemistry whose main mode of action occurs through interacting with PS I has been commercialized. These are the bipyridyl herbicides paraquat (otherwise known as methyl viologen) and diquat (Figure 2). These are fast-acting, nonselective contact herbicides. Paraquat is more active against grass species, whereas diquat has some increased activity against broadleaf weeds.[9]

Mode of Action of Bipyridyl Herbicides

Paraquat and diquat are redox active compounds which interact with PS I (Figure 1). Electrons are siphoned from one of the iron-sulfur centers, F_B,[8,10] forming a bipyridyl (BP) cation radical (Equation 1).

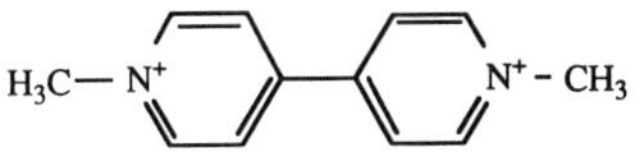

Paraquat (E_{m7} = -446 mV)

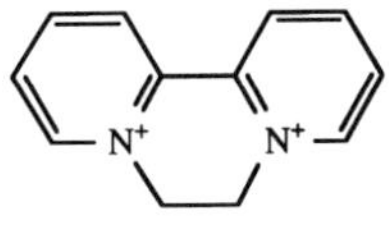

Diquat (E_{m7} = -349 mV)

Figure 2. Structure and midpoint redox potentials of the bipyridyl herbicides paraquat and diquat.

$$BP^{2+} + e^- \rightarrow BP^{\cdot+} \qquad [1]$$

The bipyridyl cation radical is unstable and will react rapidly with O_2 to form superoxide, regenerating the bipyridyl cation (Equation 2).

$$BP^{\cdot+} + O_2 \rightarrow BP^{2+} + O_2^- \qquad [2]$$

The plant is able to detoxify superoxide via superoxide dismutase (SOD) producing hydrogen peroxide and molecular oxygen (Equation 3).

$$2\, O_2^- + 2\, H^+ \rightarrow H_2O_2 + O_2 \qquad [3]$$

H_2O_2 itself is toxic and can be further detoxified by the enzymes of the ascorbate-glutathione cycle (Figure 3). The overall reaction can be described as (Equation 4):

$$H_2O_2 + NADPH + H^+ \rightarrow 2\, H_2O + NADP^+ \qquad [4]$$

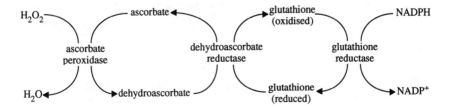

Figure 3. The ascorbate-glutathione cycle for detoxification of H_2O_2 within the chloroplast. Modified from Reference 14.

There are other reactions which can occur within the chloroplast involving H_2O_2. Trace levels of Fe^{2+} can catalyze the Fenton reaction[11] (Equation 5).

$$Fe^{2+} + H_2O_2 \rightarrow Fe^{3+} + OH^- + OH\cdot \qquad [5]$$

Other reactions are also possible (Equations 6 and 7).

$$Fe^{3+} + H_2O_2 \rightarrow Fe^{2+} + O_2^- + H^+ \qquad [6]$$

$$Fe^{3+} + O_2^- \rightarrow Fe^{2+} + O_2 \qquad [7]$$

The net result of these reactions is the production of the hydroxyl radical, $OH\cdot$. Other alternatives are the various versions of Winterbourn's reactions[12] (Equations 8 and 9).

$$BP^{\cdot+} + H_2O_2 \rightarrow BP^{2+} + OH^- + OH\cdot \qquad [8]$$

$$BP^{\cdot+} + Fe^{3+} \rightarrow BP^{2+} + Fe^{2+} \qquad [9]$$

The Fe^{2+} generated can then react through the Fenten reaction (Equation 5). The hydroxyl radical, rather than superoxide, is probably the damaging species. Babbs et al.[13] have shown that considerable quantities of hydroxyl radicals are produced in plants following paraquat application.

The active oxygen species produced following paraquat or diquat action can peroxidate lipids. The radicals attack double bonds in the fatty acid side chains of lipids,[14] with the subsequent membrane disruption resulting in death of the plant cell.

Paraquat and diquat are able to interact with other electron transfer systems and hence are toxic to animals. They can also be toxic to plants kept in the dark, but the herbicidal action is slower.[15,16] The plant mitochondrial electron transport system is possibly not the site of paraquat action in the dark as the redox potential of paraquat is too low to accept electrons from the NADH-NAD$^+$ couple.[17] The cytochrome P450 enzyme system, which is the site of paraquat action in lung tissue,[18] is a more likely site of paraquat action in the dark.

Agricultural Uses

Paraquat and diquat have found numerous agricultural and industrial uses as nonselective, rapid-action herbicides. These herbicides are tightly bound by clays and normally have no soil activity,[19] which makes them ideal for use preplanting as the crop can be sown soon after herbicide application. These herbicides are also used for postemergent weed control during the dormant stage of perennial crops such as alfalfa and peanuts, or in situations such as orchards and vineyards where they can be applied away from the foliage of the crops. Another important use is as a preharvest dessicant for crops such as potatoes. They have also been used to clear irrigation waterways of weed growth.

There is generally little selectivity of these herbicides; however, some species show more tolerance as a result of waxy cuticles or other structures which reduce herbicide penetration.[20] Perennial species are normally tolerant of these herbicides as they have the capacity to regrow from dormant structures.

RESISTANCE TO BIPYRIDYL HERBICIDES

Resistance in Weed Species

After almost two decades of use, paraquat resistance was first reported in the late 1970s and early 1980s when resistance was observed in *Conyza bonariensis* (L.) Cronq. from vineyards and citrus plantations in Egypt,[21] *Erigeron philadelphicus* L. from mulberry fields,[22] and *E. canadensis* L. from vineyards in Japan,[23] *Hordeum glaucum* Steud. from alfalfa fields in Australia,[24] and *Poa annua* L. from hop gardens in the U.K.[25] In total, resistance to paraquat or diquat (frequently both) has been reported in the field for 4 grass species and 12 broadleaf weeds (Table 1). In many cases resistance has appeared in more than one biotype of a species and, in several cases, on different continents.

Resistance to these herbicides has become apparent in the field following many exposures. Frequently resistance has appeared following 5 to 10 applications of the herbicides each year for 5 or more years.[22,32,34] In other situations consecutive once annual applications for 12 years or more has led to resistance.[16,27,28] Some paraquat-resistant biotypes of *E. canadensis*, *P. annua*, and *Epilobium ciliatum* Rafin are also resistant to the triazine herbicides following selection with both herbicides.[31,32] In all of the cases examined to date the paraquat-resistant plants have shown some resistance to diquat (Table 1), but, with the exception of the triazine-resistant biotypes mentioned above, there is no resistance to other classes of herbicides. Where reported, the level of diquat resistance is considerably less than that for paraquat, except in the case of *Vulpia bromoides* (L.) S. F. Gray.[38]

It is often difficult to compare resistance indices between biotypes because of the diversity of techniques used to measure resistance. The most relevant method to measure resistance in agricultural weeds is to spray plants grown in soil under agriculturally relevant conditions and to examine growth or survival. The rapid reaction of paraquat and diquat as bleaching herbicides has allowed laboratory techniques involving floating leaf pieces on solutions of herbicide under lights and then measuring one or more of chlorophyll bleaching, solute leakage, or pheophytin accumulation. While providing rapid and reproducible results, it is readily apparent that such measurements will not always provide a true indication of resistance at the whole plant level, and the mix of techniques for measuring resistance indices often makes comparisons difficult.

Table 1. Weed species that have developed resistance to bipyridyl herbicides

Species	Resistance index[a] Paraquat	Diquat	Occurrence	Ref.
Amaranthus lividus L.	+	nd	Malaysia	26
Arctotheca calendula (L.) Levyns	60	10	Australia	27, 28
Crassocephalum crepidioides (Benth.) S. Moore	+	nd	Malaysia	26
Conyza bonariensis (L.) Cronq.[b]	150	5	Egypt	21
	+	nd	Hungary	29
	+	nd	Kenya	29
Epilobium ciliatum Rafin	10	nd	Belgium	30
	>5	nd	U.K.	31
Erigeron canadensis L.[b]	45	50	Hungary	32, 33
	100	+	Japan	34, 35
Erigeron philadelphicus L.[b]	250	3	Japan	22
Erigeron sumatrensis Retz.[b]	400	10	Japan	34
	+	nd	Malaysia	26
Hordeum glaucum Steud.	250	>6	Australia	16
Hordeum leporinum Link	100	>6	Australia	36
Parthenium hysterophorus L.	+	nd	Kenya	29
Poa annua L.	3–4	nd	U.K.	31
Solanum americanum Mill.	12	4	U.S.A.	37
Solanum nigrum L.	+	nd	Malaysia	26
Vulpia bromoides (L.) S. F. Gray	5–6	~5	Australia	38
Youngia japonica (L.) DC.	110	nd	Japan	34

[a]Resistance index is the ratio of the concentration of paraquat which causes 50% injury (or death) to the resistant biotype divided by the concentration which causes 50% injury in the susceptible biotype. +, A report of resistance with insufficient information provided to calculate an index of resistance. nd, Resistance to this herbicide was not examined.

[b]*Conyza* and *Erigeron* are considered by some authorities to be the same genus. The biotypes here are sorted by specific names.

Table 2. Paraquat-resistant mutants obtained through laboratory selection

Species	Resistance index[a] Paraquat	Diquat	Selection system	Ref.
Ceratopteris richardii Brongn.	10–20	+	Gametophyte generation	39
Chlamydomonas reinhardtii Dangeard	~3	nd	UV mutants	40
	3	nd	Natural variation	41
Lolium perenne L.	6	+	Natural variation	42
Nicotiana tabacum L.	2–3	+	Cell culture	43, 44

[a]Resistance index is the ratio of the concentration of paraquat which causes 50% injury (or death) to the resistant biotype divided by the concentration which causes 50% injury in the susceptible biotype. +, A report of resistance with insufficient information provided to calculate an index of resistance. nd, Resistance to this herbicide was not examined.

Resistance Obtained through Laboratory Selection

Resistance to paraquat has also been deliberately selected in plants in both the laboratory and the field (Table 2). Paraquat resistance was selected in *Lolium perenne* L. by the continued selection of surviving lines.[42] Hickok and Schwarz[39] selected paraquat-resistant mutants of *Ceratopteris richardii* Brongn.

by applying the selection agent to the haploid gametophyte generation. Paraquat-resistant biotypes of the unicellular green alga *Chlamydomonas reinhardtii* Dangeard have also been selected.[40,41] Paraquat-resistant cell cultures have been obtained for several species and in at least two cases paraquat-resistant plants that breed true have been regenerated from tobacco cell cultures.[43,44]

MECHANISMS OF RESISTANCE TO BIPYRIDYL HERBICIDES

Absorption

Absorption of both paraquat and diquat through the leaf cuticle is extremely rapid despite the divalent cationic charge on these herbicides.[45] In the cases where it has been studied, absorption of ^{14}C-labeled paraquat or diquat does not differ between R and S biotypes,[28,46-48] or, in the case of *C. bonariensis*, more paraquat was absorbed by the R biotype.[21]

Translocation

Most studies examining the translocation of paraquat have been performed by applying ^{14}C-herbicide through cut leaves or petioles. It should be noted that this is an artificial method of examining translocation of these herbicides as translocation would normally be expected to occur in a basipetal direction following foliar application. Early studies on the translocation of these herbicides suggested that paraquat moved in the xylem and could be translocated by local reversals of xylem flow.[49-51]

Reduced herbicide translocation within the leaf of resistant biotypes, as observed by autoradiography, has been demonstrated for *E. philadelphicus*,[52] *E. canadensis*,[52] *H. glaucum*,[46] and *C. bonariensis*[21] following herbicide application via cut petioles or stem bases. Recently, reduced translocation has been demonstrated in R biotypes of *H. glaucum* and *H. leporinum* Link compared to the S biotypes. In these experiments, paraquat was applied to intact plants at field doses under conditions approximating field applications. Paraquat was then assayed directly and reduced translocation in a basipetal direction was observed.[53] Experiments on these R biotypes grown in the field have also observed that herbicide translocation is more rapid in the S biotypes.[54] Reduced basipetal translocation of ^{14}C-labeled paraquat applied to the leaf lamina has been demonstrated in a second R biotype of *H. leporinum*.[28] Clearly, in these two *Hordeum* species, reduced paraquat translocation occurs in the R biotypes compared to the S biotypes.

An alternative, and very useful, method of estimating paraquat and diquat translocation is to measure photosynthetic parameters. The interaction of paraquat at the active site leads to a reduction in net O_2 evolution (due to increased Mehler reaction) and hence, inhibition of CO_2 fixation, as a result of diversion of electrons from NADPH. Paraquat disruption of PS I also results in a quenching of variable fluorescence, due to the efficient accepting of electrons by the herbicide which maintains the plastoquinone pool in an oxidized state. Therefore, both photosynthetic gas exchange and fluorescence techniques can be used as *in vivo* assays for the disruptive effects of paraquat on photosynthesis. These techniques have been used to examine the effects of paraquat on leaves and such measurements can be a valuable tool for assessing the presence and effect of herbicide at the active site, information which cannot be obtained using conventional ^{14}C techniques. In almost all cases reported so far, S biotypes have shown an immediate and rapid decrease in photosynthesis following application of the herbicide. The responses of the R biotypes fall into three main patterns: no change, a delayed or reduced inhibition, and a transient inhibition which recovers.

No inhibition of CO_2 fixation was observed in an R biotype of *L. perenne* following paraquat application, whereas CO_2 fixation was completely inhibited in the S biotype when plants were illuminated after treatment.[55] In this biotype a mechanism which prevents herbicide from getting to the chloroplast appears to operate. This does not correlate with increased amounts of SOD observed in this biotype[56] (see below), but as resistance is multigenic[57] both mechanisms may operate.

CO_2-dependent O_2 evolution in leaf tissue of R biotypes of both *H. glaucum* and *H. leporinum* decreased following paraquat application to whole plants. However, this decrease was delayed in the R biotypes compared to the S biotypes.[53] These plants were kept in the dark after treatment. This data is consistent with a mechanism of reduced translocation in the resistant biotypes of these two species that has been demonstrated using other techniques.[46,53]

Using fluorescence imaging of intact leaves, reduced penetration of diquat to the active site from a single droplet applied to the leaves of resistant *Arctotheca calendula* plants was observed.[48] Diquat penetration to the active site was rapid in the S biotype, but much slower in the R biotype for plants kept in the dark. The differences observed with fluorescence imaging, which detects herbicide at the active site, were not observed when translocation was monitored by ^{14}C-labeled diquat, despite an identical application technique.[48] The differences observed with fluorescence imaging were mirrored by measurements of CO_2-dependent O_2 evolution which showed that diquat and paraquat reached the active site of the S biotype and inhibited O_2 evolution rapidly following treatment with agriculturally relevant rates of herbicide. Inhibition of O_2 evolution was not as rapid nor as pronounced in the R biotype. Preston et al.[48] concluded that there was no difference in gross translocation of herbicide in the R biotype of *A. calendula*, but movement to the active site was reduced, probably as a result of reduced penetration of herbicide into the leaf cells.

A transient inhibition of photosynthesis, measured as CO_2 fixation, O_2 evolution, or fluorescence, was observed in an R biotype of *E. canadensis* following paraquat treatment if plants were placed in the light.[58] Photosynthetic activity of the S biotype was completely inhibited. The R biotype showed signs of recovery

4 h after treatment; however, full recovery did not occur until more than 24 h had elapsed. No recovery of photosynthesis was observed in either biotype if plants were placed in the dark. These results are consistent with a hypothesis of light-driven active removal of paraquat from the chloroplasts of the resistant biotype. Paraquat transiently inhibited $^{14}CO_2$ fixation of the R biotype of *C. bonariensis* to a greater,[59] or lesser,[60] extent. In the light full recovery of the R biotype occurred in 4 h and the S biotype remained inhibited.[59] In the dark, or low light, both biotypes recovered from the effects of paraquat treatment within 3 h.[61] These results have been suggested to show that paraquat can be rapidly detoxified or removed from the chloroplasts of the R biotype in the light and both biotypes in the dark.[59,61] In direct contradiction of these results is the work of Fuerst et al.,[21] who observed inhibition of fluorescence in both biotypes following 4 h of darkness after treatment with paraquat, diquat, or triquat. In all cases the inhibition was more severe in the S biotype compared to the R biotype indicating that more herbicide was at the active site in the S biotype.

To conclude, photosynthetic measurements have proved to be a valuable tool for following the appearance of herbicides at the active site. The studies done so far clearly indicate delayed and/or reduced appearance of herbicide at the active site in R biotypes of *H. glaucum, H. leporinum, L. perenne*, and *A. calendula*. Active removal of herbicide from the chloroplast is indicated in the case of *E. canadensis*; however, the situation in *C. bonariensis* is ambiguous and needs clarification.

Sequestration

Sequestration of paraquat as a mechanism for resistance has been proposed for both *C. bonariensis*[21,62] and *H. glaucum*[63] with increased binding of herbicide to cell walls suggested as the mechanism for sequestration. Recently, Hart et al.[64] have shown in maize roots that >75% of applied paraquat becomes associated with cell walls. When the interaction of paraquat with cell walls of R and S biotypes of *C. bonariensis* was examined, no difference could be demonstrated.[21] In *A. calendula*, no differences in the binding of diquat to cell walls, as measured by either absorption or efflux, could be detected between the R and S biotypes.[48] Paraquat movement into leaf cells has been examined in R biotypes of *H. glaucum*,[53,65] *H. leporinum*,[53] *A. calendula*,[48] and *C. bonariensis*.[66] In these studies paraquat movement into cells of thin leaf slices was determined indirectly by measuring inhibition of O_2 evolution or chlorophyll destruction. The results obtained in all cases, except *H. leporinum*,[53] suggested that paraquat movement into leaf cells and to the active site was restricted in the R biotype compared to the S biotype. This reduced movement of paraquat into leaf cells has been examined in some detail for *H. glaucum*, where two components of paraquat uptake were identified. One could be competitively inhibited by the addition of the polyamine cadaverine, but not by putrescine, and was absent in the R biotype. The second component was present in both biotypes.[65] From these studies it was suggested that a change had occurred in the transporter involved in paraquat uptake into cells which resulted in reduced herbicide uptake and hence resistance in this biotype. Hart and DiTomaso,[67] in a study examining efflux of pre-loaded ^{14}C-labeled paraquat from roots of R and S biotypes of *H. glaucum,* observed slower efflux of paraquat from the vacuoles of the R biotype, and suggested that paraquat might be sequestered in the vacuoles of the R biotype. From the limited data available it appears that paraquat might be "sequestered" outside the leaf cells in R biotypes of *H. glaucum, A. calendula* and *C. bonariensis*. This sequestration is not due to increased binding of the herbicide to the cell wall, but is a result of reduced herbicide uptake into leaf cells.

Active Site Changes

In the several resistant biotypes where the active site has been examined, no differences have been observed in the interaction of paraquat, diquat, or triquat with PS I between the R and S biotypes. This has been shown for R biotypes of *H. glaucum*,[63] *H. leporinum*,[28] *C. bonariensis*,[21] *L. perenne*,[47] and *A. calendula*.[48] Paraquat and diquat have similar affinities for PS I in isolated thylakoid membranes (about 5 to 15 µM) in all R and S biotypes examined. Given the redox nature of the interaction of paraquat with PS I, resistance due to active site changes is unlikely to occur. Mutants deficient in PS I, and hence resistant to paraquat, will occur; however, in the field these will be lethal.

Metabolism of the Herbicide

Despite herbicide metabolism being a major selectivity and resistance mechanism in many plant species (see Chapters 2, 4, 5, 6, and 9), metabolism is unlikely to be responsible for resistance to paraquat. Metabolism of paraquat has been observed in a few isolates of bacteria and fungi.[68-70] The main products of metabolism observed are monoquat and *N*-methyl isonicotinic acid. Monopyridone and methylamine have also been detected as metabolites. In all cases these catabolites represented only a small percentage of the total ^{14}C-paraquat applied. No metabolism of paraquat or diquat has yet been detected for any higher plants. However, in resistance studies, metabolism of these herbicides has only been investigated in R biotypes of *L. perenne*[47] and *C. bonariensis*.[66] Neither of these biotypes showed any metabolism of paraquat.

Metabolism of Toxic Products of Herbicide Action

The herbicidal action of paraquat and diquat are not, primarily, a result of diverting electron flow from PS I, but are a result of the action of toxic O_2 species produced by the interaction of the herbicide cation radicals with O_2. There are a variety of enzymes present in plants to detoxify these toxic species and many of these enzymes become elevated with stress. With SOD, at least, the triggering factor for this increase in enzyme amount is the presence of superoxide.[71] There is considerable compartmentation of these enzymes within a plant cell; for example, different isoenzymes of SOD are localized to the chloroplast, the cytoplasm, and the mitochondrion.[72] The chloroplastic enzyme is by far the most abundant in green leaf tissue.[72] Ascorbate peroxidase activity can be observed in chloroplasts and cytosol,[73] glutathione reductase activity can be observed mainly in the chloroplast,[74] although a cytoplasmic form of the enzyme has also been observed,[75] and catalase is absent from the chloroplast.[76]

Elevated amounts of these enzymes have been proposed as the mechanism of resistance to paraquat in a number of biotypes. A resistant biotype of *C. bonariensis* was reported to have increased amounts of SOD and two new isoenzymes when compared to the S.[60] A later study was unable to observe any consistent differences in SOD isoenzymes in a population of *C. bonariensis* segregating into R and S individuals.[77] A third study was unable to observe increased amounts of SOD in whole plant extracts, but could observe a 50% increase in SOD activity in isolated chloroplasts from the R biotype.[78] In addition, amounts of ascorbate peroxidase and glutathione reductase were also increased, by 150 and 200%, respectively. A fourth study reported increased amounts of ascorbate peroxidase, by 29%, in chloroplasts of the R biotype,[66] with no increases in SOD or glutathione reductase. In addition this study observed no resistance to paraquat using isolated chloroplasts or protoplasts, but did observe resistance using thin leaf slices.[66] The levels of SOD and glutathione reductase in both R and S biotypes of *C. bonariensis* were observed to vary widely during ontogeny as did the level of paraquat resistance in the R biotype.[79] The resistance index was highest (~300-fold) when the ratio of activity of these enzymes in the R biotype compared to the S biotype was highest (~1.2- to -1.5-fold). However, for some of the developmental period there was no increased levels of these enzymes in the R biotype, yet the resistance index was still 20- to 30-fold.[79]

Paraquat resistance in *L. perenne* was also correlated with an increase in the amount of SOD by 50 to 100% in four R cultivars compared to ten S cultivars.[56] Peroxidase and catalase enzyme concentrations were also increased in some lines but not others. Glutathione reductase levels were not measured. In laboratory-developed paraquat-resistant biotypes of *C. reinhardtii*, increases of between 60 and 120% were observed in the activity of SOD,[41] and in paraquat-resistant tobacco, the concentration of this enzyme was increased by 200 to 300%.[80] In the latter case the amounts of ascorbate peroxidase and glutathione reductase were unchanged. Enzyme amounts in intact chloroplasts isolated from paraquat-resistant *E. canadensis* were also increased by between 60 and 90% for SOD, ascorbate peroxidase, monodehydroascorbate reductase, and glutathione reductase.[81] In this biotype resistance to bleaching by paraquat was reported in isolated protoplasts, but not isolated chloroplasts.

No difference in amounts of these enzymes has been observed in isolated chloroplasts from R and S biotypes of *C. richardii*,[82] in crude extracts from R and S biotypes of *H. glaucum*,[63] or from separate accessions of R and S *E. canadensis*.[83] In addition, resistance was not observed when protoplasts or chloroplasts were isolated from leaves of the R biotype of *H. glaucum*.[63] Increases in SOD or other enzymes of the ascorbate-glutathione cycle do not contribute to paraquat resistance in these species.

There has been considerable debate in the literature over the ability of increased amounts of these enzymes to contribute to paraquat resistance, particularly in *C. bonariensis*.[84] For this reason it is worthwhile examining this proposal at some length. The first point to consider is that the levels of these enzymes in the plant respond to environmental stimuli such as light, water stress, temperature, pollutants, pathogens, and xenobiotics.[71,72] In the abundant literature available, no single pattern emerges regarding enzyme response to all such variables.[85-88] It is therefore vitally important that enzyme determinations are performed on unstressed plants grown under the same conditions at the same time with leaves of the same age.[79] Secondly, increased amounts of these enzymes in themselves will not confer much resistance to paraquat as the reaction producing superoxide regenerates the paraquat cation (Equation 2). Therefore, paraquat will remain in the chloroplast generating toxic radicals, and to survive, the plant must have some method of removing the herbicide from the active site. Plants probably have a capacity for removal of compounds like paraquat to the vacuole,[66] although this has not been conclusively demonstrated. An increase in the level of detoxifying enzymes might allow an increase in the amount of herbicide removed to the chloroplast and thereby provide a modest increase in resistance.

The third point is that increased amounts of detoxifying enzymes should also give tolerance to other conditions which lead to superoxide production in the chloroplast. Where the radical-generating component is similar to paraquat, for example diquat and morfamquat, the resistance index, all other things being equal, should be similar. Conversely, cross resistance to some radical-generating conditions, such as ozone and drought, could vary as these have effects unrelated to superoxide generation in the chloroplast.[89,90] The paraquat-resistant *C. bonariensis* biotype has been reported to be resistant to a wide range of other stress conditions including diquat,[61,91,92] but is not resistant to some stresses,[62,93] most notably morfamquat. In addition, the resistance index of the paraquat-resistant biotypes of *C. bonariensis* and *E. canadiensis* to diquat is much less than that to paraquat. A paraquat-resistant biotype of *L. perenne*, in contrast, shows a fivefold increase in resistance to morfamquat.[94]

Recently, tobacco plants have been successfully transformed with genes for some of the enzymes of the ascorbate-glutathione cycle (Table 3). Transgenic tobacco plants with increased constitutive Cu/Zn SOD showed no increase in paraquat resistance.[95] In a second report, transgenic tobacco plants transformed with Mn SOD, the mitochondrial protein, targeted to the chloroplast show both elevated SOD and a twofold increase in resistance to paraquat.[96] The discrepancy between these two reports apparently lies with the fact that Cu/Zn SOD is sensitive to hydrogen peroxide, the product of the SOD reaction (Equation 3), whereas Mn SOD is not.[72] In contrast, another study which elevated Cu/Zn SOD in the chloroplast of transgenic tobacco did observe an

Table 3. Paraquat resistance obtained in transgenic tobacco plants transformed with genes encoding superoxide dismutase and glutathione reductase

Enzyme	Enzyme activity (% increase)	Targeted	Paraquat resistance index obtained	Ref.
Cu/Zn SOD	3000	Chloroplast	1	96
	100	Chloroplast	2	97
Mn SOD	300	Chloroplast	2	98
Glutathione reductase	230	Chloroplast	2–3	99
	300	Cytoplasm	2	99, 100

increase in resistance to paraquat of about twofold.[97] Transgenic tobacco plants containing increased constitutive glutathione reductase give about a twofold increase in paraquat resistance, when the gene was targeted to either the chloroplast or cytoplasm.[98,99] The paraquat-resistant laboratory strains of tobacco[44] and *C. reinhardtii*,[40] which contain elevated amounts of SOD also had two- to threefold increases in resistance.

To conclude, increased amounts of detoxifying enzymes can clearly provide resistance to paraquat; however, the paraquat-resistance index obtained appears to be small (Table 3). Resistance in plants with increased amounts of detoxifying enzymes should also be apparent at the protoplast and, probably, the chloroplast levels. In the paraquat-resistant biotypes of *C. bonariensis* and *E. canadiensis* another mechanism, other than increased enzyme activities, appears to be needed to explain the high level of resistance observed in these biotypes. This second mechanism could well involve decreased translocation of paraquat as has been reported in other studies on these biotypes.[21,52,66] Clearly other mechanisms of resistance in these biotypes need to be investigated in more detail before resistance can be attributed solely to increased constitutive oxygen detoxifying enzymes.

To date only two mechanisms providing resistance to paraquat have been observed. Reduced translocation and/or sequestration of the herbicides have been shown to be a mechanism operating in R biotypes of *H. glaucum*, *H. leporinum*, *A. calendula*, *C. bonariensis*, *E. canadensis*, and *E. philadelphicus*. Increased levels of SOD, glutathione reductase, or other protective enzymes have been unambiguously shown to confer low-level paraquat resistance in laboratory-developed strains of tobacco and *C. reinhardtii*. This mechanism has also been proposed for R biotypes of *C. bonariensis*, *L. perenne*, and *E. canadensis*; however, this remains to be comprehensively proven.

BIOLOGY OF RESISTANT BIOTYPES

Aspects of Growth and Fitness

The general biology of paraquat-resistant weed biotypes has not been investigated in great detail. Relative fitness has been examined in paraquat-resistant biotypes of *H. glaucum* and *H. leporinum* only. The R biotype of *H. glaucum* proved to have slightly reduced fitness when compared to the S biotype both in accumulation of biomass and in classical replacement series competition experiments.[100] The R biotype of *H. leporinum* showed no reduction in fitness under field conditions.[28] However, this biotype matured earlier than the S biotype. Growth of the R biotype of *E. canadensis* was less vigorous than that of the S biotype, suggesting the R biotype would be less fit.[34] Further details on fitness of R biotypes can be found in Chapter 11.

The relative resistance index for paraquat was observed to change with season in an R biotype of *H. leporinum*. This biotype showed a high resistance index when grown in the winter, but in summer the resistance index was dramatically decreased.[101] This change in the level of resistance is a result of increased temperature and not of increased light intensity.[101] This decrease in resistance level is due to increased translocation of herbicide in the R biotype at high temperatures.[28]

Inheritance of Bipyridyl Resistance

Where studied, most cases of paraquat and diquat resistance can be attributed to a single major gene (see Chapter 10). Paraquat resistance in *C. bonariensis* was shown to be a result of a single dominant nuclear-encoded gene.[102] Moreover, the increased levels of oxygen detoxifying enzymes segregated in the F2 population. Paraquat resistance in both *E. philadelphicus*[103] and *E. canadensis*[35] is also the result of a single dominant, nuclear-encoded gene. Paraquat and/or diquat resistance in *H. glaucum*,[104] *H. leporinum*,[105] and *A. calendula*[105] has been shown to be a result of a single, nuclear-encoded, incompletely dominant gene. In these three species the F2 individuals segregate into R, S, and intermediate individuals. Paraquat resistance in *C. richardii* is the result of a nuclear-encoded recessive gene with two separate alleles which gave different resistance indices.[106] A second, modifying gene which gave a higher resistance index in combination with the resistance gene, but no resistance on its own, was also identified.[107] Paraquat resistance in cultivars of *L. perenne*, in contrast to the above studies, was observed to be inherited as the result of a number of genes.[57] In this study no direct comparison was made between a single R and a single S biotype, so the number of genes involved in any one R biotype is unknown. More information on the inheritance of resistance genes can be found in Chapter 10.

In most of the above cases, no correlation has been attempted between the genetics and the mechanism of resistance, largely because the mechanisms are unknown. As the two are closely linked, i.e., a single gene encodes a single protein, such comparisons must be made. It is difficult to reconcile the mechanism of resistance in *C. bonariensis*, due to increased activities of three enzymes,[102] with a single gene change, unless that gene is a regulatory gene. A single operon regulates nine genes involved in the detoxification of active oxygen species in bacteria;[108] however, the genetic regulation of oxygen detoxifying enzymes in higher plants is not understood,[72] so it is difficult to properly interpret these results. In the case of *L. perenne*, where resistance may be due to more than one gene,[57] it is easier to envisage increases in activity of more than one enzyme.

CONCLUSIONS

Resistance to paraquat is largely a minor nuisance although it has appeared in a significant number of species. With few exceptions these resistant biotypes have not developed resistance to other herbicides and so their control has not, so far, been a serious problem. The long time scale for the development of paraquat resistance and widespread usage due to the inexpensive and efficacious nature of these herbicides means that new cases of resistance can be expected.

In general there has been relatively little progress in understanding the mechanisms of paraquat resistance at the molecular level. In a number of biotypes, resistance is due to decreased translocation of the herbicide, which in the cases of *H. glaucum*,[53] *C. bonariensis*,[66] and *A. calendula*[48] appears to be a result of decreased herbicide movement into leaf mesophyll cells. The molecular basis for this decreased movement is not known, largely because of the paucity of information available on the mechanism of paraquat uptake into plant leaf cells. Paraquat uptake into plant root cells, at least in maize roots, has been suggested to occur on a polyamine carrier.[109] An alteration in a membrane transporter of this type could easily account for resistance to paraquat in *H. glaucum*. Unfortunately the polyamine carrier in maize roots transports putrescine, cadaverine, and paraquat,[109] whereas only cadaverine seems to inhibit paraquat uptake into *H. glaucum* leaf cells.[65] This discrepancy remains to be investigated.

Resistance to paraquat has been a particularly intriguing subject because of the nature of the herbicide. The perception that resistance to such nonselective, rapid-action herbicides would be rare has, in part, been confirmed when compared to some herbicides with other modes of action. In addition, from the resistance point of view, paraquat has the advantages of not having the two most common pathways to resistance, metabolism of the herbicide and target site changes, available. Hence the mechanisms of resistance to the bipyridyl herbicides, when established, are likely to be present at very low frequencies in wild, unselected populations and be of considerable scientific interest. Clearly, more effort will be required before the mechanistic basis of paraquat resistance in the field is fully understood.

ACKNOWLEDGMENTS

I wish to thank Dr. E. P. Fuerst, Professor J. Gressel, Dr. S. B. Powles and Dr. J. M. DiTomaso for allowing me to quote from unpublished material. Thanks also to Dr. E. P. Fuerst for his constructive comments on the manuscript.

REFERENCES

1. Golbeck, J. H. "Structure and Function of Photosystem I," *Annu. Rev. Plant Physiol. Plant Mol. Biol.* 43:293-324 (1992).
2. Mansfield, R. W., and M. C. W. Evans. "EPR Characteristics of the Electron Acceptors A_0, A_1, and (Iron-sulfur)$_X$ in Digitonin and Triton X-100 Solubilized Pea Photosystem I," *Isr. J. Chem.* 28:97-102 (1988).
3. Malkin, R. "On the Function of Two Vitamin K_1 Molecules in the PS I Electron Acceptor Complex," *FEBS Lett.* 208:343-346 (1986).
4. Izawa, S. "Acceptors and Donors for Chloroplast Electron Transport," *Methods Enzymol.* 69:413-434 (1980).
5. Graan, T., D. R. Ort, and R. C. Prince. "Methyl Purple, an Exceptionally Sensitive Monitor of Chloroplast Photosystem I Turnover: Physical Properties and Synthesis," *Anal. Biochem.* 144:193-198 (1985).
6. Trebst, A. "Measurements of Hill Reactions and Photoreduction," *Methods Enzymol.* 24:146-165 (1972).
7. Mathis, P., and A. W. Rutherford. "The Primary Reactions of Photosystems I and II of Algae and Higher Plants," in *Photosynthesis*, J. Amesz, Ed. (Amsterdam: Elsevier, 1987), pp. 63-96.
8. Fujii, T., E. Yokoyama, K. Inoue, and H. Sakurai. "The Sites of Electron Donation of Photosystem I to Methyl Viologen," *Biochim. Biophys. Acta* 1015:41-48 (1990).
9. Calderbank, A., and P. Slade. "Diquat and Paraquat," in *Herbicides: Chemistry, Degradation and Mode of Action*, P. C. Kearney and D. D. Kaufman, Eds. (New York: Marcel Dekker Inc., 1976), pp. 501-540.
10. Golbeck, J. H., and J. M. Cornelius. "Photosystem I Charge Separation in the Absence of Centers A and B. I. Optical Characterization of Centre A_2 and Evidence for its Association with a 64-kDa Protein," *Biochim. Biophys. Acta* 849:16-24 (1986).
11. Fridovich, I. "Superoxide Radical: an Endogenous Toxicant," *Annu. Rev. Pharmacol. Toxicol.* 23:239-257 (1983).
12. Winterbourn, C. C. "Production of Hydroxyl Radicals by Paraquat Radicals and H_2O_2," *FEBS Lett.* 128:2339-2342 (1981).
13. Babbs, C. F., J. A. Pham, and R. C. Coolbaugh. "Lethal Hydroxyl Radical Production in Paraquat-treated Plants," *Plant Physiol.* 90:1267-1270 (1989).
14. Kunert, K. J., and A. D. Dodge. "Herbicide-induced Radical Damage and Antioxidative Systems," in *Target Sites of Herbicide Action*, P. Böger, and G. Sandmann, Eds. (Boca Raton, FL: CRC Press Inc., 1989), pp. 45-63.
15. Faulkner, J. S., and B. M. R. Harvey. "Paraquat Tolerant *Lolium perenne* L.: Effects of Paraquat on Germinating Seedlings," *Weed Res.* 21:29-36 (1981).
16. Powles, S. B. "Appearance of a Biotype of the Weed, *Hordeum glaucum* Steud., Resistant to the Herbicide Paraquat," *Weed Res.* 26:167-172 (1986).
17. Wikström, M., and M. Saraste. "The Mitochondrial Respiratory Chain," in *Bioenergetics*, L. Ernster, Ed. (Amsterdam: Elsevier, 1984), pp 49-94.
18. Kappus, H. "Overview of Enzyme Systems Involved in Bio-reduction of Drugs and in Redox Cycling," *Biochem. Pharmacol.* 35:1-6 (1986).
19. Weber, J. B., and S. B. Weed. "Adsorption and Desorption of Diquat, Paraquat and Prometone by Montmorillonitic and Kaolinitic Clay Minerals," *Soil Sci. Soc. Am. Proc.* 32:485-487 (1968).

20. Thrower, S. L., N. D. Hallam, and L. B. Thrower. "Movement of Diquat, 1,1'-ethylene-2,2'-bipyridylium Dibromide in Leguminous Plants," *Ann. Appl. Biol.* 55:253-260 (1965).
21. Fuerst, E. P., H. Y. Nakatani, A. D. Dodge, D. Penner, and C. J. Arntzen. "Paraquat Resistance in *Conyza*," *Plant Physiol.* 77:984-989 (1985).
22. Watanabe, Y., T. Honma, K. Ito, and M. Miyahara. "Paraquat Resistance in *Erigeron philadelphicus* L.," *Weed Res. (Japan)* 27:49-54 (1982).
23. Kato, A., and Y. Okuda. "Paraquat Resistance in *Erigeron canadensis* L.," *Weed Res. (Japan)* 28:54-56 (1983).
24. Warner, R. B., and W. B. C. Mackie. "A Barley Grass *Hordeum leporinum* ssp. *glaucum* Steud. Population Tolerant to Paraquat (Gramoxone)," *Aust. Weed Res. Newsl.* 31:16 (1983).
25. Putwain, P. D. "Herbicide Resistance in Weeds — an Inevitable Consequence of Herbicide Use?," *Br. Crop Prot. Conf. — Weeds 1982*:719-728 (1982).
26. Itoh, K., M. Azmi, and A. Ahmad. "Paraquat Resistance in *Solanum nigrum*, *Crassocephalum crepidioides*, *Amaranthus lividus* and *Conyza sumatrensis* in Malaysia," in *Proceedings of the First International Weed Control Congress*, Vol. 2, R. G. Richardson, Ed. (Melbourne: Weed Science Society of Victoria, 1992), pp. 224-228.
27. Powles S. B., E. S. Tucker, and T. R. Morgan. "A Capeweed (*Arctotheca calendula*) Biotype in Australia Resistant to Bipyridyl Herbicides," *Weed Sci.* 37:60-62 (1989).
28. Purba, E., C. Preston, and S. B. Powles. Unpublished results (1993).
29. LeBaron, H. M. "Distribution and Seriousness of Herbicide-resistant Weed Infestations Worldwide," in *Herbicide Resistance in Weeds and Crops*, J. C. Caseley, G. W. Cussans, and R. K. Atkin, Eds. (Oxford: Butterworth-Heinemann, 1991), pp. 27-43.
30. Bulke, R., F. Verstraete, M. Van Himme, and J. Stryckers. "Biology and Control of *Epilobium ciliatum* Rafin. (Syn.: *E. adenocaulon* Hausskn.)," in *Weed Control on Vine and Soft Fruits*, R. Carvolloro and D. W. Robinson, Eds. (Rotterdam: A. A. Balkema, 1987), pp. 57-67.
31. Clay, D. V. "New Developments in Triazine and Paraquat Resistance and Coresistance in Weed Species in England," *Br. Crop Prot. Conf. — Weeds 1989*, 317-324 (1989).
32. Pölös, E., J. Mikulàs, Z. Szigeti, G. Laskay, and E. Lehoczki. "Cross-resistance to Paraquat and Atrazine in *Conyza canadensis*," *Br. Crop Prot. Conf. — Weeds 1987*, 909-916 (1987).
33. Pölös, E., Z. Szigeti, G. Váradi, and E. Lehoczki. "Diquat Resistance in Paraquat/atrazine Coresistant *Conyza canadensis*," in *Herbicide Resistance in Weeds and Crops*, J. C. Caseley, G. W. Cussans, and R. K. Atkin, Eds. (Oxford: Butterworth-Heinemann, 1991), pp. 468-469.
34. Itoh, K., and S. Matsunaka. "Parapatric Differentiation of Paraquat Resistant Biotypes in some Compositae Species," in *Biological Approaches and Evolutionary Trends in Plants*, S. Kawano, Ed. (London: Academic Press, 1990), pp. 33-49.
35. Yamasue, Y., K. Kamiyama, Y. Hanioka, and T. Kusanagi. "Paraquat Resistance and its Inheritance in Seed Germination of the Foliar-resistant Biotypes of *Erigeron canadensis* L. and *E. sumatrensis* Retz.," *Pestic. Biochem. Physiol.* 44:21-27 (1992).
36. Tucker, E. S., and S. B. Powles. "A Biotype of Hare Barley (*Hordeum leporinum*) Resistant to Paraquat and Diquat," *Weed Sci.* 39:159-162 (1991).

37. Bewick, T. A., W. M. Stall, S. R. Kostewicz, and K. Smith. "Alternatives for Control of Paraquat Tolerant American Black Nightshade (*Solanum americanum*)," *Weed Technol.* 5:61-65 (1991).
38. Purba, E., C. Preston, and S. B. Powles. "Paraquat Resistance in a Biotype of *Vulpia bromoides* (L.) S. F. Gray," *Weed Res.* 33:409-413 (1993).
39. Hickok, L. G., and O. J. Schwarz. "An *in vitro* Whole Plant Selection System: Paraquat Tolerant Mutants in the Fern *Ceratopteris*," *Theor. Appl. Genet.* 72:302-306 (1986).
40. Kitayama, K., and R. K. Togasaki. "Characterization of Paraquat Resistant Mutants of *Chlamydomonas reinhardtii*," in *Research in Photosynthesis,* Vol. IV, N. Murata, Ed. (Dordrecht, The Netherlands: Kluwer Academic Publishers, 1992), pp. 487-490.
41. Bray, D. F, J. R. Bagu, and Nakamura, K. "Ultrastructure of *Chlamydomonas reinhardtii* Following Exposure to Paraquat: Comparison of Wild Type and a Paraquat-resistant Mutant," *Can. J. Bot.* 71:174-182 (1993).
42. Faulkner, J. S. "A Paraquat Resistant Variety of *Lolium perenne* Under Field Conditions," *Br. Crop Prot. Conf. — Weeds 1976*, 485-490 (1976).
43. Hughes, K. W., D. Negretto, M. E. Daub, and R. L. Meeusen. "Free-radical Stress Response in Paraquat-sensitive and Resistant Tobacco Plants," *Environ. Exp. Bot.* 24:151-157 (1984).
44. Furusawa, I., K. Tanaka, P. Thanutong, A. Mizuguichi, M. Yazaki, and K. Asada. "Paraquat Resistant Tobacco Cell Calluses with Enhanced Superoxide Dismutase Activity," *Plant Cell Physiol.* 25:1247-1254 (1984).
45. Brian, R. C. "The Uptake and Adsorption of Diquat and Paraquat by Tomato, Sugar Beet and Cocksfoot," *Ann. Appl. Biol.* 59:91-99 (1967).
46. Bishop, T., S. B. Powles, and G. Cornic. "Mechanism of Paraquat Resistance in *Hordeum glaucum*. II. Paraquat Uptake and Translocation," *Aust. J. Plant Physiol.* 14:539-547 (1987).
47. Harvey, B. M. R., J. Muldoon, and D. B. Harper. "Mechanism of Paraquat Tolerance in Perennial Ryegrass. I. Uptake, Metabolism and Translocation of Paraquat," *Plant Cell Environ.* 1:203-209 (1978).
48. Preston, C., S. Balachandran, and S. B. Powles. "Mechanisms of Resistance to Bipyridyl Herbicides in *Arctotheca calendula* (L.) Levyns," *Plant Cell Environ.* in press (1994).
49. Slade, P., and E. G. Bell. "The Movement of Paraquat in Plants," *Weed Res.* 6:267-274 (1966).
50. Baldwin, B. S. "Translocation of Diquat in Plants," *Nature* 198:872-873 (1963).
51. Smith, J. M., and G. R. Sagar. "A Re-examination of the Influence of Light and Darkness on the Long-distance Transport of Diquat in *Lycopersicon esculentum* Mill.," *Weed Res.* 6:314-321 (1966).
52. Tanaka, Y., H. Chisaka, and H. Saka. "Movement of Paraquat in Resistant and Susceptible Biotypes of *Erigeron philadelphicus* and *E. canadensis*," *Physiol. Plant.* 66:605-608 (1986).
53. Preston, C., J. A. M. Holtum, and S. B. Powles. "On the Mechanism of Resistance to Paraquat in *Hordeum glaucum* and *H. leporinum*. Delayed Inhibition of Photosynthetic O_2 Evolution after Paraquat Application," *Plant Physiol.* 100:630-636 (1992).
54. Preston, C., and S. B. Powles. Unpublished results (1993).
55. Harvey, B. M. R., and T. W. Fraser. "Paraquat Tolerant and Susceptible Perennial Ryegrasses: Effects of Paraquat Treatment on Carbon Dioxide Uptake and Ultrastructure of Photosynthetic Cells," *Plant Cell Environ.* 3:107-117 (1980).

56. Harper, D. B., and B. M. R. Harvey. "Mechanism of Paraquat Tolerance in Perennial Ryegrass. II. Role of Superoxide Dismutase, Catalase and Peroxidase," *Plant Cell Environ.* 1:211-215 (1978).
57. Faulkner, J. S. "Heritability of Paraquat Tolerance in *Lolium perenne* L.," *Euphytica* 23:281-288 (1974).
58. Lehoczki, E., G. Laskay, I. Gaál, and Z. Szigeti. "Mode of Action of Paraquat in Leaves of Paraquat-resistant *Conyza canadensis* (L.) Cronq.," *Plant Cell Environ.* 15:531-539 (1992).
59. Shaaltiel, Y., and J. Gressel. "Kinetic Analysis of Resistance to Paraquat in *Conyza*. Evidence that Paraquat Transiently Inhibits Leaf Chloroplast Reactions in Resistant Plants," *Plant Physiol.* 85:869-871 (1987).
60. Youngman, R. J., and A. D. Dodge "On the Mechanism of Paraquat Resistance in *Conyza* sp.," in *Photosynthesis VI. Photosynthesis and Productivity, Photosynthesis and Environment*, G. Akoyunoglou, Ed. (Philadelphia: Balaban International Science Services, 1981), pp. 537-544.
61. Jansen, M. A. K., C. Malan, Y. Shaaltiel, and J. Gressel. "Mode of Evolved Photooxidant Resistance to Herbicides and Xenobiotics," *Z. Naturforsch.* 45c:463-469 (1990).
62. Vaughn, K. C., M. A. Vaughan, and P. Camilleri. "Lack of Cross-tolerance of Paraquat-resistant Hairy Fleabane (*Conyza bonariensis*) to Other Toxic Oxygen Generators Indicates Enzymatic Protection is not the Resistance Mechanism," *Weed Sci.* 37:5-11 (1989).
63. Powles, S. B., and G. Cornic. "Mechanism of Paraquat Resistance in *Hordeum glaucum*. I. Studies with Isolated Organelles and Enzymes," *Aust. J. Plant Physiol.* 14:81-89 (1987).
64. Hart, J. J., J. M. DiTomaso, D. L. Linscott, and L. V. Kochian. "Characterization of the Transport and Cellular Compartmentation of Paraquat in Roots of Intact Maize Seedlings," *Pestic. Biochem. Physiol.* 43:212-222 (1992).
65. Preston, C., J. A. M. Holtum, and S. B. Powles. "Do Polyamines Contribute to Paraquat Resistance in *Hordeum glaucum*?," in *Research in Photosynthesis*, Vol. III, N. Murata, Ed. (Dordrecht, The Netherlands: Kluwer Academic Publishers, 1992), pp. 571-574.
66. Norman, M. A., E. P. Fuerst, R. J. Smeda, and K. C. Vaughn. "Evaluation of Paraquat Resistance Mechanisms in *Conyza*," *Pestic. Biochem. Physiol.* 46:236-249 (1993).
67. Hart, J. J., and J. M. DiTomaso. Unpublished results (1993).
68. Imai, Y., and S. Kuwatsuka. "Characteristics of Paraquat-degrading Microbes," *J. Pestic. Sci.* 14:475-480 (1989).
69. Funderbunk, H. H., and G. A. Bozarth. "Review of the Metabolism and Decomposition of Diquat and Paraquat," *J. Agric. Food Chem.* 15:563-567 (1967).
70. Carr, R. J. G., R. F. Bilton, and T. Atkinson. "Mechanism of Biodegradation of Paraquat by *Lipomyces starkeyi*," *Appl. Environ. Microbiol.* 49:1290-1294 (1985).
71. Bowler, C., T. Alliotte, M. De Loose, M. Van Montagu, and D. Inzé. "The Induction of Manganese Superoxide Dismutase in Response to Stress in *Nicotiana plumbaginifolia*," *EMBO J.* 8:31-38 (1989).
72. Bowler, C., M. Van Montagu, and D. Inzé. "Superoxide Dismutase and Stress Tolerance," *Annu. Rev. Plant Physiol. Plant Mol. Biol.* 43:83-116 (1992).
73. Asada, K. "Ascorbate Peroxidase — a Hydrogen Peroxide-scavenging Enzyme in Plants," *Physiol. Plant.* 85:235-241 (1992).

74. Foyer, C. H., and B. Halliwell. "The Presence of Glutathione and Glutathione Reductase in Chloroplasts: Proposed Role in Ascorbic Acid Metabolism," *Planta* 133:21-25 (1976).
75. Drumm-Herrel, H., U. Gerhäußer, and H. Mohr. "Differential Regulation by Phytochrome of the Appearance of Plastidic and Cytoplasmic Isoforms of Glutathione Reductase in Mustard (*Sinapis alba* L.) Cotyledons," *Planta* 178:103-109 (1989).
76. Tolbert, N. E. "Microbodies — Peroxisomes and Glyoxysomes," *Annu. Rev. Plant Physiol.* 22:45-74 (1971).
77. Vaughn, K. C., and E. P. Fuerst. "Structural and Physiological Studies of Paraquat-resistant *Conyza*," *Pestic. Biochem. Physiol.* 24:86-94 (1985).
78. Shaaltiel, Y., and J. Gressel. "Multienzyme Oxygen Radical Detoxifying System Correlated with Paraquat Resistance in *Conyza bonariensis*," *Pestic. Biochem. Physiol.* 26:22-28 (1986).
79. Amsellem, Z., M. A. K. Jansen, A. R. J. Driesenaar, and J. Gressel. Unpublished results (1993).
80. Tanaka, K., I. Furusawa, N. Kondo, and K. Tanaka. "SO_2 Tolerance of Tobacco Plants Regenerated from Paraquat-tolerant Cells," *Plant Cell Physiol.* 29:743-746 (1988).
81. Matsunaka, S., and K. Ito. "Paraquat Resistance in Japan," in *Herbicide Resistance in Weeds and Crops*, J. C. Caseley, G. W. Cussans, and R. K. Atkin, Eds. (Oxford: Butterworth-Heinemann, 1991), pp. 77-86.
82. Carroll, E. W., O. J. Schwarz, and L. G. Hickok. "Biochemical Studies of Paraquat-tolerant Mutants of the Fern *Ceratopteris richardii*," *Plant Physiol.* 87:651-654 (1988).
83. Pölös, E., J. Mikulàs, Z. Szigeti, B. Matkovics, D. Q. Hai, À. Pàrducz, and E. Lehoczki. "Paraquat and Atrazine Co-resistance in *Conyza canadensis* (L.) Cronq.," *Pestic. Biochem. Physiol.* 30:142-154 (1988).
84. Fuerst, E. P., and K. C. Vaughn. "Mechanisms of Paraquat Resistance," *Weed Technol.* 4:150-156 (1990).
85. Schöner, S., and G. H. Krause. "Protective Systems Against Active Oxygen Species in Spinach: Response to Cold Acclimation in Excess Light," *Planta* 180:383-389 (1990).
86. Tsang, E. W. T., C. Bowler, D. Hérouart, W. Van Camp, R. Villarroel, C. Genetello, M. Van Montagu, and D. Inzé. "Differential Regulation of Superoxide Dismutases in Plants Exposed to Environmental Stress," *Plant Cell* 3:783-792 (1991).
87. Osswald, W. F., R. Kraus, S. Hippeli, B. Benz, R. Volpert, and E. F. Elstner. "Comparison of the Enzymatic Activities of Dehydroascorbic Acid Reductase, Glutathione Reductase, Catalase, Peroxidase and Superoxide Dismutase of Healthy and Damaged Spruce Needles (*Picea abies* (L.) Karst.)," *J. Plant Physiol.* 139:742-748 (1992).
88. Wingsle, G., A. Mattson, A. Ekblad, J.-E. Hällgren, and E. Selstem. "Activities of Glutathione Reductase and Superoxide Dismutase in Relation to Changes of Lipids and Pigments due to Ozone in Seedlings of *Pinus sylvestris* (L.)," *Plant Sci.* 82:167-178 (1992).
89. Mehlhorn, H., B. J. Tabner, and A. R. Wellburn. "Electron Spin Resonance Evidence for the Formation of Free Radicals in Plants Exposed to Ozone," *Physiol. Plant.* 79:377-383 (1990).

90. Burke, J. J., P. E. Gamble, J. L. Hatfield, and J. E. Quisenberry. "Plant Morphological and Biochemical Responses to Field Water Deficits. I. Responses of Glutathione Reductase Activity and Paraquat Sensitivity," *Plant Physiol.* 79:415-419 (1985).
91. Shaaltiel, Y., A. Glazer, P. F. Bocion, and J. Gressel. "Cross Tolerance to Herbicidal and Environmental Oxidants of Plant Biotypes Tolerant to Paraquat, Sulfur Dioxide and Ozone," *Pestic. Biochem. Physiol.* 31:13-23 (1988).
92. Jansen, M. A. K., Y. Shaaltiel, D. Kazzes, O. Canaani, S. Malkin, and J. Gressel. "Increased Tolerance to Photoinhibitory Light in Paraquat-resistant *Conyza bonariensis* Measured by Photoacoustic Spectroscopy and $^{14}CO_2$-fixation," *Plant Physiol.* 91:1174-1178 (1989).
93. Preston, C., J. A. M. Holtum, and S. B. Powles. "Resistance to the Herbicide Paraquat and Increased Tolerance to Photoinhibition are not Correlated in Several Weed Species," *Plant Physiol.* 96:314-318 (1991).
94. Harvey, B. M. R., and D. B. Harper. "Tolerance to Bipyridylium Herbicides," in *Herbicide Resistance in Plants*, H. M. LeBaron and J. Gressel, Eds. (New York: John Wiley & Sons, 1982), pp. 215-233.
95. Tepperman, J. M., and P. Dunsmuir. "Transformed Plants with Elevated Levels of Chloroplastic SOD are not More Resistant to Superoxide Toxicity," *Plant Mol. Biol.* 14:501-511 (1990).
96. Bowler, C., L. Slooten, S. Vandenbranden, R. De Rycke, J. Botterman, C. Sybesma, M. Van Montagu, and D. Inzé. "Manganese Superoxide Dismutase can Reduce Cellular Damage Mediated by Oxygen Radicals in Transgenic Plants," *EMBO J.* 10:1723-1732 (1991).
97. Sen Gupta, A., J. L. Heinen, A. S. Holladay, J. J. Burke, and R. D. Allen. "Increased Resistance to Oxidative Stress in Transgenic Plants that Overexpress Chloroplastic Cu/Zn Superoxide Dismutase," *Proc. Natl. Acad. Sci. U.S.A.* 90:1629-1633 (1993).
98. Aono, M., A. Kubo, H. Saji, K. Tanaka, and N. Kondo. "Enhanced Tolerance to Photooxidative Stress of Transgenic *Nicotiana tabacum* with High Chloroplastic Glutathione Reductase Activity," *Plant Cell Physiol.* 34:129-135 (1993).
99. Aono, M., A. Kubo, H. Saji, T. Natori, K. Tanaka, and N. Kondo. "Resistance to Active Oxygen Toxicity of Transgenic *Nicotiana tabacum* that Expresses the Gene for Glutathione Reductase from *Escherichia coli*," *Plant Cell Physiol.* 32:691-697 (1991).
100. Tucker, E. S., and S. B. Powles. Unpublished results (1989).
101. Purba, E., C. Preston, and S. B. Powles. "Temperature Influences on the Level of Resistance to Paraquat in a Biotype of *Hordeum leporinum* Link," in *Proceedings of the First International Weed Control Congress,* Vol. 2, R. G. Richardson, Ed. (Melbourne: Weed Science Society of Victoria, 1992), pp. 421-423.
102. Shaaltiel, Y., N.-H. Chua, S. Gepatein, and J. Gressel. "Dominant Pleiotropy Controls Enzymes Co-segregating with Paraquat Resistance in *Conyza bonariensis*," *Theor. Appl. Genet.* 75:850-856 (1988).
103. Itoh, K., and M. Miyahara. "Inheritance of Paraquat Resistance in *Erigeron philadelphicus* L.," *Weed Res. (Japan)* 29:301-307 (1984).
104. Islam, A. K. M. R., and S. B. Powles. "Inheritance of Resistance to Paraquat in Barley Grass *Hordeum glaucum* Steud.," *Weed Res.* 28:393-397 (1988).
105. Purba, E., C. Preston, and S. B. Powles. "Inheritance of Bipyridyl Herbicide Resistance in *Arctotheca calendula* and *Hordeum leporinum*," *Theor. Appl. Genet.* 87:598-602 (1993).

106. Hickok, L. G., and O. J. Schwarz. "Paraquat Tolerant Mutants in *Ceratopteris*: Genetic Characterization and Reselection for Enhanced Tolerance," *Plant Sci.* 47:153-158 (1986).
107. Hickok, L. G., and O. J. Schwarz. "Genetic Characterization of a Mutation that Enhances Paraquat Tolerance in the Fern *Ceratopteris richardii*," *Theor. Appl. Genet.* 77:200-204 (1989).
108. Greenberg, J. T., P. Monach, J. H. Chou, P. D. Josephy, and B. Demple. "Positive Control of a Global Oxidant Defence Regulon Activated by Superoxide-generating Agents in *Escherichia coli*," *Proc. Natl. Acad. Sci. U.S.A.* 87:6181-6185 (1990).
109. Hart, J. J., J. M. DiTomaso, D. L. Linscott, and L. V. Kochian. "Transport Interactions Between Paraquat and Polyamines in Roots of Intact Maize Seedlings," *Plant Physiol.* 99:1400-1405 (1992).

CHAPTER 4

Resistance To Acetolactate Synthase Inhibiting Herbicides

L. L. Saari, J. C. Cotterman, and D. C. Thill

INTRODUCTION

One of the most significant occurrences in herbicide resistance has been the advent of weeds resistant to herbicides that inhibit acetolactate synthase (ALS; also called acetohydroxyacid synthase or AHAS; EC 4.1.3.18). This is because ALS inhibitor herbicides have become extremely important new tools in agricultural production, and any development which might limit their utility is regarded as serious. The first introduction of an ALS inhibitor herbicide was that of the sulfonylurea, chlorsulfuron, in 1982 for use in cereals. Since then, ALS inhibitor herbicides have been registered for use in several crops as well as for vegetation management (i.e., roadsides, railways, and industrial areas). The use of two major classes of ALS inhibitor herbicides, sulfonylureas and imidazolinones, alone, has grown to a 1991 market value of $1.3 billion.[1] This popularity is due to relatively low use rates, sound environmental properties, low mammalian toxicity, wide crop selectivity, and high efficacy. Five years after the initial use of an ALS inhibitor herbicide, the first resistant weeds appeared, and their incidence has steadily increased both in number of sites and species. Ironically, it is the high efficacy that quickly selects for the resistant phenotype, the same characteristic that enables ALS inhibitor herbicides to be used at very low rates.

In this chapter, we review the status of ALS inhibitor resistance in plants for dissimilar herbicides, especially as the phenomenon relates to resistance in weeds, and discuss the mechanisms and associated characteristics that are responsible for resistance. A number of recent reviews have covered the mode of action of ALS inhibitor herbicides[2-4] and resistance to ALS inhibitor herbicides.[5-12] Several informative reviews on ALS inhibitors, branched-chain amino acid metabolism, and ALS itself are contained in two monographs.[13,14]

CHEMISTRY OF THE HERBICIDE GROUP AND USE IN AGRICULTURE

Currently, four chemical classes of ALS inhibitor herbicides are commercialized or in development: sulfonylureas,[15-17] imidazolinones,[18] triazolopyrimidines,[19,20] and pyrimidinyl thiobenzoates[21] (Figure 1). In addition to these four classes, eleven other chemical classes of ALS inhibitors have been described: carbamoylpyrazolines, sulfonyliminotriazinyl heteroazoles, N-protected valylanilides, sulfonylamide azines, pyrimidyl mandelic acids, benzenesulfonyl carboxamide compounds, substituted sulfonyldiamides, ubiquinone-0,[22] carbonyl sulfonamides,[23] triazole sulfonanilides,[24] and gliotoxin,[25] but it is unknown whether any are headed for commercialization.

Commercial ALS inhibitor herbicides are comprised of 30 active ingredients for selective use in at least 10 crops (Table 1). Sulfonylurea herbicides are the most numerous on an active ingredient basis (22 active ingredients) followed

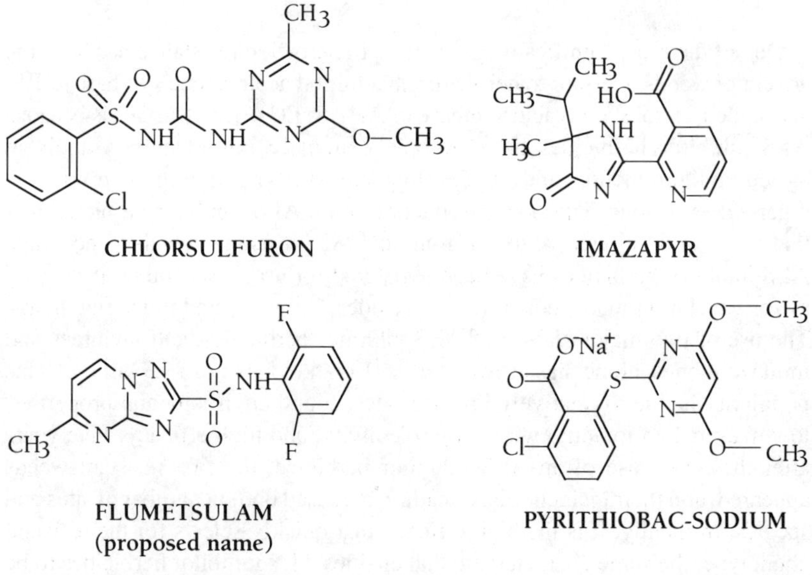

Figure 1. Initial herbicide products from each ALS inhibitor class include chlorsulfuron (sulfonylurea), imazapyr (imidazolinone), flumetsulam (triazolopyrimidine), and pyrithiobac-sodium (pyrimidinyl thiobenzoate).

Table 1. ALS-inhibitor herbicides — commercial and developmental

Common name	Code number	Chemical class	Company	Crop	Year
Chlorsulfuron	DPX-W4189	Sulfonylurea	DuPont	Cereals	1982
Metsulfuron methyl	DPX-T6376	Sulfonylurea	DuPont	Cereals	1986
Sulfometuron methyl	DPX-T5648	Sulfonylurea	DuPont	Vegetation management	1982
Chlorimuron ethyl	DPX-F6025	Sulfonylurea	DuPont	Soybeans	1986
Thifensulfuron methyl	DPX-M6316	Sulfonylurea	DuPont	Cereals, soybeans, maize	1988
Tribenuron methyl	DPX-L5300	Sulfonylurea	DuPont	Cereals	1989
Bensulfuron methyl	DPX-F5384	Sulfonylurea	DuPont	Rice	1987
Nicosulfuron	DPX-V9360/SL-950	Sulfonylurea	DuPont/ISK	Maize	1991
Ethametsulfuron methyl	DPX-A7881	Sulfonylurea	DuPont	Canola	1989
Rimsulfuron	DPX-E9636	Sulfonylurea	DuPont	Maize, other	1991
Triflusulfuron methyl	DPX-66037	Sulfonylurea	DuPont	Sugar beets	1995
Triasulfuron	CGA-131036	Sulfonylurea	Ciba-Geigy	Cereals	1992
Primisulfuron methyl	CGA-136872	Sulfonylurea	Ciba-Geigy	Maize	1990
Cinosulfuron	CGA-142464/BAS-9133	Sulfonylurea	Ciba-Geigy/BASF	Rice	1995
—	CGA-152005	Sulfonylurea	Ciba-Geigy	Maize	1995
Amidosulfuron	HOE-075032	Sulfonylurea	Hoechst	Cereals	1990
Fluzasulfuron	SL-160	Sulfonylurea	ISK	Turf	1990
Imazosulfuron	TH-913	Sulfonylurea	Takeda	Rice	1993
Pyrazosulfuron ethyl	NC-311	Sulfonylurea	Nissan	Rice	1989
Halosulfuron[a]	MON-12000/NC-319	Sulfonylurea	Monsanto/Nissan	Maize	1995
—	NC-330	Sulfonylurea	Nissan	Wheat	
Imazapyr	AC 322 140	Imidazolinone	American Cyanamid	Vegetation management	Late 1990s
Imazamethabenz methyl	AC 243 977	Imidazolinone	American Cyanamid	Cereals	1986
Imazethapyr	AC 222 293	Imidazolinone	American Cyanamid	Soybeans	1987
Imazaquin	AC 263 499	Imidazolinone	American Cyanamid	Soybeans	1986
—	AC 252 214	Imidazolinone	American Cyanamid	Peanuts	1995
Flumetsulam[a]	AC 263 222	Triazolopyrimidine	DowElanco	Soybeans, maize	1995
Metosulam[a]	XRD 498	Triazolopyrimidine	DowElanco	Cereals	1993
Pyrithiobac sodium	XRD 511				
	KIH-2031/DPX-PE350	Pyrimidinylthiobenzoate	Kumiai/DuPont	Cotton	1995

[a]Proposed common name

by imidazolinones (five active ingredients), triazolopyrimidines (two active ingredients), and pyrimidinyl thiobenzoates (one active ingredient). New commercial compounds will likely be found among these herbicide classes and possibly in other ALS inhibitor classes also. For all known cases, crop selectivity is due to metabolism and is discussed in detail below.

MODE OF ACTION

ALS inhibitor herbicides are so named because they inhibit acetolactate synthase (EC 4.1.3.18), the first enzyme common to the biosynthesis of the branched-chain amino acids, valine, leucine, and isoleucine (Figure 2). Evidence for the mode of action of the ALS inhibitor herbicides comes from several scientific disciplines. From an enzymatic point of view, sulfonylureas,[26-28] imidazolinones,[29] triazolopyrimidines,[30] and pyrimidinyl thio-[21] and oxy-[31,32] benzoates directly inhibit ALS activity. Furthermore, sulfonylureas[26,33] and imidazolinones[34] inhibit the growth of plant roots or cells in culture, and this inhibition can be reversed via the addition of branched-chain amino acids, implicating this pathway in sulfonylurea and imidazolinone function.

Genetic evidence indicating that sulfonylurea and imidazolinone herbicides act upon ALS includes the observation that insensitivity of the enzyme to ALS

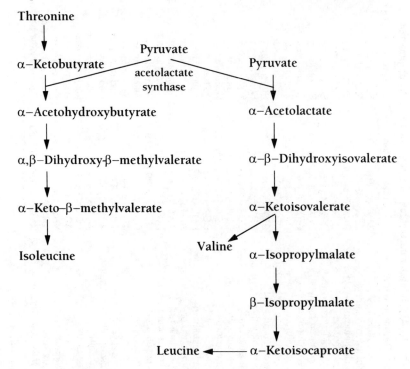

Figure 2. Biosynthetic pathway for branched-chain amino acids.

inhibitor herbicides cosegregates with the herbicide resistance phenotype.[28,35-37] Gene reintroduction experiments (i.e., transformation of sensitive plants with an ALS gene coding for an ALS inhibitor-insensitive ALS) further show that the herbicide resistance phenotypes are due to altered, insensitive ALS enzymes.[38-40] Other evidence comes from demonstrations that single nucleotide changes in the ALS gene are associated with sulfonylurea[38] and imidazolinone[40] resistance.

Finally, plants altered only by the inclusion of an insensitive ALS gene that are subjected to high rates of ALS inhibitor herbicides appear normal (i.e., no symptomology as a result of herbicide application). This is a strong argument for there not being a second site at which ALS inhibitors act in plants.

EVALUATION OF RESISTANCE

There are several ways to evaluate herbicide resistance. Unfortunately, different conclusions may be drawn for the same plant, or other organism, depending on what type of methodology is used to study the resistance. In order to better assess the comparisons made in the following sections pertaining to ALS inhibitor-resistant organisms, it is important to first understand how determinations of resistance are made, and what their relative strengths and weaknesses are.

Evaluating resistance is often performed using plants, cultured cells, or bacteria. In addition, the determination of enzymic sensitivities, metabolism rates, and mutations conferring resistance can be used. However, the most straightforward assessment of resistance is obtained by the application of herbicides to whole plants. Although target site cross resistance to herbicides of the same chemical class as the selection agent is anticipated, it is not necessarily guaranteed. Of interest is to determine whether and how much resistance exists to structurally similar herbicides as well as target site cross resistance to chemical classes of different structure but with the same mode of action. For target site resistance, it is expected that herbicides with alternate modes of action will be effective. For example, *Kochia scoparia* (L.) Schrad. selected with and resistant to chlorsulfuron is not resistant to other mode of action herbicides such as bromoxynil.[41] One must be careful in comparing resistance patterns especially when a herbicide is not labeled for use on a particular weed. For such a comparison to be valid, control of the susceptible biotype by the herbicide must be demonstrated.

A second useful method, especially for target site resistance, is the determination of "cross insensitivity" patterns at the enzyme level. That is, do ALS inhibitors of different chemistries inhibit the ALS isolated from a herbicide-resistant organism? To extrapolate enzymic data to whole plant response requires that the herbicide used for I_{50} determinations also be herbicidal toward the plant or weed in question. It is also of interest, however, to understand whether other chemical classes of inhibitors of the same mode of action are potentially affected by the resistance mutation, regardless of plant sensitivity. Enzymic analysis allows this determination because differences in uptake, translocation, metabolism, and sequestration are avoided.

The interpretation of enzymic results may have to be reevaluated if the quaternary structure of eukaryotic ALS is shown to be the same as prokaryotic ALS. In enteric bacteria, ALS is a tetramer of two large and two small subunits,[42-44] and the kinetic properties of the large subunits are affected by the small subunits.[45,46] While arguments exist for the possibility of an ALS small subunit in plants,[47] there is no direct evidence for this type of quaternary structure. With regard to ALS inhibitor resistance, enzymic analysis has been performed assuming that the small subunit is not present in plants and not affecting ALS inhibition by ALS inhibitor herbicides.

A useful measurement for comparing resistant organisms is the resistance factor, which is a ratio of the response, organismic or enzymic, of resistant to susceptible isolates. For whole plant data, this would be the fold increase in herbicide rates required to inhibit some growth parameter (e.g., LD_{50} or GR_{50}) as measured by changes in fresh weight, dry weight, or other means. For enzymic data, the resistance factor is the ratio of the resistant ALS I_{50} value to the susceptible ALS I_{50} value for a given herbicide (binding constants could also be used but are generally more difficult to obtain).

The protocol used in determining target site cross resistance is important, especially for ALS inhibitors. Many evaluations of the growth of plants, cells, or bacteria, or the assay of ALS, for example, have been performed at one or at most, a few herbicide concentrations. Ideally, several herbicide concentrations should be used so that an accurate interpolation, rather than an extrapolation, of the GR_{50} or ALS I_{50} can be made. It is imperative that susceptible plants be evaluated in the same test. Generally, it is inadequate to claim cross resistance based on determinations with only one herbicide concentration unless preliminary testing has been performed to determine a discriminating concentration. Also, the response to ALS inhibitor herbicides is environment and time dependent, and for this reason, it is important to consider differences in how and when plants or cultures were evaluated when comparing studies.

A third method is to compare the mutation(s) in the target site that confers resistance. An identical mutation in the ALS gene would be expected to result in similar cross resistance patterns to ALS inhibitor herbicides among different plants assuming that uptake, translocation, sequestration, and metabolism do not interfere. If many mutations are identified in the laboratory, as they have been for ALS inhibitor resistance, then many mutations may also be present in agronomic settings. However, this is not always the case as evidenced by triazine resistance (see Chapter 2). While several mutations in the *psbA* gene resulting in triazine resistance have been identified, only one mutation (ser264 to gly) exists in higher plants.[48] Thus, laboratory-generated mutants may have a limited usefulness in studying herbicide resistance in weeds, and the predictions of cross resistance patterns in the field based on laboratory mutants must be done cautiously.

To accurately assess target site cross resistance, several experimental details must be addressed. Is cross resistance based on whole plant analysis, bacterial disc diffusion assays, or enzymic data? Were herbicide concentrations appropriate to perform accurate interpolations of GR_{50} values for both susceptible and resistant plants? At what physiological age were injury determinations assessed? How large a change in resistance factor connotes cross resistance? What mutations are responsible, if known? Which specific herbicides, not herbicide classes, display the phenomenon of cross resistance? More precise information regarding questions of cross resistance will contribute to a better overall understanding of resistance and cross resistance.

EVOLUTION AND DISTRIBUTION

Resistance in weeds due to treatment with ALS inhibitor herbicides was first observed in *Lactuca serriola* L. in 1987,[49] only five years after the commercial introduction of chlorsulfuron. Soon thereafter, resistant *K. scoparia* was identified,[41] and since that time, resistance to ALS inhibitors has been reported in 12 additional weed species: 8 dicots, 3 grasses, and 1 sedge. A listing of resistant species documented to date is presented in Table 2. At least one resistant biotype of one or more of these species has been confirmed in 15 states in the United States and 3 Canadian provinces, as well as in Australia, Denmark, England and Israel (Table 3). In addition, reports of resistance to ALS inhibitors in *Ioxophorus unisetus*[50] and *Eleusine indica* (L.) Gaertn.[51] exist, but these cases have not been documented.

Resistance to ALS inhibitors has evolved relatively quickly in both crop and non-crop (including roadsides and railroad rights-of-way) situations, with the first resistant biotypes of several species becoming apparent after four to seven applications of the selection agent(s).[52-59] Reliance on a single herbicide mode of action in monoculture or non-crop areas has been associated with most cases of resistance.[41,49,52-57,60] The long residual activity of many of the selection agents has also contributed to the rapid development of resistance by increasing the effective kill and thus the selection pressure.[61] In some cases, despite the use of herbicide mixtures with different modes of action, the selection pressure for resistance to the ALS inhibitor remained relatively high because effective control of susceptibles by the ALS inhibitor often continued beyond the efficacious period of the shorter residual component of the mixture.

Although evidence exists for the independent development of resistance in agricultural fields through treatment of crops with selective herbicides, there has been an ongoing debate about whether the use of long residual, nonselective herbicides in non-crop areas has been an important source of resistant weeds in adjacent areas. Particularly with highly mobile weeds such as *K. scoparia*[62] and *Salsola iberica*,[63] the potential for distribution of seeds to previously uninfested sites is relatively high. Spreading resistance by pollen may also occur.[63,64] The possibility remains that non-crop use has been responsible for some of the observed resistance in crop sites although no conclusive evidence exists to support this contention.

Examination of Table 3 reveals that since the initial observations of resistance in 1987, there has been a steady increase in the number of sites, the

Table 2. Weeds resistant to ALS-inhibitor herbicides

Latin name	Common name	Selection agent(s)	Resistance mechanism[a]	Ref.
Alopecurus myosuroides	Blackgrass	Chlorotoluron/isoproturon	Unknown	147
Amaranthus blitoides	Prostrate pigweed	Sulfometuron-methyl[b]	Unknown	198
Amaranthus retroflexus	Redroot pigweed	Sulfometuron-methyl[b]	Insensitive ALS	52,198
Cyperus difformis	Smallflower umbrella sedge	Bensulfuron-methyl	Insensitive ALS	53
Kochia scoparia	Kochia	Chlorsulfuron Sulfometuron-methyl	Insensitive ALS	41,77
Lactuca serriola	Prickly lettuce	Chlorsulfuron/metsulfuron-methyl	Insensitive ALS	49,199
Lolium perenne	Perennial ryegrass	Sulfometuron-methyl/chlorsulfuron	Insensitive ALS	55,58
Lolium rigidum	Rigid ryegrass	Chlorsulfuron	Insensitive ALS	54
Lolium rigidum	Rigid ryegrass	Diclofop-methyl	Increased metabolism	54,131,146
Sagittaria montevidensis	California arrowhead	Bensulfuron-methyl	Insensitive ALS	53
Salsola iberica	Russian thistle	Sulfometuron-methyl	Insensitive ALS	55
Sisymbrium orientale	Oriental mustard	Chlorsulfuron/triasulfuron	Unknown	200
Sonchus oleraceus	Annual sowthistle	Chlorsulfuron	Unknown	200
Stellaria media	Common chickweed	Chlorsulfuron	Insensitive ALS	55,56
Xanthium strumarium	Common cocklebur	Imazaquin	Insensitive ALS	57,69,79

[a]Resistance mechanism to ALS-inhibitor herbicides only.
[b]Sulfometuron methyl and simazine were applied simultaneously but simazine probably dissipated before germination of Amaranthus spp.

Table 3. Confirmed cases of weeds with evolved resistance to ALS-inhibitor herbicides

State/Province/Country	Weed	Crop			Non-crop		
		Number of sites	Year 1st occurred[a]	Ref.	Number of sites	Year 1st occurred[a]	Ref.
Australia							
South Australia	*Lolium rigidum*	~100[b]	1989	54,201	—		
	Sisymbrium orientale	8	1991	200	—		
New South Wales	*Sonchus oleraceus*	1	1991	200	—		
Canada							
Alberta	*Kochia scoparia*	1	1990	59,202	—		
	Stellaria media	9	1988	55,59,202,203	—		
Manitoba	*Kochia scoparia*	1	1991	202	2	1988	59,202,206
Saskatchewan	*Kochia scoparia*	42	1988	59,202	2	1989	59,202
	Salsola iberica	1	1989	59,202	—		
	Stellaria media	1	1991	59,204	—		
Denmark							
England	*Alopecurus myosuroides*[c]	1	1984	205	—		
Israel	*Amaranthus blitoides*	—			1	1991	198
	Amaranthus retroflexus	—			1	1991	52
United States							
California	*Lolium perenne*	—			1	1989	55
	Salsola iberica	—			4	1989	59
	Cyperus difformis	2	1992	53	—		
	Sagittaria montevidensis	2	1992	53	—		
Colorado	*Kochia scoparia*	60	1988	59,68	7	1987	59
Idaho	*Kochia scoparia*	55	1991	59,65	30	1990	59,65
	Lactuca serriola	10	1987	49,195,199	2	1988	49,195
Kansas	*Kochia scoparia*	51	1987	41,59	7	1988	59
	Salsola iberica	1	1988	59	—		
Mississippi	*Lolium perenne*	—			1	1992	59
	Xanthium strumarium	2	1991	57,69	—		
Missouri	*Xanthium strumarium*	1	1992	57	—		
Montana	*Kochia scoparia*	448	1988	59,67	—		
	Salsola iberica	2	1988	59	—		

Table 3 (continued). Confirmed cases of weeds with evolved resistance to ALS-inhibitor herbicides

State/Province/Country	Weed	Crop			Non-crop		
		Number of sites	Year 1st occurred[a]	Ref.	Number of sites	Year 1st occurred[a]	Ref.
Nebraska	*Kochia scoparia*	3	1992	59			
New Mexico	*Kochia scoparia*	—			3	1988	59
North Dakota	*Kochia scoparia*	59	1987	59			
	Salsola iberica	1	1988	59	—		
Oklahoma	*Kochia scoparia*	2	1991	59	5	1991	59
South Dakota	*Kochia scoparia*	36	1988	59	—		
Tennessee	*Xanthium strumarium*	3	1992	57			
Texas	*Kochia scoparia*	13	1988	59	2	1988	59
	Lolium perenne	5	1991	58	—		
Washington	*Kochia scoparia*	—			3	1988	59
	Salsola iberica	42	1988	55,59,66	22	1991	66

[a]Whenever possible, date is the year the first resistant sample was collected.
[b]Confirmed sites throughout Australia. Estimates suggest there may be up to 1000; sites with cross resistance to ALS inhibitors due to a non-ALS mechanism may increase this total. S. Powles pers. comm. 1992; Reference 201; AVCA Survey 1992.
[c]Resistant to chlorotoluron and isoproturon; cross resistant to chlorsulfuron.

diversity of species, and the geographical areas affected. For the purposes of Table 3, a site is defined as a single agricultural field or non-crop area managed according to the same practices. Exceptions to this definition are most of the *K. scoparia* sites in Idaho, most of the *S. iberica* sites in Washington, approximately one third of the *K. scoparia* sites in Montana, and nine *K. scoparia* sites in Colorado. In these cases, a site was tabulated if at least one resistant plant was found in a 156-km^2 (Idaho),[65] a 197-km^2 (Washington),[66] a 164-km^2 (Montana),[67] or a 92-km^2 (Colorado)[68] sector. Numbering 832 of the 1056 total, *K. scoparia* accounts for by far the greatest number of resistant sites (Table 3). It is important to note that sampling programs specifically for *K. scoparia* were conducted in several states; hence, other species may be somewhat underrepresented relative to *K. scoparia* in the existing data.

Most of the known resistance cases can be understood on the basis of herbicide selection pressure and biological attributes of the weeds (see sections on "Modeling and Resistance" and "Biology and Genetics"). A puzzling question is why resistance has not occurred in certain areas where weed species have been exposed to high selection pressure. Examples are *S. iberica* in cereals in Montana where resistant *K. scoparia* is widespread, and *Amaranthus* spp. in soybeans (*Glycine max* (L.) Merrill) in the southern United States where resistant *Xanthium strumarium* L. was recently confirmed.[57,69] A corollary to this question is why there are so many more resistant *K. scoparia* sites than any other species. Mobility (gene flow) may be a partial explanation for *K. scoparia*, but *S. iberica* can also move long distances. The main explanation for these apparent inconsistencies probably lies in our incomplete knowledge of weed biology, the initial frequencies of resistant genes in diverse environments, and the interactions of biological traits with cultural practices and the factors that promote resistance. Consequently, although parameters that affect the development of resistance have been described (see below), prediction of which species will be the next to become resistant can be tenuous. A related and critical question for the future, as new ALS inhibitor herbicides are developed in an increasing number of crops, is whether the consistent use of ALS inhibitors in crop rotations and/or the frequent use of short residual ALS inhibitors will result in equivalent selection pressure (and therefore resistance cases) to the current monoculture/long residual scenarios.

MODELING AND RESISTANCE

Resistance to ALS inhibitor herbicides has developed quickly, possibly faster than for several other mode of action herbicides. The parameters important to the development of herbicide resistance are described in mathematical models[61,70-72] and include the selection pressure imposed by the herbicides, the absolute and relative fitness of resistant biotypes, the initial frequency of resistant genes, the average lifetime in the soil seed bank, and gene flow (see Chapter 1).

Selection pressure is cited as one of the most important determinants in the evolution of resistance.[61] The importance and degree of selection pressure in the advent of ALS inhibitor resistance can be appreciated by comparing the commercial herbicide rates with the rates required to control weeds that have become resistant to ALS inhibitor herbicides. With the exception of *Lolium rigidum* Gaudin and *Lolium perenne* L., all of the susceptible counterparts of the resistant weeds listed in Table 2 (selected with ALS inhibitor herbicides) are controlled by one half or less of the commercial use rate. The commercial rate is often determined by the amount necessary to control the least sensitive weeds. Thus, the selection pressure imposed on the weeds controlled at fractional rates was very high relative to the selection pressure on weeds which required full rates for control. Herbicidal residual activity, a component of selection pressure, is also a major factor in the development of ALS inhibitor resistance. The herbicide selection agents listed in Table 2 include chlorsulfuron, sulfometuron-methyl, and imazaquin, all of which have relatively long residual activity. In fact, all of the ALS inhibitors listed as selection agents have some residual activity.

An intriguing question is whether the initial frequency of target site ALS inhibitor resistance is higher than the 10^{-5} to 10^{-6} frequency quoted for single-gene dominant mutations.[61] Initial frequency is a function of the number of genes involved, the dominance, and the ploidy.[61] As discussed later, target site ALS inhibitor resistance is inherited in a dominant or semidominant manner, and several ALS inhibitor resistant weeds thus far identified are diploid.[6]

Rigorous studies have not been performed on the initial frequency of ALS inhibitor resistance mutations in weeds, but model studies suggest that the frequency is around 10^{-6} or less. Haughn and Somerville[73] screened *Arabidopsis thaliana* (L.) Heynh. M2 seed for resistance to chlorsulfuron and estimated that the frequency of spontaneous chlorsulfuron-resistant ALS mutations to be approximately 10^{-9} assuming a 5000-fold enhancement in frequency due to the mutagen. A similar frequency was observed for imidazolinone-resistant *A. thaliana* mutants using imazapyr as the selection agent.[37] The mutation rate per cell generation was calculated to be 2.7×10^{-8} for non mutagenized tobacco (*Nicotiana tabacum* L.) cell cultures selected with primisulfuron.[74]

Two lines out of 20 million germinated alfalfa (*Medicago sativa* L.) seeds selected with chlorsulfuron resulted in plants with ALS that is less sensitive to chlorsulfuron inhibition.[75] The frequency of sulfonylurea-resistant tobacco leaf protoplasts selected on sulfonylurea impregnated media is approximately 10^{-5} to 10^{-6}, but protoplast medium itself may induce mutations and thereby increase the frequency observed.[76] Inconsistencies between the appearance of resistance and lack of selection pressure do exist (e.g., southern Idaho), but the results of these studies suggest that the rapid appearance of resistance to sulfonylurea and imidazolinone herbicides probably is not a function of an unusually high frequency of ALS mutations (i.e., greater than 10^{-6} frequency ascribed to mono-gene, dominant mutations) but rather due to the high selection pressure of these herbicides combined with other characteristics favoring resistance, such as high seed production (at least for several weeds), rapid and frequent seed germination, efficient seed and pollen distribution systems, and high levels of fitness (see Chapter 1). These latter traits of gene flow, seed biology, and fitness, which also contribute to the appearance of resistance, are discussed later in the section on "Biology and Genetics".

CROSS RESISTANCE

All of the weed biotypes resistant to ALS-inhibitor herbicides are also resistant to other ALS-inhibitor herbicides. A biotype of resistant *L. serriola* selected with chlorsulfuron/metsulfuron-methyl is also resistant to eight other sulfonylurea herbicides, imazapyr, imazethapyr, but not imazaquin.[49] Likewise, a *K. scoparia* biotype is resistant to five sulfonylurea herbicides and one imidazolinone herbicide, imazapyr.[41,77] While resistant to ALS inhibitor herbicides, these weed biotypes are not resistant to herbicides with alternate modes of action.[41,49] A *Stellaria media* (L.) Vill. biotype, selected with and resistant to chlorsulfuron, shows target site cross resistance to DE-489,[56] an ALS inhibiting triazolopyrimidine herbicide, and another *S. media* biotype, also selected with chlorsulfuron, is resistant to sulfometuron-methyl, triasulfuron, and imazapyr.[55] Imazamethabenz-methyl was also used to evaluate *S. media*,[56] but since control of the S biotype by this compound was not well established, it remains difficult to interpret similarities or differences in the control of S and R *S. media* biotypes by imazamethabenz-methyl.

L. perenne and *S. iberica* biotypes are resistant to three sulfonylurea and one imidazolinone herbicide(s).[55] Similarly, a target site mutant of *L. rigidum*, selected with chlorsulfuron, is cross resistant to triasulfuron, metsulfuron-methyl, sulfometuron-methyl, imazapyr, and imazamethabenz-methyl.[54]

A notable exception in cross resistance patterns is seen with *X. strumarium* selected with imazaquin. The R biotype is resistant to several imidazolinone herbicides but not chlorimuron-ethyl, a sulfonylurea herbicide used in soybeans.[57,69] Because of the significantly different cross resistance pattern, this biotype probably contains a different ALS mutation than those selected with sulfonylurea herbicides. This biotype is controlled by herbicides with alternate modes of action.

Cross resistance studies for ALS inhibitor resistant weeds are rather limited. However, for those studies that have been done, weeds selected with sulfonylurea herbicides are cross resistant to virtually all other sulfonylureas tested. Also, cross resistance is generally seen to imazapyr, an imidazolinone herbicide, but to a lesser degree. Other ALS inhibitors have been shown to be affected by resistance mutations (e.g., the triazolopyrimidine, DE-489, and the imidazolinone, imazamethabenz) but the ubiquity of the cross resistance to these compounds is unknown.

TARGET SITE RESISTANCE

Weeds

As the name suggests, target site resistance involves a decrease in the sensitivity of the herbicide target site to inhibition by the herbicides. The first target site resistance to ALS inhibitor herbicides was identified in *K. scoparia*.[77] Subsequently, an ALS with decreased sensitivity to ALS inhibitor herbicides has been reported in nine other weed species (Table 2). In the cases of *K. scoparia*,[77] *S. media*, *S. iberica*, and *L. perenne*,[55] not only was the resistance mechanism identified as an insensitive ALS, but increased metabolism by the R biotypes was excluded as a mechanistic possibility. Differential uptake and translocation were also excluded in *K. scoparia*.[77]

It is interesting to contrast the predominant type of resistance found in North America versus the predominant type found in Australia. In North America, selection with sulfonylurea herbicides has resulted in target site resistance with no other concurrently selected resistance mechanisms. In Australia, *L. rigidum* selected with diclofop-methyl results in multiple resistance mechanisms (see Chapter 9) including the possibility of non-target site, metabolism-based resistance mechanisms to ALS inhibitors.[54] This phenomenon occurs, in part, because *L. rigidum* is an obligate cross-pollinated species, and resistant genomes can readily exchange among the relatively few plants remaining after a herbicide application.[78] In areas where ALS inhibitor herbicides are used, multiple resistance mechanisms may be less likely because the high selection pressure of ALS inhibitor herbicides overwhelms other mechanisms except target site resistance.

Several herbicides have been evaluated for their ability to inhibit ALS activity isolated from resistant weeds. ALS I_{50} ratios (R ALS I_{50}/S ALS I_{50}) using published data on ALS inhibitor resistant weeds have been calculated and reported in Table 4. As expected, the ALS I_{50} ratio is relatively high for the selection agent. All of the resistant weeds selected with a sulfonylurea herbicide are also cross insensitive at the enzyme level to all other sulfonylurea herbicides tested. Triazolopyrimidine herbicides appear to be similarly affected by the mutation conferring resistance to the sulfonylureas. Another general trend is that there is a lower but consistent cross insensitivity to imidazolinone herbicides, especially imazapyr, for weeds selected with sulfonylureas.

In contrast, a *X. strumarium* biotype selected with imazaquin[57,69] displays high ALS insensitivity to imazaquin but not the sulfonylurea, chlorimuron-ethyl, or the triazolopyrimidine, flumetsulam[79] (also known as XRD 498, not to be confused with DE-489).

The effect of mutations in weed ALS genes on the inhibition of ALS by pyrimidinyl oxybenzoates (same general ALS inhibitor class as the pyrimidinyl thiobenzoates) has not been reported. However, Subramanian et al.[32] showed that for ALS isolated from resistant tobacco or cotton (*Gossypium hirsutum* L.) selected with the triazolopyrimidine herbicide, DE-489, there was a high degree of cross insensitivity to pyrimidinyl oxybenzoates. The ALS from these

RESISTANCE TO ACETOLACTATE SYNTHASE HERBICIDES

Table 4. I$_{50}$ ratios (resistant/susceptible) for various ALS inhibitor herbicides using ALS isolated from different weeds

		Ratio of ALS I$_{50}$ values (resistant/susceptible)[a]										
Weed species	Selection agent	Chlor-sulfuron	Chlori-muron ethyl	Triasulf-uron	Metsulfuron methyl	Sulfome-turon methyl	Thifensulf-uron methyl	Triazolo-pyrimidine[b]	Imazapyr	Imaze-thapyr	Imazame-thabenz	Imazaquin	Ref.
Kochia scoparia	Chlorsulfuron	20	20	10	5	30	10	20	6	2	3	ND[c]	55
Kochia scoparia	Chlorsulfuron	30	ND	4	3	ND	4	ND	ND	ND	ND	ND	202
Lolium perenne	Sulfometuron-methyl	40	ND	20	10	50	20	>20	7	ND	ND	ND	55
Lolium perenne	Chlorsulfuron	20	ND	20	8	80	ND	>9[e]	>2	ND	>20	ND	58
Lolium rigidum	Chlorsulfuron	ND	ND	ND	ND	>70	ND	ND	8	5	2[d]	ND	54
Salsola iberica	Sulfometuron-methyl	8	ND	8	4	20	9	8	4	ND	ND	ND	55
Stellaria media	Chlorsulfuron	10	ND	7	9	9	20	9	2	ND	ND	ND	55
Stellaria media	Chlorsulfuron	30	ND	10	80	100	ND	200	ND	ND	4	ND	203
Stellaria media	Chlorsulfuron	>200	ND	20	10	100	ND	>300	ND	ND	7	ND	203
Xanthium strumarium	Imazaquin	ND	1	ND	ND	ND	ND	1[e]	ND	ND	ND	>40	79

[a]Calculated to one significant figure.
[b]Triazolopyrimidine used was 1,2,4-triazolo-(1,5-a)-2,4-dimethyl-3-(N-sulfonyl-(2,6-dichloroanilide))-1,5-pyrimidine; known as DE-489.
[c]ND = not determined.
[d]Imazamethabenz-methyl, not imazamethabenz, was used.
[e]Flumetsulam (XRD 498) used.

resistant cotton and tobacco lines also display cross insensitivity to imazethapyr, chlorsulfuron, and nicosulfuron. In another study, an *A. thaliana* line selected for resistance to the triazolopyrimidine, DE-489, was cross resistant to chlorsulfuron but not imazapyr or a pyrimidinyl oxybenzoate herbicide.[80]

The cross resistance discussed above and cross insensitivity patterns reported in Table 4 come from a rather limited set of biotypes. A more extensive survey, nearly epidemiological in scope, is represented by the data presented in Figures 3 to 6.[59] *K. scoparia* biotypes suspected of resistance to sulfonylureas were collected from 1988 through 1991 in six states and three Canadian provinces, covering an area from Texas to Alberta and Washington to the Dakotas. The I_{50} ratios for ALS isolated from these biotypes were calculated for chlorsulfuron. Of the sampled biotypes, 298 exhibited target site insensitivity (the criterion was that the ratio of R ALS I_{50}/S ALS I_{50} for chlorsulfuron was at least three). I_{50} ratios were also determined for other ALS inhibitors using ALS isolated from some or all of these 298 biotypes: metsulfuron-methyl (179 biotypes), sulfometuron-methyl (298 biotypes), triasulfuron (298 biotypes), and imazapyr (158 biotypes). Biotypes were considered cross insensitive to these other ALS inhibitors when the I_{50} was at least twofold greater than the I_{50} obtained using susceptible *K. scoparia* ALS. The results were sorted in descending order of chlorsulfuron ALS I_{50} ratios, and the I_{50} ratios for sulfometuron-methyl, triasulfuron, metsulfuron-methyl, and imazapyr for each corresponding *K. scoparia* biotype were plotted in Figures 3, 4, 5, and 6, respectively.

For the sulfonylurea herbicides, a general correlation exists between ALS insensitivity to chlorsulfuron and a lessened sensitivity to metsulfuron-methyl, sulfometuron-methyl, and triasulfuron. In fact, 179 of 179, 298 of 298, and 297 of 298 biotypes that were less sensitive to inhibition by chlorsulfuron were also less sensitive (at least twofold) to inhibition by metsulfuron-methyl, sulfometuron-methyl, and triasulfuron, respectively.

The data for imazapyr were somewhat different (Figure 6). Of 158 biotypes having ALS with lessened sensitivity to chlorsulfuron inhibition, ALS from 145 biotypes was also less sensitive to inhibition by imazapyr. However, the imazapyr I_{50} ratios were generally less than for triasulfuron, sulfometuron-methyl, and metsulfuron-methyl, remaining at 5 to 6. Resistance to both chlorsulfuron and imazapyr was evident in some of these biotypes at the whole plant level.[41] The imazapyr ratios did not track the chlorsulfuron ratios as closely as the ratios for the other ALS inhibitors studied. This was due, in part, to the high percentage of indefinite ALS I_{50} ratios (i.e., the highest imazapyr concentration in the ALS assay was 38 μM, and the R ALS I_{50} for imazapyr was >38 μM; the ratio is therefore >6.4, indicated by a solid triangle) in the first 40 biotypes having a high resistance factor (Figure 6). In light of these results where 145 of 158 *K. scoparia* biotypes had lessened ALS sensitivity to both chlorsulfuron and imazapyr, it is surprising that four (of four reported) *K. scoparia* biotypes resistant to chlorsulfuron were completely controlled by imazapyr in another study.[8]

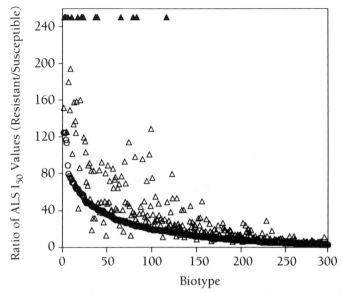

Figure 3. Ratio of I_{50} values (resistant/susceptible) for chlorsulfuron (○) and sulfometuron-methyl (△) using ALS isolated from 298 *K. scoparia* biotypes. For some resistant ALS I_{50} determinations, the highest concentration of herbicide used in the assay was below the I_{50} for sulfometuron-methyl and the subsequent ratio (▲) is a minimum rather than finite value.

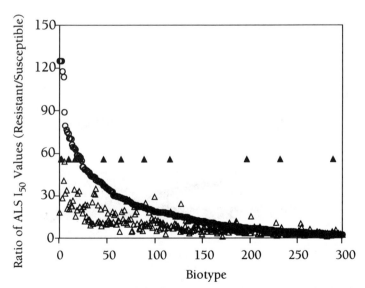

Figure 4. Ratio of I_{50} values (resistant/susceptible) for chlorsulfuron (○) and triasulfuron (△) using ALS isolated from 298 *K. scoparia* biotypes. For some resistant ALS I_{50} determinations, the highest concentration of herbicide used in the assay was below the I_{50} for triasulfuron and the subsequent ratio (▲) is a minimum rather than a finite value.

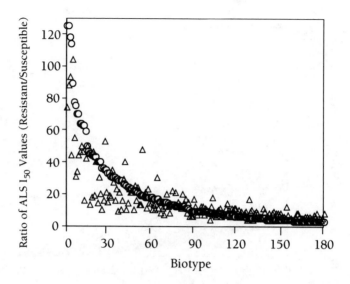

Figure 5. Ratio of I_{50} values (resistant/susceptible) for chlorsulfuron (○) and metsulfuron-methyl (△) using ALS isolated from 179 *K. scoparia* biotypes.

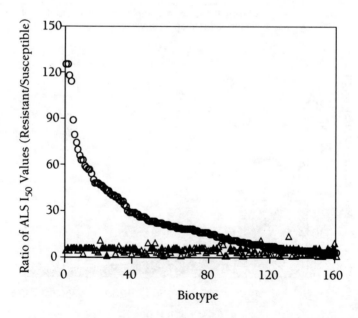

Figure 6. Ratio of I_{50} values (resistant/susceptible) for chlorsulfuron (○) and imazapyr (△) using ALS isolated from 158 *K. scoparia* biotypes. For some resistant ALS I_{50} determinations, the highest concentration of herbicide used in the assay was below the I_{50} for imazapyr and the subsequent ratio (▲) is a minimum rather than a finite value.

While the genotypes of these R *K. scoparia* isolates are unknown, this collection was obtained over a large area of North America over a relatively short time frame and is representative of ALS inhibitor resistance in *K. scoparia* in North America. Many of these biotypes (Figures 3 to 6) have not been tested at the whole plant level for target site cross resistance; however, where both whole plant and enzymic data are available, no discrepancies have been found.

Since the results presented in Figures 3 to 6 are from field populations of *K. scoparia*, several explanations are available for apparent differences in ALS I_{50} ratios among the *K. scoparia* biotypes. First, some *K. scoparia* samples may have been a mixture of susceptible and resistant plants resulting in a mixture of sensitive and resistant forms of ALS upon extraction. It is also possible that heterozygous plants were present which expressed both sensitive and resistant ALS. Finally, different mutations in the ALS gene, if present, would be expected to give different phenotypes. For these field samples, all three situations may be represented.

In spite of these limitations to interpretation, there exist some trends in cross insensitivity (Figures 3 to 6) that mirror those seen with whole plant analyses previously described. The enzymic data from this study indicate that a lessened sensitivity to all sulfonylurea herbicides is expected if the selection agent is a sulfonylurea herbicide. Also, a general trend of lesser cross insensitivity at the enzyme level to imazapyr is seen, again similar to the whole plant scenario. As more ALS inhibitor-resistant weeds become available for study, it will be interesting to see whether these trends continue.

Laboratory Mutants

Several mutants resistant to ALS inhibitors have been selected or generated in laboratory or greenhouse settings. In most cases, the organisms selected are identified as, or probably are, target site mutants, and include tobacco;[32,38,74,81-85] *Escherichia coli*;[86] yeast (*Saccharomyces cerevisiae*);[35,86] *Synechococcus*;[87,88] soybeans;[89] maize (*Zea mays* L.);[90] canola (*Brassica napus* L.);[91] *Chlamydomonas reinhardtii*;[92,93] cotton;[32,83] *Datura innoxia* Mill.;[94-96] *Salmonella typhimurium*;[27] *A. thaliana*;[36,39,40,80] birdsfoot trefoil (*Lotus corniculatus* L.);[97] wheat (*Triticum aestivum* L.);[98] and sugarbeet (*Beta vulgaris* L.).[99,100] The nucleotide changes in the ALS gene responsible for resistance have not been elucidated in most of the above-mentioned organisms; however, mutations have been identified in tobacco,[38] *Synechococcus*,[88] yeast,[86] *E. coli*,[86] and *A. thaliana*.[39,40]

Among the more interesting resistant organisms selected in the laboratory are those that are hypersensitive to one class of herbicide upon selection with a different class of ALS inhibitor. For example, a *C. reinhardtii* mutant selected with imazaquin has a higher sensitivity to chlorsulfuron[93] and a *D. innoxia* mutant resistant to sulfonylureas is more sensitive to imidazolinones.[94] If resistant weeds were more susceptible to other classes of herbicides, then this hypersensitivity could be used to selectively reduce the resistant population via the use of the alternate herbicide, as has been suggested for the management of triazine-resistant weeds.[101] Unfortunately, there is no evidence of a hypersensitive phenomenon occurring among ALS inhibitor-resistant weeds selected in the field, either to other classes of ALS inhibitors or other mode of action herbicides.[41,49,54-56,77]

From these studies and from the knowledge that many ALS mutations confer resistance,[5] it is clear that several different profiles of resistance to ALS inhibitor herbicides are possible in model or laboratory systems. However, the selection pressure in the field is much more demanding than in the laboratory, and it is unknown how many of these mutations (and consequent phenotypes) will be common in agronomic settings.

ALS Mutations Resulting in Herbicide Resistance

Resistant Plants Selected with Sulfonylureas

There is a great deal of information regarding mutations in the ALS gene from bacteria, yeast, and plants that confer resistance to ALS inhibitor herbicides. This is, in part, due to the fact that resistant mutants are easily selected in the laboratory. In addition, site-directed mutagenesis of the ALS gene has been performed. While several mutations resulting in ALS inhibitor herbicide resistance have been identified in model studies, the germane question here is, "what mutations are responsible for ALS inhibitor resistance in agronomic settings?".

First, mutations identified in laboratory experiments are discussed. Two mutations in yeast ALS and *E. coli* ALS II that confer sulfonylurea resistance are a pro192 to ser change and an ala26 to val mutation, respectively.[86] Several mutations conferring ALS inhibitor herbicide resistance have been identified in yeast;[5] in fact, 24 mutations at 10 different sites have been described. The following ALS mutations resulting in a sulfonylurea-resistant phenotype have been identified in plants (weeds not included): a pro197 to ser mutation in *A. thaliana* line GH50,[39] a pro196 to gln mutation in tobacco,[38] and a pro196 to ala coupled with a trp573 to leu mutation in tobacco[38] which is resistant to both sulfonylurea and imidazolinone herbicides[5] and enhances the degree of resistance to sulfonylureas over the single mutant.[85] Tobacco protoplasts were transformed using mutant chimeric tobacco ALS genes coding for either a trp568 to leu or ala199 to asp change, and these mutations by themselves confer resistance.[9] Resistance to both sulfonylurea and imidazolinone herbicides was achieved in tobacco by cloning a chimeric ALS gene composed of two individual mutations (pro to ser and ser to asn), each of which independently confer resistance to either sulfonylureas or imidazolinones.[102]

Guttieri et al.[103] are the first to determine the identity of specific ALS mutations in weeds that have become resistant because of sulfonylurea herbicide use. A pro173 to his change was identified in resistant *L. serriola* and a pro173 to thr change in resistant *K. scoparia*. Unfortunately, cross resistance patterns for these specific biotypes are not available.

Possible Role of Proline Mutations

The previous studies indicate that pro mutations in the ALS gene may be more prevalent than other mutations resulting in ALS inhibitor resistance in plants. Of the many resistant *K. scoparia* biotypes examined (Figures 3 to 6), are pro mutations involved? The answer to this question requires an analysis of the whole plant cross resistance and enzymic cross insensitivity patterns, where available, of the mutants described above. By comparing resistance and insensitivity patterns of known mutants to the enzymic patterns observed for *K. scoparia*, an extrapolation to the type of ALS mutations in *K. scoparia*, and perhaps for other sulfonylurea selected weeds, may be possible. In addition, information is now available on the deduced amino acid sequence of a highly conserved region of the ALS, domain A, which includes the residues found at position 173 (pro in the S biotype), for ten different resistant *K. scoparia* biotypes.[104]

Two single point mutations in the ALS gene are thought to be selectively resistant. That is, the previously mentioned pro197 mutation in *A. thaliana*[39] and the pro196 mutation in tobacco[38] are claimed to result in sulfonylurea but not imidazolinone resistance.[5,36,38,80] In contrast, the survey of resistant *K. scoparia* patterns (Figures 3 to 6) indicates a high and slight degree of ALS insensitivity to chlorsulfuron and imazapyr, respectively. At first inspection, the data would seem to suggest that pro mutations are not responsible for the resistance observed in the *K. scoparia* collection since some ALS insensitivity to imazapyr is observed. However, closer examination of the studies using tobacco and *A. thaliana* can be interpreted to include the possibility that pro mutations confer a slight degree of cross resistance to imazapyr and thus allow the prospect that pro mutations are prevalent in resistant *K. scoparia* and possibly other weeds.

The *A. thaliana* mutant line (GH50) is described as resistant to chlorsulfuron and sulfometuron-methyl but not imazapyr.[36] This line is undoubtedly resistant to the sulfonylurea herbicides; however, the I_{50} for imazapyr increased approximately fourfold using ALS isolated from line GH50 relative to wild-type ALS, and it is possible that a slight degree of resistance to imazapyr would not have been apparent relative to the large change in sensitivity to sulfonylurea herbicides. Closer inspection of *A. thaliana* plants containing ALS with pro197 to ser changes (line GH50) may well show low levels of imazapyr resistance. A similar study was conducted using transgenic tobacco that only expresses an ALS containing a pro196 to ser mutation.[105] ALS isolated from this tobacco line is highly insensitive to inhibition by all sulfonylureas tested and shows a low-level cross insensitivity to imazapyr.

The same pro mutation in either *A. thaliana* (line GH50) or transgenic tobacco yields ALS that is insensitive to inhibition by chlorsulfuron, thifensulfuron-methyl, ethametsulfuron-methyl, and tribenuron-methyl, but not the imidazolinones, imazaquin, imazethapyr, or imazapyr.[106] Again on closer review, there exists a slightly lower sensitivity of either *A. thaliana* GH50 or transgenic tobacco ALS to inhibition by imazapyr relative to the wild-type ALS. Seedling growth was not evaluated using imazapyr, only imazaquin.

Other plant studies on resistance to various ALS inhibitor herbicides have been done using well-characterized mutants. Hattori et al.[102] cloned the *A. thaliana* ALS gene carrying the pro to ser mutation into tobacco. While the interpretation was that the transformants were sensitive to imazapyr, the growth of the transformed seedlings was not directly compared to sensitive lines, and ALS from the transformed lines appeared somewhat less sensitive to imazapyr than ALS from untransformed controls. Charest et al.[107] introduced the ALS gene from sulfonylurea-resistant *A. thaliana* line GH50 into tobacco. As before, high levels of resistance to chlorsulfuron occur, and transformed plantlets display a slight cross resistance to imazethapyr. Smith et al.[108] saw differences in zones of growth inhibition using chlorsulfuron, sulfometuron-methyl, bensulfuron-methyl, but not imazaquin for *E. coli* expressing plant ALS containing a pro to ser mutation. Transformed canola containing the *csr*1 gene from *A. thaliana* (i.e., pro to ser change in ALS) is resistant to chlorsulfuron but not imazethapyr.[109] Wiersma et al.[110] introduced, via site-directed mutagenesis, a ser for pro173 into canola ALS; *S. typhimurium* expressing this canola ALS displays high levels of resistance to chlorsulfuron and imazamethabenz-methyl and lesser resistance to imazethapyr in disc diffusion assays.

The imazapyr concentration required for the control of *A. thaliana* mutant line *csr*1 (pro197 to ser in ALS) as measured on a seedling fresh weight basis was marginally higher relative to nearly isogenic wild-type plants.[80] Correspondingly, ALS isolated from this line was inhibited slightly less by imazapyr relative to the inhibition of wild-type ALS. As in previous studies, these authors did not conclude that pro to ser changes in ALS conferred any increased resistance to imazapyr, presumably because the differences were very small.

There are two studies that are not wholly consistent with the foregoing observations. Gabard et al.[106] in studying *A. thaliana* GH50 (pro to ser) and transgenic tobacco containing the gene from line GH50, noted that the ALS isolated from these plants is highly inhibited by triasulfuron. Our experience is that pro to ser mutations, in addition to lessening ALS sensitivity to chlorsulfuron, decreases sensitivity to triasulfuron (by at least 14-fold) as well as many other sulfonylureas.[105] The other contradictory result was with transformed tobacco which apparently contains an ALS gene with a pro196 to ser change.[111] This construct yields ALS that is insensitive to several sulfonylureas as expected, but is also highly cross insensitive to imazaquin (150-fold). For other organisms with pro to ser mutations, low but not high levels of cross insensitivity to imidazolinones are observed; further clarification is required regarding this point.

Small differences in imazapyr I_{50} values[36,80,102,106] may be overlooked relative to the very large differences in I_{50} values (sometimes >300-fold) seen using sulfonylurea herbicides. Closer evaluation of the enzymic data using ALS from plants with pro mutations shows that resistance to chlorsulfuron and other sulfonylureas along with a slight degree of resistance to imazapyr is probable. ALS from resistant *K. scoparia* (Figures 3 to 6) displays this inhibition pattern, and therefore, the possibility exists that pro mutations are responsible for the ALS inhibitor resistance observed in this very geographically diverse collection.

Direct evidence now exists to support this possibility. Of ten *K. scoparia* biotypes isolated from different North American locations, seven biotypes have pro changes in the highly conserved domain A region of the ALS gene while three biotypes have domain A sequences identical to the wild-type gene (i.e., not pro mutations).[104] While pro changes may be among the more common ALS mutations observed in resistant weeds selected by sulfonylurea herbicides, other ALS mutations apparently are also available for selection by ALS inhibitor herbicides in the field.

Resistant Weeds Selected with Imidazolinones

Since there exists only one well-described case of a weed developing resistance through selection with an imidazolinone herbicide, generalizations regarding possible mutations are not possible. However, it is interesting that the *X. strumarium* selected with imazaquin seemingly has the same whole plant cross resistance and ALS insensitivity profile to imidazolinone and sulfonylurea herbicides,[57,69,79] as does a well-characterized mutant of *A. thaliana*. That is, an *A. thaliana* GH90 line contains an ALS with a ser653 to asn alteration,[40] and this line is selectively resistant to imidazolinones but not sulfonylureas or triazolopyrimidines.[37,40,80] Cloning of this gene into tobacco gave similar results with regard to the selective imidazolinone resistance.[102] Similarly, the imidazolinone resistant *X. strumarium* is resistant to imazaquin but not cross resistant to chlorimuron-ethyl,[57,69] and the ALS from this resistant weed is insensitive to inhibition by imidazolinones but not chlorimuron-ethyl or flumetsulam.[79] It will be interesting to see whether the resistant *X. strumarium* harbors this amino acid change (i.e., ser653 to asn), and whether future imidazolinone-resistant weeds are target site mutants with predominantly ser mutations in the ALS.

Other Target Site-Related Mechanisms: ALS Overexpression and Double Mutants

There are no known cases of ALS inhibitor resistance in weeds due to the overexpression of wild-type ALS. In fact, evidence exists in several resistant weeds that ALS expression (as measured by specific enzyme activity) is similar in both the S and R biotypes.[54,55,77,112] However, plant mutants owing their resistance to ALS overexpression have been isolated in the laboratory, and their study may be instructive in appreciating whether overexpression of ALS is a future possibility in agronomic settings.

Harms et al.[111] obtained a tobacco cell line that was resistant to both sulfonylurea and imidazolinone herbicides by selecting first with a lethal level of cinosulfuron and subsequently with stepwise, increasing concentrations of primisulfuron. The tobacco cell line acquired resistance via two mechanisms: one was a herbicide-insensitive ALS and the other was an amplification of ALS. After attaining a target site change in ALS, it appears that stepwise selection probably favored the amplification of ALS genes, and this ALS amplification, at least in one case, decreased in the absence of selection pressure. ALS gene copy number was linearly correlated with ALS activity, suggesting that there were no limitations to overexpression of the enzyme in these cell lines. In contrast, Odell et al.[113] used the cauliflower mosaic virus 35S promoter to elevate ALS mRNA levels by 25-fold and saw only a 2-fold increase in ALS-specific activity. In this case, posttranscriptional control of mRNA may have kept enzyme levels low. The difference possibly results from the use of cell lines[111] vs. plants,[113] since the phenomenon of ALS amplification has yet to be demonstrated at the whole plant level.

There are other reports of increased ALS expression possibly influencing resistance in plants. Different tobacco clones had variable degrees of sulfonylurea resistance apparently due to differences in the expression level of mutant genes.[38] Naturally occurring differences in ALS levels in the roots of maize inbred lines contributed to observed chlorsulfuron tolerance.[114] Slightly higher total protein levels were seen in R birdsfoot trefoil selected with stepwise, increasing concentrations of thifensulfuron-methyl.[97] While ALS amplification may have played a role in the resistance observed in birdsfoot trefoil, ALS insensitivity to thifensulfuron-methyl was probably the predominant mechanism of resistance.

The stepwise selection for resistance may favor the appearance of gene amplification, as has been suggested previously.[115] While this phenomenon has not yet been demonstrated as a resistance mechanism in ALS inhibitor-resistant weeds, a route to the appearance of these weeds can be postulated. That is, increasing usage of ALS inhibitor herbicides with time may promote the appearance of resistant weeds with amplified, resistant ALS genes.

In the laboratory, a double mutant rather than a gene amplification mechanism was obtained using a two-step selection process.[85] Reselection of sulfonylurea-resistant tobacco cells that contained a single mutation (S4) with a second, higher level of sulfometuron-methyl resulted in the isolation of the double mutant, HRA, which has an increased resistance of approximately fivefold to ALS inhibitors over the single mutant. Higher ALS expression was not observed. Obtaining a double mutant weed biotype upon further selection with ALS inhibitors in agronomic settings seems unlikely because many, if not most, ALS inhibitor-resistant weeds are cross resistant to other ALS inhibitor herbicides. This cross resistance would likely preclude the use of other ALS inhibitors on the previously selected weed. However, selection of double mutants is possible if an ALS inhibitor resistant weed is not cross resistant to a different chemical class of ALS inhibitor. Double mutants may be very resistant to several or all ALS inhibitors, similar to that seen with other plant ALS double mutants.[5,102] Although plants containing the HRA genes (double mutant) perform well under cultivated conditions,[9] the fitness of weeds with a second ALS mutation under competitive conditions is unknown.

ALS Binding Domains and Relation to Resistance

ALS binds many inhibitors of diverse structure, and an explanation for the variety of binding structures accommodated by ALS may result from the relationship between ALS and pyruvate oxidase. The herbicide binding domain in ALS is hypothesized to be a vestigial site of quinone binding in pyruvate oxidase.[116] The sequence of the *pox*B gene coding for pyruvate oxidase shows substantial sequence homology with ALS,[117] suggesting an evolutionary relationship.[116] ALS probably evolved from pyruvate oxidase by losing the ability to bind quinones with long isoprene tails in order to facilitate the catalysis of pyruvate to acetolactate or acetohydroxybutyrate rather than acetate, the product of pyruvate oxidase. Although quinones with long isoprene tails are not bound by ALS, quinones with no or short isoprene tails such as ubiquinone-0 or ubiquinone-5 are bound by ALS and like sulfonylureas, imidazolinones, and triazolopyrimidines, inhibit ALS.

It is of interest to know whether these structurally diverse ALS inhibitors have overlapping, nonoverlapping, or both overlapping and nonoverlapping binding sites in the ALS enzyme. If ALS inhibitor herbicides bind to the same domain, then cross resistance would be expected. Alternatively, if the binding domains are unique, then cross resistance would not be anticipated (except in the unusual case where one mutation denies access of inhibitors to two different binding domains). This expectation can be analyzed by determining whether different ALS inhibitor herbicides compete with the binding of the substrate, pyruvate, or with other ALS inhibitors. If both substrate and inhibitor can bind to the enzyme simultaneously, as is the case for purely noncompetitive and uncompetitive inhibition, then their binding domains must be unique, whereas competitive binding suggests mutually exclusive binding.

Unfortunately, results from kinetic studies do not give a uniform answer. With respect to pyruvate binding, sulfometuron-methyl binding is competitive using *S. typhimurium* ALS,[118] noncompetitive using pea (*Pisum sativum* L.) ALS,[119] and uncompetitive using *Methanococcus* ALS.[120] Chlorsulfuron binds noncompetitively with pyruvate to barley (*Hordeum vulgare* L.) ALS.[121] Uncompetitive binding is seen for imazaquin versus pyruvate using plant ALS,[22,29,121] but noncompetitive binding has also been reported for the imidazolinones.[119] A triazolopyrimidine herbicide[22,122] and a pyrimidinyl oxybenzoate herbicide[22] show mixed-type inhibition with respect to pyruvate. Some of the discrepancies may be due to the complicated time-dependent behavior of ALS inhibition and the apparent inactivation of ALS in the presence of imidazolinones.[123]

Competitive binding studies are clearer. ^{14}C-sulfometuron-methyl bound to *S. typhimurium* ALS II can be displaced by imazaquin, 2-NO$_2$-6-methylsulfonanilide (the triazolopyrimidine, DE-489), and ubiquinone-0.[116] Similarly, ^{14}C-chlorsulfuron bound to maize ALS is replaced with increasing concentrations of imazaquin.[121] These studies indicate that the binding of these inhibitors is at least partially overlapping and in effect, mutually exclusive.

The evidence for non-shared binding domains within ALS for ALS inhibitor herbicides derives from studies with resistant mutants. There are several examples of mutant cell lines or plants selected with either sulfonylureas, imidazolinones, or triazolopyrimidines which show little or no cross resistance to other ALS inhibitor herbicide classes (see previous sections). If a plant is resistant to sulfonylurea but not imidazolinone herbicides, this is evidence for distinct binding domains in that the mutation resulting in resistance occurred in a region important to sulfonylurea but not imidazolinone binding. Alternatively, mutations in ALS that affect both classes of herbicides provide circumstantial evidence of shared binding domains within ALS. Mutations affecting the binding of extraneous site inhibitors[124] (i.e., inhibitors that bind either entirely or partially to a site outside of the enzyme active site), such as the ALS inhibitor herbicides, may be more abundant because this protein domain probably no longer has a vital catalytic function in ALS.

An interpretation consistent with both observations (i.e., mutually exclusive binding of various ALS inhibitors and resistance to one but not other ALS inhibitor classes) is that there exist both overlapping and nonoverlapping binding domains of ALS inhibitor herbicides in the ALS molecule. Thus, ALS mutations have the potential of affecting the binding of all, some combination, or single ALS inhibitor herbicide(s), and the consequent cross resistance patterns, in theory, should mirror these possibilities.

NON-TARGET SITE RESISTANCE

Uptake and Translocation

Limitations in uptake of a herbicide or its movement to the site of action are often cited as possible mechanisms of resistance. Nevertheless, few examples exist for ALS inhibitors in which a lack of herbicide absorption and/or translocation to the site of action contributes to herbicide resistance. Some that do exist will be discussed here.

One such case in crops is the selectivity of triasulfuron between wheat (resistant) and *L. perenne* (susceptible). There is little difference between wheat and *L. perenne* in the percentage of foliarly applied ^{14}C-triasulfuron that is absorbed, and 1.5% or less of the absorbed ^{14}C is translocated out of the treated leaf in either species.[125] However, following ^{14}C-triasulfuron application to the roots, more is absorbed and translocated to various portions of the shoot in *L. perenne* than in wheat, on a dry weight basis. After 1 and 6 days of exposure, *L. perenne* roots and shoot fractions contain 1.3- to 2.4-fold and 3.3- to 13-fold, respectively, more ^{14}C/mg dry weight than wheat.[125] Data for the foliar application were not expressed on a dry weight basis, so it is not clear whether or not preferential accumulation in *L. perenne* also occurs when triasulfuron is applied to the leaf. Evidence suggests that reduced uptake and translocation play a role in wheat resistance to preemergence (but not postemergence) applications of triasulfuron, but rapid metabolic inactivation by wheat is probably the major selectivity mechanism for either pre- or postemergence applications.[125,126]

In another example, comparison is made of two maize inbreds with different sensitivities to thifensulfuron-methyl. No significant difference exists between inbreds in the amount of ^{14}C-thifensulfuron-methyl absorbed from a foliar application; however, slightly more of the absorbed ^{14}C is translocated out of the treated leaf in susceptible inbred A619 (13.1%) than resistant inbred A671 (5.8%).[127] The observed differences in translocation are probably too small to account for the large differences in sensitivity, and more rapid metabolism of thifensulfuron-methyl in inbred A671 appears to be the major mechanism conferring resistance. In fact, the reduced translocation in A671 may result from resistance rather than contribute to it. Thifensulfuron-methyl metabolites, which are formed more rapidly in A671 than A619, may be less readily translocated than the parent herbicide.[127]

The preceding examples attribute at least a contributory role to limited uptake and/or translocation in resistance, whereas the following reports suggest virtually no involvement of these processes. Like maize and cereals, soybeans exhibit a level of resistance to thifensulfuron-methyl (Table 5). There is no correlation between the percent taken up and sensitivity of soybeans and four weed species to a foliar application of thifensulfuron-methyl.[128] One and three days after application, soybeans take up nearly as much or more ^{14}C-thifensulfuron-methyl as susceptible *Amaranthus retroflexus* L., *Chenopodium album* L., and *Abutilon theophrasti* Medicus, or moderately susceptible *Ipomoea purpurea* (L.) Roth. Metabolism of thifensulfuron-methyl to herbicidally inactive products occurs much more rapidly in the resistant than the susceptible species and therefore metabolic inactivation is, again, the probable mechanism of selectivity.

Similarly, uptake and distribution of foliarly applied ^{14}C-ethametsulfuron-methyl is nearly identical in commercial brown mustard (*Brassica juncea* (L.) Czern.) (resistant) and wild mustard (*Sinapis arvensis* L.) (susceptible) between 8 and 72 h after application.[129] Resistance is likely due to the fact that ethametsulfuron-methyl is converted to two major metabolites more rapidly in commercial brown mustard than wild mustard. In some of the earliest work on ALS inhibitor selectivity, Sweetser et al.[130] observed a tendency for a smaller percentage of applied ^{14}C-chlorsulfuron to penetrate leaves of resistant species such as wheat, barley, and wild oat (*Avena fatua* L.) than susceptible species such as sugarbeet, soybean, cotton, and wild mustard. The small difference, however, does not account for the up to 4000-fold difference in sensitivity to chlorsulfuron observed among the species tested. Furthermore, there is no clear correlation between the percentage of applied ^{14}C that is translocated out of the treated leaf and sensitivity of the species.[130]

Now considering weeds, the involvement of differential uptake and/or translocation has been investigated in two species that have evolved resistance to ALS inhibitors, though in neither instance does uptake or translocation make

Table 5. Metabolism-based crop selectivity demonstrated for ALS-inhibitor herbicides

Herbicide	Crop[a]	Ref.	Sensitive ALS documented
Bensulfuron-methyl	Rice	207	Yes (Ref. 135)
Chlorimuron-ethyl	Soybean	139	Yes
Chlorsulfuron	Wheat, barley	130	Yes (Ref. 135)
Chlorsulfuron	Flax	208	
Ethametsulfuron-methyl	Canola	134	
Ethametsulfuron-methyl	Brown mustard	129	Yes
Flumetsulam[b]	Maize, soybean	20, 141	Yes (Ref. 141)
Imazamethabenz-methyl	Maize, wheat	143	Yes (Ref. 142)
Imazapyr	Conifers, rubber trees	142	
Imazaquin	Soybean	142	
Imazethapyr	Legumes	142	Yes
Metsulfuron-methyl	Wheat, barley	209	
MON-120000 or NC-319	Maize	210, 211	Yes
Nicosulfuron	Maize	134	Yes
Primisulfuron	Maize	140, 174	Yes
Rimsulfuron	Maize	212	
Rimsulfuron	Potato	213	
Thifensulfuron-methyl	Wheat	214	Yes
Thifensulfuron-methyl	Maize	127	Yes
Thifensulfuron-methyl	Soybean	128	Yes
Triasulfuron	Wheat	125, 126	
Triflusulfuron-methyl	Sugarbeet	215	

[a]Crops that are naturally resistant to the specified ALS inhibitor due to metabolism.
[b]Proposed common name.

a significant contribution to resistance. In *L. rigidum* biotype SR4/84, which is resistant to chlorsulfuron due to rapid conversion of chlorsulfuron to inactive metabolites (Table 6), the percentage of ^{14}C-chlorsulfuron absorbed from a foliar application is not significantly different than that in SRS2, a susceptible biotype, up to 48 h after application.[131] Although there is a tendency for an approximately twofold greater percentage of the absorbed ^{14}C to be translocated out of the treated leaf in SRS2 than in SR4/84, this difference may result from reduced mobility of the metabolites (formed more rapidly in SR4/84) compared to the parent herbicide,[131] as in the case of maize inbreds above.[127]

Hundreds of biotypes of *K. scoparia* are resistant to ALS inhibitors due to reduced sensitivity of the target site (Tables 3 and 4; Figures 3 to 6). Uptake and translocation of ^{14}C-chlorsulfuron applied to a selected leaf was examined in a representative biotype, and the results demonstrate that during most of the period from 1 to 48 h after application, there is no significant difference in the percentage of applied ^{14}C absorbed, or of the absorbed ^{14}C translocated out of the treated leaf, between the susceptible and resistant biotypes.[77] When there is a significant difference, it is a slightly greater percentage absorbed or translocated in the resistant biotype, the opposite of what would be expected if uptake or translocation was a factor in producing resistance.

Reduced sensitivity of plants to several herbicides is associated with lower rates of either uptake or translocation from the site of uptake to the site of action.[132,133] However, in the cases of resistance to ALS inhibitors where these

Table 6. Documented weeds with naturally occurring (nonselected) metabolism-based resistance to ALS inhibitors

Herbicide	Common name	Species	Ref.
Bensulfuron-methyl	Barnyardgrass	Echinochloa crus-galli	207
Chlorsulfuron	Wild oats	Avena fatua	130
Chlorsulfuron	Annual bluegrass	Poa annua	130
Chlorsulfuron	Johnsongrass	Sorghum halepense	130
Chlorsulfuron	Giant foxtail	Setaria faberi	130
Chlorsulfuron	Black nightshade	Solanum nigrum	208
Chlorsulfuron	Leafy spurge	Euphorbia esula	216
Imazaquin	Florida beggarweed[a]	Desmodium tortuosum	217
Imazaquin	Sicklepod[a]	Cassia obtusifolia	217
Primisulfuron	Barnyardgrass	Echinochloa crus-galli	140

[a]Fresh weight reduction observed only at the highest rate tested.

potential mechanisms have been investigated, they play only a secondary role (or no role at all) compared to the primary mechanism of either metabolic inactivation or target site insensitivity.

Metabolism Prior to the Target Site

Naturally Occurring Resistant Crops

The single most important mechanism of naturally occurring (as opposed to evolved) resistance to ALS inhibitors is metabolic alteration of the active herbicide. A wealth of information exists on the mechanism of naturally occurring resistance in crops (i.e., crop selectivity) to sulfonylurea, triazolopyrimidine, and imidazolinone products that either have been commercialized or are being developed (Table 5). In every case for which the mechanism has been determined, inherent crop selectivity of a particular ALS inhibitor in a given crop is based on the crop's ability to metabolize the herbicide to nonphytotoxic compounds rapidly enough to prevent lethal herbicide levels from reaching ALS. There exists one possible exception, imazamethabenz-methyl selectivity in maize and wheat, which will be discussed at the end of this section.

The metabolic reactions involved in crop selectivity to a broad range of sulfonylureas were recently reviewed.[134] The authors cite references to nine distinct metabolic reactions at numerous positions on the sulfonylurea molecule that lead to inactivation. Among the more common reactions are aryl and aliphatic hydroxylation, O-dealkylation, and deesterification. Cytochrome P450 (Cyt P450) monooxygenase systems have been implicated in some of the hydroxylation reactions.[126,135-138] As will become apparent in the discussion below, several of the reactions listed are important in the selectivity of the triazolopyrimidines and the imidazolinones as well. An additional reaction, homoglutathione conjugation, has only been reported for soybean inactivation of chlorimuron-ethyl.[139]

In addition to the sulfonylureas previously reviewed, metabolism of primisulfuron in maize[136,140] and triasulfuron in wheat[125,126] have also been described. Interestingly, primisulfuron is metabolized in maize by apparently analogous reactions to those reported for nicosulfuron, another maize selective sulfonylurea. Nicosulfuron is rapidly metabolized to 5-hydroxypyrimidinyl nicosulfuron, a herbicidally inactive derivative which is then conjugated to glucose.[134] Maize microsomes hydroxylate primisulfuron *in vitro* on either the phenyl or the pyrimidine ring, catalyzed by a Cyt P450 system.[136] In maize seedlings, the major metabolites are glucose conjugates of the same ring hydroxylated intermediates.

Similarly, triasulfuron metabolism in wheat resembles that of the closely related chlorsulfuron, which is rapidly hydroxylated at the 5 position on the phenyl ring, followed by conjugation to glucose.[130] Although the hydroxylated intermediate is active against ALS, it does not accumulate in wheat because of rapid conversion to the glucose conjugate.[135] Frear et al.[126] demonstrated that microsomes from wheat seedlings metabolize triasulfuron to its 5-hydroxyphenyl derivative *in vitro*, apparently via a Cyt P450 monooxygenase. Further, studies with intact wheat plants showed that the hydroxylated triasulfuron derivative is conjugated to a sugar (probably glucose) *in vivo*.[125]

Flumetsulam, a triazolopyrimidine that is selective in cereals, maize, and soybeans, also owes its selectivity to metabolic detoxification. Tolerant plants oxidize flumetsulam to one or more hydroxylated metabolites, and soybeans also form a pyrimidine ring opened metabolite.[141] Each of these metabolites is at least fivefold less active than flumetsulam on ALS.

Although differential metabolism is also the common thread in the selectivity of the four commercial imidazolinones, there are distinct nuances to the involvement of metabolic inactivation for each product. For imazaquin and imazethapyr, rapid metabolic inactivation is clearly the basis for selectivity in soybeans. Imazaquin is initially metabolized to two herbicidally inactive compounds that are produced by ring opening or ring closure.[142] These are further metabolized to numerous other breakdown products. Different reactions are involved in imazethapyr selectivity in soybeans, where the first step in imazethapyr inactivation is the hydroxylation of the ethyl substituent on the pyridine ring.[142] This intermediate is only 2.5-fold less inhibitory to ALS than the parent herbicide, so the actual inactivation occurs upon subsequent conjugation to glucose. It is interesting to note that maize, a species with intermediate sensitivity to imazethapyr, also produces the moderately active hydroxylated metabolite; however, maize does not further modify the metabolite by conjugation to glucose. This supports the conclusion that glucose conjugation is required to fully inactivate imazethapyr. Highly sensitive species such as *Xanthium* spp. exhibit very little degradation of imazethapyr.[142]

Differential metabolism is responsible for imazapyr selectivity through a less common mechanism. Imazapyr is a broad-spectrum herbicide that is generally used in nonselective applications. It is, however, used selectively in some crops such as conifers and rubber trees, where rapid metabolism results in the formation of an active, but immobile, metabolite. The herbicide is not actually detoxified, but it is rendered nonherbicidal because phytotoxic quantities are prevented from reaching the growing points.[142]

The final case to be considered is imazamethabenz-methyl. This imidazolinone is unique among commercial ALS inhibitors because selectivity is based on metabolic activation by the sensitive weed, wild oat, and metabolic inactivation by resistant crops such as maize and wheat. The commercial form of the herbicide is a methyl ester that is inactive against ALS; the active form of the herbicide is imazamethabenz, the free acid.[142] Wild oat deesterifies imazamethabenz-methyl to the free acid which accumulates following treatment. In contrast, little or no free acid was detected in maize or wheat.[143] Instead, maize and wheat hydroxylate the methyl substituent on the benzene ring forming an inactive alcohol. An alcohol derivative of imazamethabenz was also observed in maize and wheat, suggesting that both the parent and the acid may be hydroxylated. Therefore, maize and wheat may produce the active acid, but metabolic inactivation occurs so rapidly that the acid never accumulates to an appreciable extent. Alternatively, deesterification in maize and wheat may occur after hydroxylation of the parent. *Avena fatua* also produces the same alcohol and glucoside metabolites of imazamethabenz-methyl as maize and wheat.[143] Hence, the selectivity of imazamethabenz-methyl results from the production and accumulation of acid in wild oat and not in maize and wheat. As an additional note, for sulfonylureas that have not been commercialized, activation of pro herbicides (i.e., an inactive form of the herbicide) by sensitive species has also been reported.[144]

The foregoing examples highlight two key reactions by which ALS inhibitors are inactivated by crops: aryl and aliphatic hydroxylation, both of which are often followed by glucose conjugation. These reactions, which are likely catalyzed by Cyt P450, have been reported to be responsible for detoxification of herbicides in such diverse crops as wheat, barley, maize, flax (*Linum usitatissimum* L.), and soybeans. Additional reactions[134] are known to function for other selective herbicides in these and other crops (Table 5). The conclusion that metabolic inactivation is the key determinant in crop selectivity of ALS inhibitors is further supported by the demonstration that ALS from several naturally resistant crops is sensitive to inhibition by these herbicides[145] (Table 5).

Naturally Occurring Resistant Weeds

The mechanism of resistance has been investigated in a few key weed species that are inherently resistant to particular ALS inhibitors. In each herbicide/resistant weed combination that has been reported, resistance is based on the ability of the weed to metabolize the herbicide to inactive products (Table 6). Sensitive weeds that have been studied typically metabolize an active herbicide very slowly, if at all.[135] Therefore, just as metabolic inactivation is the mechanism by which crops are intrinsically resistant to specific ALS inhibitors, so it also appears to be a mechanism responsible for poor or variable control of some weeds by certain ALS inhibitor herbicides.

Evolved Resistance in Weeds

While resistance to ALS inhibitors has evolved in at least 14 weed species (Table 2), rapid metabolic inactivation has so far proven to be the basis of resistance only in *L. rigidum*.[131,146] In addition, *Alopecurus myosuroides* Huds. that is resistant to chlorotoluron and isoproturon due to rapid degradation of those herbicides shows non-target site cross resistance to chlorsulfuron.[147] However, the involvement of sulfonylurea herbicide metabolism has not yet been demonstrated.

Two *L. rigidum* biotypes (SR4/84 and SLR 31) that developed resistance to diclofop-methyl through its continuous use in wheat culture shows non-target site cross resistance to chlorsulfuron and other wheat selective ALS inhibitors (Table 7).[112,146] Chlorsulfuron is metabolized fourfold more rapidly in roots[131] and two- to fourfold more rapidly in shoots[54,131,146] of these biotypes than in susceptible *L. rigidum* (Table 7). Although chlorsulfuron metabolism is faster in both the shoots and roots of the resistant biotype, the rate of metabolism by roots may, in fact, play the critical role in determining the sensitivity of *L. rigidum* to chlorsulfuron since the rate of root metabolism is more limiting than shoot metabolism. The half life of chlorsulfuron is approximately three times longer in roots than shoots of each *L. rigidum* biotype. In sensitive shoots, the half life is only twice as long as in wheat where resistance is due to metabolic inactivation (Table 7). Sensitivity to chlorsulfuron in selected grasses, including *Setaria viridis* (L.) Beauv. and a few of the more sensitive varieties of wheat and barley, was related to how slowly chlorsulfuron was degraded in the roots but was independent of the degradation rate in the shoots.[148]

The major chlorsulfuron metabolite in R and S *L. rigidum* is the phenyl-*o*-glucoside,[131] the same metabolite produced in wheat,[130] and minor metabolites are the triazine amine and sulfonamide. All of these metabolites are inactive against ALS. There was no significant difference in the uptake or translocation

Table 7. Characteristics of *Lolium rigidum* with metabolism-based cross resistance to chlorsulfuron[a]

Characteristic	Susceptible[b]	Resistant[c]
GR_{50} (diclofop-methyl) (g ai/ha)	250	>4,000
GR_{50} (chlorsulfuron) (g ai/ha)	7	160
Half-life (chlorsulfuron)[d] (h):		
Shoots	4	1
Roots	13	3
ALS I_{50} (chlorsulfuron and metabolites) (nM)		
Chlorsulfuron	15	14
Glucose conjugate	540	820
Triazine amine	>18,000	>18,000
Sulfonamide	>18,000	>18,000

[a]Compiled from Reference 131.
[b]Susceptible = biotype SRS2.
[c]Resistant = biotype SR4/84.
[d]Half-life for chlorsulfuron in wheat = 2 h in shoots and 3 h in roots.

of foliarly applied chlorsulfuron in SR4/84 compared to a susceptible biotype,[131] nor was there any difference in sensitivity to chlorsulfuron for ALS isolated from SR4/84 or SLR 31 and a susceptible biotype.[54,112,131] Taken together, these results confirm that the basis for non-target site cross resistance to chlorsulfuron in these diclofop-methyl resistant biotypes is rapid metabolic detoxification (see Chapter 9).

In contrast to biotypes SR4/84 and SLR 31, *L. rigidum* biotype WLR 1 developed resistance to nonselective and wheat selective ALS inhibitors following seven consecutive annual treatments with chlorsulfuron. ALS isolated from WLR 1 is less sensitive than ALS from a susceptible biotype to inhibition by a variety of ALS inhibitors,[54] and target site insensitivity appears to be the main mechanism of resistance. In addition, metabolism of chlorsulfuron is somewhat enhanced in WLR 1, though it is not as rapid as in SLR 31, and differential metabolism is probably a secondary mechanism of resistance in WLR 1.[54]

These results demonstrate that it is possible for more than one resistance mechanism to exist in the same biotype. Nevertheless, when other resistant weeds that have insensitive ALS have been examined for additional resistance mechanisms, none has been found. That is, there is no difference in metabolism (Table 8) or uptake and translocation[77,131] of sulfonylureas between susceptible and resistant biotypes. Additional species with metabolism-based resistance, or even multiple resistance mechanisms, may be discovered if sufficient numbers of biotypes are studied. As an example, whereas the preponderance of cases of triazine resistance are due to a reduction in target site sensitivity, a biotype of velvetleaf was discovered that is resistant because of increased atrazine inactivation by two glutathione *S*-transferase isozymes.[149]

Metabolism and Non-Target Site Cross Resistance

Does the occurrence of metabolism-based resistance increase the potential for non-target site cross resistance to herbicides with different modes of action? Opinions on this question vary. Metabolic inactivation is the basis of resistance to sulfonylureas in some resistant *L. rigidum*[54,131] and phenylureas in resistant *Alopecurus myosuroides*.[147] These biotypes are also cross resistant in varying

Table 8. Metabolic half-life of sulfonylurea herbicides in shoots of weeds with target site-based resistance

		Half-life (h)		
Weed	Herbicide	Susceptible	Resistant	Ref.
Stellaria media	Chlorsulfuron	>24	>24	55
Kochia scoparia	Chlorsulfuron	>21	>21	77
Lolium perenne	Sulfometuron-methyl	>24	>24	55
Lactuca serriola	Metsulfuron-methyl	>20	>20	218
Salsola iberica	Chlorsulfuron	>24	>24	55
Wheat[a]	Chlorsulfuron	—	2	131

[a]Wheat, which has metabolism-based resistance to chlorsulfuron, included for comparison.

degrees to a range of herbicide classes with different modes of action.[147,150] It is hypothesized that these weeds mimic wheat's ability to metabolize wheat selective herbicides with a single enzyme system,[151] but little direct evidence exists in support of this idea. The basis for resistance to diclofop-methyl in the cross resistant *L. rigidum* biotypes has not been conclusively determined, and the metabolism of diclofop-methyl to inactive products appears to play only a minor role.[152,153] Additionally, the basis for resistance to sulfonylureas or other herbicide classes in cross resistant *A. myosuroides* has not been established.

An alternative hypothesis is that wide-ranging non-target site cross resistance (see Chapter 9), which threatens to severely limit the usefulness of several wheat selective herbicides for control of *L. rigidum*[154] or *A. myosuroides*,[155] results from the existence of multiple mechanisms of resistance.[78,154] Through extensive outcrossing, more than one mechanism may occur in the same individual or population, resulting in multiple resistance.[78,154] Hence, further study is required to determine whether or not metabolic inactivation inherently presents a greater potential than other resistance mechanisms to confer broad non-target site cross resistance to herbicides with different modes of action. But regardless of the basis of cross resistance and multiple resistance, these biotypes[154,155] pose a difficult challenge in weed control to the agricultural community (see Chapter 12).

HERBICIDE-RESISTANT CROPS

The topic of herbicide-resistant crops in agriculture has been reviewed by Mazur and Falco.[5] Since their review, development of crops resistant to ALS inhibitor herbicides has continued, and resistance has been incorporated into a number of crops that are not naturally resistant to these herbicides. Reports of ALS inhibitor-resistant crops include tobacco,[32,81,111] flax,[156-159] soybeans,[89] maize,[90] canola,[91,160] sugarbeets,[99,100,161,162] wheat,[98] birdsfoot trefoil,[97] chicory (*Cichorium intybus* L.),[163] and rice (*Oryza sativa* L.).[164,165] In all of these cases, the resistance apparently is due to an ALS that is less sensitive to ALS-inhibitor herbicides.

While metabolic inactivation (or lack of activation) is the predominant mechanism for naturally occurring resistance to ALS-inhibitor herbicides, none of the ALS inhibitor-resistant crops scheduled for commercialization uses this resistance mechanism. Nevertheless, the introduction of genes into plants for herbicide-metabolizing enzymes has been achieved. For example, Cyt P450 genes from *Streptomyces griseolus*[137] function in plants to increase sulfonylurea herbicide metabolism.[166] Similarly, the incorporation of bacterial genes encoding specific metabolizing enzymes has resulted in the successful development of plants that are resistant to the commercial herbicides bromoxynil[167,168] (*bxn* gene) and phosphinothricin[169,170] (*bar* and *pat* genes) due to metabolic inactivation.

In agronomic situations where few herbicide options exist, the introduction of herbicide-resistant biotypes may be the only way to introduce herbicides with different modes of action.[171] On the other hand, some agronomic sectors currently benefit (from a resistance point of view) from the use of several herbicides with different modes of action coupled with crop rotation. The increased use of ALS inhibitor herbicides and the introduction of ALS inhibitor-resistant crops could limit such herbicide diversity and hence increase the potential for resistance. For example, the increased use of new ALS inhibitor herbicides in soybeans and maize (see Table 1) as well as the introduction of ALS inhibitor-resistant maize and soybeans will increase the selection pressure imposed by ALS inhibitor herbicides in these crops.

BIOLOGY AND GENETICS

Knowledge of weed biology and ecology is essential to the development of effective herbicide-resistant weed management strategies (see Chapters 10, 11, and 12). The biology of several weeds (*S. media*, *K. scoparia*, *L. serriola*, and *S. iberica*) resistant to sulfonylurea herbicides has been reviewed.[6] In general, however, knowledge of basic weed biology is lacking. Little is known about intra- and interspecific competition, environmental influences on most physiological processes, and genetics of weeds. This section will review information on inheritance, fitness, and gene flow of ALS inhibitor-resistant plants (Table 9) and present recommendations for prevention and management of herbicide-resistant weeds based on current biology/ecology findings.

Inheritance of Resistance

As fully discussed in Chapter 1, a two-part theoretical model was developed recently to predict the evolution of herbicide resistance in weed populations.[71] The plant population model simulates the evolution and spread of resistant weeds, while the inheritance model forecasts the proportion of resistant and susceptible genotypes in future generations. Thus, to predict how resistance to ALS inhibitor herbicides will be passed onto future generations it is necessary to understand how the trait is inherited. Studies on the inheritance of ALS inhibitor resistance in crops and weeds have provided this information (see also Chapter 10).

Crops and Microbes

Inheritance of sulfonylurea and/or imidazolinone resistance in bacteria (*S. typhimurium*),[172] yeast (*S. cerevisiae*),[35] green alga (*C. reinhardtii*),[92,93] flax,[159] and canola[91,109] is due to a dominant or semidominant mutation. Sulfonylurea resistance in selected tobacco plants is due to a single, semidominant nuclear mutation,[81] while in transgenic tobacco plants resistance is due to a dominant mutation.[9,39] Double mutations resulting in enhanced resistance in tobacco ALS also are inherited as single semidominant traits.[85] In *A. thaliana*, sulfonylurea[36] and imidazolinone[37] resistance is due to a single, dominant nuclear mutation. Sulfonylurea herbicide resistance in soybean mutants has

Table 9. Components affecting the inheritance, fitness, and gene flow of ALS inhibitor-resistant weeds

Trait	Weed species	Observation[a]	Ref.
Inheritance	Lactuca serriola	Target-site resistance inherited	177
	Kochia scoparia	dominantly	178,179
Seed germination	Lactuca serriola	R[1] seed germinated faster than S[1] seed	186
	Kochia scoparia	R seed germinated faster at cool temperatures than S seed, but not at warm temperatures	178,187
Competitiveness	Lactuca serriola	R and S biotypes equal	188
	Kochia scoparia	R and S biotypes equal	178
Growth	Lactuca serriola	S grew faster and larger than R	188
	Kochia scoparia	R and S growth was similar	189,191
Catalytic competency	Kochia scoparia	Specific activity and K_m for pyruvate were similar for ALS from R and S biotypes	77
	Salsola iberica		55
	Stellaria media		55
	Lolium perenne		55
Seed production	Lactuca serriola	R and S biotypes produced equal number of seeds	186
	Kochia scoparia		189
Seed longevity	Lactuca serriola	R and S seed viability was equal in the field	186
Cross pollination	Kochia scoparia	1 to 13% in the field between R and S plants	63
Seed dispersal	Salsola iberica	Wind moved plants over 5 km in 1 week	63

[a] R and S refer to resistant and susceptible, respectively.

been reported to be a recessive trait[173] when the resistance was not due to an altered ALS and a semidominant trait when resistance was due to a sulfonylurea-insensitive ALS.[89] In maize, resistance to primisulfuron is metabolism based and is inherited as a single, dominant trait.[174] Imidazolinone resistance in maize is inherited as a semidominant trait when ALS is insensitive to the herbicide[90,175] and as a recessive trait when ALS is sensitive to the herbicide.[176] Hence, to date, resistance due to an altered ALS is always associated with dominant or semidominant inheritance.

Weeds

Inheritance of sulfonylurea resistance in weeds has been investigated in *L. serriola*[177] and *K. scoparia*.[178,179] Both weeds are diploid, as are *S. media* and *S. iberica*.[6] One sulfonylurea-resistant biotype of *S. iberica* was reported recently to be a polyploid.[103] Mulugeta et al.[179] reported that resistance appeared to be a dominant trait in *K. scoparia*. Greenhouse studies conducted by Thompson and Thill[178] confirmed that sulfonylurea resistance in *K. scoparia* is inherited as a dominant, nuclear trait. However, it was impossible to separate homozygous and heterozygous F_2 phenotypes for the resistance trait.

Sulfonylurea herbicide resistance in *Lactuca* spp. is controlled by a single nuclear gene with incomplete dominance.[177] The best fit for Chi Square analysis of the F_2 generation of both the susceptible X resistant *L. serriola* and Bibb lettuce (*Lactuca sativa* L.) X resistant *L. serriola* crosses was a 1:2:1 ratio indicating the trait was controlled by a single nuclear gene with incomplete dominance. Evaluation of F_3 plants grown from seed collected from intermediate and resistant F_2 plants confirmed F_2 generation findings. Resistant F_3 plants did not segregate, while intermediate F_3 Bibb and *L. serriola* plants segregated 1:2:1 and 3:1, respectively. It was impossible to differentiate between homozygous and heterozygous resistant *L. serriola* phenotypes in the F_3 generation.

Selection for an ALS gene insensitive to ALS inhibitor herbicides appears to be relatively easy based on the number of resistant microbe[172] and plant[6,36,39,81,89,92] biotypes selected in laboratory experiments and through commercial use of the herbicides in field situations. In most cases, the ALS inhibitor herbicide resistance trait is controlled by a single nuclear gene with incomplete dominance[9,35,36,39,65,81,85,89,92,93,172,178,179] that resulted in survival of both the homozygous resistant and the heterozygous intermediate plants treated with herbicide applied at typical field rates. Immigration of resistant pollen or seed from fields or non-crop sites infested with ALS inhibitor-resistant weeds into previously susceptible weed population could increase the proportion of resistant plants, even in the absence of any selection pressure, and the subsequent use of an ALS inhibitor herbicide would accelerate selection of R biotypes. Thus, spread of resistant weeds must be prevented (see gene flow section for more discussion).

Fitness

The theoretical plant population model constructed to predict herbicide resistance identified two sets of biological processes as major factors in the evolution and dynamics of herbicide-resistant weed populations, that is, ecological fitness and gene flow.[71] Knowledge about both factors is necessary to develop effective management strategies for herbicide-resistant weeds (see Chapter 11).

Fitness is a measure of survival and ability to produce viable offspring.[180] Under conditions of natural selection, genotypes with greater fitness produce, on the average, more offspring than less fit genotypes. Competition, growth rate, and physiological studies between triazine-resistant and -susceptible weed biotypes showed that triazine resistance often has negative physiological consequences for the plants.[181,182] For example, studies have shown that susceptible biotypes produce up to 64% more biomass than their resistant counterparts in noncompetitive environments, and under different growing temperatures.[183-185] Because many triazine-resistant biotypes are less fit than their susceptible counterpart, it was widely believed that plants resistant to other classes of herbicides also would be less fit than susceptible biotypes.

Crop studies indicate that plant fitness may be unaffected by resistance resulting from changes in ALS sensitivity to ALS inhibitor herbicides. Agronomic traits, including quality and yield, of flax resistant to sulfonylurea herbicides was not different compared to standard cultivars.[157,158] Seed yield, maturity, and disease tolerance of canola resistant to imidazolinone herbicides were similar to the susceptible biotype.[91] No deleterious effects attributed to the imidazolinone-resistant allele in maize have been observed.[90] Likewise, seedling shoot heights of sulfonylurea- and imidazolinone-resistant maize mutants were similar to the susceptible biotypes.[175] Growth of sulfonylurea-resistant and -susceptible tobacco cells in callus culture were not different.[81]

Components of fitness have been compared between sulfonylurea herbicide-susceptible and -resistant weeds. Parameters measured included seed germination,[178,186,187] growth and competition,[188-191] seed production,[188,189] seed longevity in soil,[188] and catalytic competency (discussed below). Species examined include *L. serriola*, *K. scoparia*, *S. iberica*, *L. perenne*, and *S. media*.

Seed Germination

Seeds from R and S biotypes of *L. serriola* were grown in the field.[186] Beginning at bud stage, ripe seed were collected for a 53-day period and germination rates (24°C, 12 h light) were compared. R and S seed lots reached nearly 100% germination within 7 days. However, the R seeds always germinated as fast or faster than susceptible seeds. Germination of R and S near-isogenic lines of Bibb lettuce were compared at 8, 18, and 28°C.[190] In all cases, the R line germinated faster than the S line.

Germination data for R and S biotypes of *K. scoparia* are similar. At 8°C, R and S *K. scoparia* seeds (Kansas biotypes, fourth-generation selfed seed produced in the greenhouse) attained maximum germination by 324 and 624 h, respectively.[178] At 18°C, maximum germination was attained at 184 h for the R biotype and at 284 h for the S biotype. Both biotypes germinated at the same rate at 28°C. Resistant *K. scoparia* biotypes from Montana germinated faster than S biotypes at 4.6 and 7.2°C, but germination rate was equal between biotypes at 10.5 and 16.8°C.[187]

Resistant lines of prickly and Bibb lettuce[186,190] always germinated faster compared to S lines, while R *K. scoparia* biotypes[178,187] germinated faster only at cool temperatures. Germination was equal between *K. scoparia* biotypes at warmer temperatures. *K. scoparia* cohorts that germinate at cooler temperatures may be more susceptible to early season tillage or herbicide applications. This could allow for some selective control of R biotypes, which may reduce their proportion in the population if the use of the selecting herbicide is discontinued.

Growth and Competition

The relative competitiveness and growth rate of R and S *L. serriola* biotypes were compared in the greenhouse using an addition series design with four densities of each biotype grown to the floral bud stage.[188] A reciprocal yield analysis regression procedure was used to develop an equation to describe the relationship between plant biomass and plant density.[192]

Relative competitive ability was calculated for each biotype.[192,193] On average, the S biotype produced 31% more biomass than the R biotype. The ratio of R to S biomass ranged from 0.7 where mixed densities were 150 to 200 plants per m^2 to 0.97 where there were 300 plants per m^2 in mixed density. The R to S ratio was 0.54 under noncompetitive conditions. The relative competitive ability of the S biotype was 0.91, which indicated that one S biotype was equivalent to 0.91 R biotype plants in terms of the S plant's yield. The relative competitive ability of the R biotype was 0.94. The ratio of both biotypes was close to one, which indicates that whether the biotype was adjacent to a growing R or S biotype was unimportant. Thus, the S biotype was superior to the R biotype in biomass production, but competitiveness was equal for both biotypes. In similar experiments, the relative competitive abilities for R and S *K. scoparia* were 0.75 and 0.85, respectively.[178] Again, both biotypes were equally competitive.

The relative growth rate of R and S biotypes of *L. serriola* was determined in greenhouse experiments from 10 days after germination until bud stage.[188] The S biotype produced 24% more biomass and accumulated biomass 52% faster than the R biotype. In similar greenhouse experiments, R biotypes of *K. scoparia* (from North Dakota and Kansas) tended to have equal or greater leaf area, stem dry weight, shoot height, and stem and shoot diameter than their S biotype counterparts 13 weeks after emergence.[189] Greenhouse and field studies conducted in Colorado also showed little or no difference in biomass and leaf area between R and S *K. scoparia* biotypes.[191]

Catalytic Competency

Molecular studies corroborate the findings at the whole plant level. That is, the catalytic competency of ALS isolated from S and R weeds appears identical. The K_m values for pyruvate using ALS isolated from *K. scoparia*,[77] *S. iberica*, *S. media*, and *L. perenne*,[55] regardless of resistance phenotype, ranged from 1.7 to 4.8 mM, which is close to previously reported K_m values for plant ALS.[119,194] Also, the expression of ALS from R and S weeds was similar assuming that *in vitro* enzyme assays reflected *in vivo* enzyme levels.[55,77] Using a triazolopyrimidine herbicide to select for mutants, Subramanian et al.[32] found that the specific activity and K_m values (pyruvate and thiamine pyrophosphate) for ALS isolated from wild-type and mutant tobacco and cotton also were comparable. Variation in the K_m for pyruvate but not V_{max} were seen using ALS from *Datura innoxia* mutants selected with either chlorsulfuron or sulfometuron-methyl.[96]

Seed Production

Seed production of R and S biotypes of *L. serriola* was compared.[186] During the 53-day sampling period, R and S biotypes of *L. serriola* produced 4870 and 4160 seeds per plant, respectively. Greenhouse-grown R *K. scoparia* produced 12,900 seeds per plant, while S *K. scoparia* produced 11,150 seeds per plant.[189]

Seed Longevity

R (Lewiston, ID) and S (Troy, ID) *L. serriola* seeds were collected in the field during late summer and were buried during September in fields near Lewiston (adjacent to where resistance was discovered) and Moscow, ID.[186] Seeds left on the soil surface and buried 7.5 and 15 cm were exhumed every 6 months for about 3 years. *L. serriola* seed viability in the field was the same for resistant and susceptible biotypes at all seed burial positions and at both locations.

Fitness-related studies showed that relative competitiveness and seed output of sulfonylurea R and S biotypes of *L. serriola*[186,188] and *K. scoparia*[189,190] were nearly equal. Also, seed longevity of R and S biotypes of *L. serriola* was the same.[186] A recently published model predicts that with reduced fitness of the R biotype, the S biotype will replace the R biotype over time after the herbicide selection pressure is abandoned.[71] If the R biotype is as competitive as the S biotype, resistance would decline slowly, if at all.[71] In a field situation, however, once sulfonylurea herbicide use was discontinued, the proportion of R *L. serriola* decreased 25 to 86% between 1988 and 1990, but its range increased.[195] Resistant biotypes likely will persist in infested fields for many years, even in the absence of any additional selection pressure.

Gene Flow

Genes can immigrate into a weed population from two sources: seed and pollen (not considered here is the spread of vegetative propagules via cultivation). Plants with special dispersal mechanisms can distribute seed long distances from the mother plant (e.g., the pappus on *L. serriola* seed aids wind dispersal) or from the original site of the mother plant (e.g., tumbling of *S. iberica* or *K. scoparia*). Pollen movement from R to S plants also may disperse the resistant trait, especially in weeds with a high percentage of cross pollination (e.g., *K. scoparia*[63] and *L. rigidum*[196]). Conversely, according to the model of Maxwell et al.,[71] pollen movement from susceptible to resistant plants may dilute the resistant trait in a population. This has yet to be demonstrated.

Cross pollination,[63,64,179,197] pollen dispersal,[63,64,197] pollen viability,[63,64] and seed dispersal[62,197] have been investigated in R and S biotypes of *K. scoparia*. Seed dispersal of *S. iberica* also has been studied.[63] In the greenhouse, branches of R and S *K. scoparia* were bagged just prior to flowering.[179] Seedlings grown from cross-pollinated S plants were sprayed with chlorsulfuron and 19% survived. Cross pollination between R and S *K. scoparia* in field experiments ranged from 1.5 to 13%.[63,64] At 30 m, up to 16 crosses per plant were observed. Pollen grains have been collected up to 62 m from the closest pollen source.[63,64] Peak pollen collection (18 grains cm^{-2}) occurred between 6 and 9 p.m., while only 4 grains/cm^2 were collected between 3 and 6 a.m.[63] In the laboratory, *K. scoparia* pollen grains were 30% viable after 5 days when stored at 4°C and high relative humidity.[64] Pollen grains remained viable for 2 days stored at

28°C and 33% relative humidity (simulating summer conditions during flowering). Fresh pollen collected from anthers in the field during late afternoon in August (24°C and 18% relative humidity) was 68% viable.[63]

Cross pollination in *K. scoparia* appears to be obligatory.[178,197] Observations indicated that male and female floral parts mature sequentially. Stigmas appeared receptive to pollen several days before pollen grains were shed from anthers in the same flower. Usually these stigmas had desiccated before pollen was shed by the anthers. Thus, pollen grains deposited on stigmas likely came from other flowers on the same plant (*K. scoparia* flowers indeterminately) or from flowers on neighboring plants.

In the field, tumbling *K. scoparia* and *S. iberica* plants can disperse seed long distances. *K. scoparia* seed dispersal is reported to decrease exponentially with distance from the point of mother plant attachment.[62] *S. iberica* plants in a winter wheat field and in a stubble field traveled about 5.5 km in 45 days.[63] Movement was in the direction of the prevailing winds and most long-distance movement was associated with heavy winds. For example, 35 km/h average wind speed (gusts up to 80 km/h) moved plants over 1.6 km in a 24-h period. Some *S. iberica* plants moved over 5.0 km within 1 week.

Gene flow through pollen and seed movement has most likely aided in the dispersal of the ALS inhibitor resistance trait, especially for *K. scoparia*[63,64] and *S. iberica*[63] and to a limited extent for *L. serriola*.[195] For *K. scoparia*, up to 13% outcrossing between R and S plants occurred,[63,64] with some hybridization occurring 30 m from the nearest R plant.[63] *K. scoparia* pollen moved at least 60 m and remained viable under warm, dry summer conditions for about 2 days.[63,64] *S. iberica* plants can move over 5 km in about 1 week showing long-distance spread potential.[63] Sulfonylurea-resistant biotypes of *S. iberica*[66] and *K. scoparia*[65] now infest much of the dry land wheat-growing region of Eastern Washington and much of the irrigated farm land in Southern Idaho, respectively. It is not known what proportion of the resistant population is associated with direct selection with herbicides or with spread of resistant pollen or seed. Most likely both factors have contributed to the current distribution of sulfonylurea herbicide-resistant weeds. Even in the absence of selection pressure, sulfonylurea herbicide-resistant weeds will continue to spread due to gene flow from infested to uninfested sites. Resistant weeds along rights-of-way and field borders, and in crop land easily can invade surrounding areas. Thus, abating the spread of ALS inhibitor-resistant weeds is imperative if the utility of these herbicides is to be maintained.

As discussed in Chapter 12, even a small proportion of resistant plants in a population will require altered weed management strategies, which include monitoring fields for weed escapes, preventing spread to surrounding areas, changing crops and tillage systems, and changing herbicides. Fields and vegetation management sites should be examined to insure that weed species known to be herbicide-resistant have been controlled. Escapes should be removed before flowering stage to prevent spread, because herbicide-resistant weeds left uncontrolled will expand into uninfested areas. Machinery used in fields or areas with known infestations of herbicide-resistant weeds must be cleaned thoroughly before moving to other areas to prevent the spread of resistant weeds. Different crop rotations or tillage practices favor different weeds. Changing one or both of these factors will prevent resistant weeds from completing their life cycle. Also, the use of herbicides with different modes of action will control ALS inhibitor-resistant weeds.

SUMMARY

Sulfonylureas, imidazolinones, triazolopyrimidines, and pyrimidinyl thiobenzoates comprise the commercial ALS inhibitor herbicides currently or soon to be available. Thirteen weed species have become resistant to ALS inhibitor herbicides as a result of the agronomic or industrial use of these herbicides. Additionally, two weed species have non-target site cross resistance to ALS-inhibitor herbicides after selection with other mode of action herbicides. For resistant weeds selected with ALS inhibitor herbicides, the mechanism conferring resistance is target based; that is, the target site, ALS, is less sensitive to inhibition by ALS inhibitor herbicides. Metabolism-based resistance is the primary basis for crop selectivity among ALS inhibitor herbicides. The only known evolved resistance due to rapid herbicide metabolism occurred in *L. rigidum* that was selected with diclofop-methyl and cross resistant to wheat selective ALS inhibitors. *Lolium* spp., and possibly grasses in general, may have numerous resistance mechanisms. However, target site resistance predominates in weeds that are extremely sensitive to ALS inhibitor herbicides, suggesting that other mechanisms cannot overcome relatively large herbicide doses as effectively as target site mutants. Extrapolation from known resistance cases suggests that the weeds most likely to become resistant to ALS inhibitor herbicides because of target site resistance are those controlled by fractions of the commercial use rates.

Laboratory research has identified several ALS mutations resulting in herbicide resistance, but the fitness cost of surviving in nature probably reduces the number of mutations available in the field. Preliminary evidence suggests that pro mutations in a highly conserved region of ALS are frequently present in weeds selected with sulfonylurea herbicides. Several ALS inhibitor-resistant crops, many of which have pro mutations, appear as agronomically viable as original cultivars. Plants resistant because of pro changes in ALS are characterized by target site cross resistance to most or all sulfonylurea herbicides, probably most triazolopyrimidines, and low-level resistance to some imidazolinones. Other classes of ALS inhibitor herbicides may select for resistant weeds with different ALS mutations and phenotypes.

Models for herbicide resistance are useful in appreciating the factors important to resistance development, but the complex inputs required limit the utility of models for prediction. The contributions of inheritance, gene flow, seed biology, and fitness to the occurrence of ALS inhibitor resistance are being evaluated. A large factor in the appearance of resistance is the high selection pressure imposed by ALS inhibitor herbicides on very sensitive weed species. The occurrence of target site, ALS inhibitor resistance has most frequently resulted from the selection pressure associated with long residual herbicides and monoculture or near monoculture conditions. Once established, gene flow via seed distribution has probably contributed to the spread of resistant *K. scoparia* and *S. iberica*.

More complex strategies for weed control are required in order to minimize the effects of herbicide resistance on ALS inhibitor use. Accordingly, resistant weed management strategies should include the use of crop rotations, herbicide mixtures or rotations, tillage, and integrated pest management techniques where possible. Unfortunately, economics and/or governmental regulations often limit the implementation of these strategies. Maize/soybean rotations provide a guide for the diversity of herbicides and practices required to allay resistance as evidenced by the lack of triazine resistance observed in areas where triazines are rotated or mixed with other herbicides.[101] In contrast, the predominant use of one herbicide in monoculture as practiced in the northern Great Plains (chlorsulfuron use on wheat) is among the least appropriate from a resistance management view. Many agronomic practices are between the extremes of monoculture and complex rotations and include both monoculture with several mode of action herbicides and rotational culture with a single mode of action herbicides. Even these more integrated markets, however, are tending towards less herbicide diversity and an increased resistance potential due to the introduction of new ALS-inhibitor herbicides and new ALS inhibitor-resistant crops.

ALS-inhibitor herbicides have become important tools in agriculture because of their effectiveness as well as their environmentally and toxicologically desirable features. Unfortunately, resistance to ALS-inhibitor herbicides will increase concomitantly with the increased use of this herbicide group. As evidenced by this review, a great deal of research on ALS inhibitor resistance has been performed, but in order to develop better resistance management tools, both fundamental and practical research efforts need to continue. Currently, organizations such as the Weed Science Societies of America (WSSA) and Australia, the ALS Inhibitor Resistance Working Group (AIRWG), and the International Organization of Pest Resistance Management (IOPRM) are working toward developing newer and more encompassing tactics for the control of resistant weeds. Hopefully, the efforts of these organizations along with the cooperation of the agricultural community will preserve the utility of ALS inhibitor herbicides far into the future.

ACKNOWLEDGMENTS

We thank Drs. J. L. Saladini, H. M. Brown, A. R. Rendina, J. V. Hay, and H. P. Hershey for reviewing the manuscript.

REFERENCES

1. County NatWest WoodMac (Wood MacKenzie), *The NatWest Investment Bank Group*, London, 1991.
2. Ray, T. B. "Herbicides as Inhibitors of Amino Acid Biosynthesis," in *Target Sites of Herbicide Action*, P. Böger and G. Sandmann, Eds. (Boca Raton, FL: CRC Press, Inc., 1989), pp. 105-125.
3. Schloss, J. V. "Acetolactate Synthase, Mechanism of Action and Its Herbicide Binding Site," *Pestic. Sci.* 29:283-292 (1990).
4. Stidham, M. A. "Herbicides that Inhibit Acetohydroxyacid Synthase," *Weed Sci.* 39:428-434 (1991).
5. Mazur, B. J., and S. C. Falco. "The Development of Herbicide Resistant Crops," *Annu. Rev. Plant Physiol. Plant Mol. Biol.* 40:441-470 (1989).
6. Thill, D. C., C. A. Mallory-Smith, L. L. Saari, J. C. Cotterman, M. M. Primiani, and J. L. Saladini. "Sulfonylurea Herbicide Resistant Weeds: Discovery, Distribution, Biology, Mechanism, and Management," in *Herbicide Resistance in Weeds and Crops*, J. C. Caseley, G. W. Cussans, and R. K. Atkin, Eds. (Oxford, England: Butterworth-Heinemann Ltd., 1991), pp. 115-128.
7. Reed, W. T., J. L. Saladini, J. C. Cotterman, M. M. Primiani, and L. L. Saari. "Resistance in Weeds to Sulfonylurea Herbicides," in *Brighton Crop Protection Conference — Weeds*, (Farnham, U.K.: The British Crop Protection Council, 1989), pp. 295-300.
8. Shaner, D. L. "Mechanisms of Resistance to Acetolactate Synthase/ Acetohydroxyacid Synthase Inhibitors," in *Herbicide Resistance in Weeds and Crops*, J. C. Caseley, G. W. Cussans, and R. K. Atkin, Eds. (Oxford, England: Butterworth-Heinemann Ltd., 1991), pp. 187-198.
9. Hartnett, M. E., C. Chui, C. J. Mauvais, R. E. McDevitt, S. Knowlton, J. K. Smith, S. C. Falco, and B. J. Mazur. "Herbicide-Resistant Plants Carrying Mutated Acetolactate Synthase Genes," in *Managing Resistance to Agrochemicals. From Fundamental Research to Practical Strategies*, M. B. Green, H. M. LeBaron, and W. K. Moberg, Eds. (Washington, D.C.: American Chemical Society, 1990), pp. 459-473.
10. Hartnett, M. E., C.-F. Chui, S. C. Falco, S. Knowlton, C. J. Mauvais, and B. J. Mazur. "Molecular Analysis of Sulfonylurea Herbicide-Resistant ALS Genes," in *Herbicide Resistance in Weeds and Crops*, J. C. Caseley, G. W. Cussans, and R. K. Atkin, Eds. (Oxford, England: Butterworth-Heinemann Ltd., 1991), pp. 343-353.
11. Newhouse, K. E., D. L. Shaner, T. Wang, and R. Fincher. "Genetic Modification of Crop Responses to Imidazolinone Herbicides," in *Managing Resistance to Agrochemicals. From Fundamental Research to Practical Strategies*, M. B. Green, H. M. LeBaron, and W. K. Moberg, Eds. (Washington, D.C.: American Chemical Society, 1990), pp. 474-481.
12. Newhouse, K. E., T. Wang, and P. C. Anderson. "Imidazolinone-Resistant Crops," in *The Imidazolinone Herbicides*, D. L. Shaner, and S. L. O'Connor, Eds. (Boca Raton, FL: CRC Press, Inc., 1991), pp. 139-150.
13. Gammon, D. W., and R. E. Ford, Eds. Pesticide Science, Vol. 29, Special Issue: American Chemical Society Symposium, *Herbicides Inhibiting Branched-Chain Amino Acid Biosynthesis*, (London: Elsevier Applied Science, 1990), 378 pp.
14. Barak, Z., D. M. Chipman, and J. V. Schloss, Eds. *Biosynthesis of Branched Chain Amino Acids*, (Weinheim, Federal Republic of Germany: VCH Verlagsgesellschaft, 1990), 530 pp.

15. Sauers, R. F., and G. Levitt. *Pesticide Synthesis Through Rational Approaches*, ACS Symposium Series No. 255, P. S. Magee, G. K. Kohn, and J. J. Menn, Eds. (Washington, D.C.: American Chemical Society, 1984), pp. 21-28.
16. Beyer, E. M., M. J. Duffy, J. V. Hay, and D. D. Schlueter. "Sulfonylureas," in *Herbicides: Chemistry, Degradation, and Mode of Action*, Vol. 3, P. C. Kearney, and D. D. Kaufman, Eds. (New York: Dekker, 1988), pp. 117-189.
17. Blair, A. M., and T. D. Martin. "A Review of the Activity, Fate and Mode of Action of Sulfonylurea Herbicides," *Pestic. Sci.* 22:195-219 (1988).
18. Los, M. *Pesticide Synthesis Through Rational Approaches*, ACS Symposium Series No. 255, P. S. Magee, G. K. Kohn, and J. J. Menn, Eds. (Washington, D.C.: American Chemical Society, 1984), pp. 29-44.
19. Kleschick, W. A., M. J. Costales, J. E. Dunbar, R. W. Meikle, W. T. Monte, N. R. Pearson, S. W. Snider, and A. P. Vinogradoff. "New Herbicidal Derivatives of 1,2,4-Triazolo[1,5-a]pyrimidine," *Pestic. Sci.* 29:341-355 (1990).
20. Kleschick, W. A., B. C. Gerwick, C. M. Carson, W. T. Monte, and S. W. Snider. "DE-498, A New Acetolactate Synthase Inhibiting Herbicide with Multicrop Selectivity," *J. Agric. Food Chem.* 40:1083-1085 (1992).
21. Takahashi, S., S. Shigematsu, and A. Morita. "KIH-2031, A New Herbicide for Cotton," in *Brighton Crop Protection Conference — Weeds*, (Farnham, U.K.: The British Crop Protection Council, 1991), pp. 57-62.
22. Babczinski, P., and T. Zelinski. "Mode of Action of Herbicidal ALS-Inhibitors on Acetolactate Synthase from Green Plant Cell Cultures, Yeast, and *Escherichia coli*," *Pestic. Sci.* 31:305-323 (1991).
23. Tseng, C. P., B. L. Finkelstein, J. A. Kerschen, and G. E. Scheiders. "Synthesis and Herbicidal Activity of Pyrazolo[1,5-a]Pyrimidine-3-Carbonylsulfonamides and Related Heterocyclic Carbonylsulfonamides," in *Proceedings of the Seventh International Congress of Pesticide Chemistry*, Vol. 1, H. Frehse, E. Kesseler-Schmitz, and S. Conway, Eds. (Hamburg: Gesellschaft Deutscher Chemiker, 1990) p. 87.
24. Head, J. C., and G. Anderson-Taylor. "1-Heteroaryl-1,2,4-Triazole-3-Sulphonanilides — A New Class of Acetolactate Synthase Inhibitors," in *Proceedings of the Seventh International Congress of Pesticide Chemistry, Vol. 1*, H. Frehse, E. Kesseler-Schmitz, and S. Conway, Eds. (Hamburg: Gesellschaft Deutscher Chemiker, 1990) p. 57.
25. Haraguchi, H., Y. Hamatani, K. Shibata, and K. Hashimoto. "An Inhibitor of Acetolactate Synthase from a Microbe," *Biosci. Biotechnol. Biochem.* 56:2085-2086 (1992).
26. Ray, T. B. "Site of Action of Chlorsulfuron. Inhibition of Valine and Isoleucine Biosynthesis in Plants," *Plant Physiol.* 75:827-831 (1984).
27. LaRossa, R. A., and J. V. Schloss. "The Sulfonylurea Herbicide Sulfometuron Methyl Is an Extremely Potent and Selective Inhibitor of Acetolactate Synthase in *Salmonella typhimurium*," *J. Biol. Chem.* 259:8753-8757 (1984).
28. Chaleff, R. S., and C. J. Mauvais. "Acetolactate Synthase Is the Site of Action of Two Sulfonylurea Herbicides in Higher Plants," *Science* 224:1443-1445 (1984).
29. Shaner, D. L., P. C. Anderson, and M. A. Stidham. "Imidazolinones. Potent Inhibitors of Acetohydroxyacid Synthase," *Plant Physiol.* 76:545-546 (1984).
30. Gerwick, B. C., M. V. Subramanian, V. I. Loney-Gallant, and D. P. Chandler. "Mechanism of Action of the 1,2,4-Triazolo[1,5-a]Pyrimidines," *Pestic. Sci.* 29:357-364 (1990).

31. Hawkes, T. R. "Studies of Herbicides Which Inhibit Branched Chain Amino Acid Biosynthesis," in *Prospects for Amino Acid Biosynthesis Inhibitors in Crop Protection and Pharmaceutical Chemistry*, L. G. Copping, J. Dalziel, and A. D. Dodge, Eds. (Farnham, U.K.: The British Crop Protection Council, 1989), pp. 131-138.
32. Subramanian, M. V., H. Hung, J. M. Dias, V. W. Miner, J. H. Butler, and J. J. Jachetta. "Properties of Mutant Acetolactate Synthases Resistant to Triazolopyrimidine Sulfonanilide," *Plant Physiol.* 94:239-244 (1990).
33. Scheel, D., and J. E. Casida. "Sulfonylurea Herbicides: Growth Inhibition in Soybean Cell Suspension Cultures and in Bacteria Correlated with Block in Biosynthesis of Valine, Leucine, or Isoleucine," *Pestic. Biochem. Physiol.* 23:398-412 (1985).
34. Anderson, P. C., and K. A. Hibberd. "Evidence for the Interaction of an Imidazolinone Herbicide with Leucine, Valine, and Isoleucine Metabolism," *Weed Sci.* 33:479-483 (1985).
35. Falco, S. C., and K. S. Dumas. "Genetic Analysis of Mutants of *Saccharomyces cerevisiae* Resistant to the Herbicide Sulfometuron Methyl," *Genetics* 109:21-35 (1985).
36. Haughn, G. W., and C. Somerville. "Sulfonylurea-Resistant Mutants of *Arabidopsis thaliana*," *Mol. Gen. Genet.* 204:430-434 (1986).
37. Haughn, G. W., and C. R. Somerville. "A Mutation Causing Imidazolinone Resistance Maps to the Csr1 Locus of *Arabidopsis thaliana*," *Plant Physiol.* 92:1081-1085 (1990).
38. Lee, K. Y., J. Townsend, J. Tepperman, M. Black, C. F. Chui, B. Mazur, P. Dunsmuir, and J. Bedbrook. "The Molecular Basis of Sulfonylurea Herbicide Resistance in Tobacco," *EMBO J.* 7:1241-1248 (1988).
39. Haughn, G. W., J. Smith, B. Mazur, and C. Somerville. "Transformation with a Mutant *Arabidopsis* Acetolactate Synthase Gene Renders Tobacco Resistant to Sulfonylurea Herbicides," *Mol. Gen. Genet.* 211:266-271 (1988).
40. Sathasivan, K., G. W. Haughn, and N. Murai. "Molecular Basis of Imidazolinone Herbicide Resistance in *Arabidopsis thaliana* var Columbia," *Plant Physiol.* 97:1044-1050 (1991).
41. Primiani, M. M., J. C. Cotterman, and L. L. Saari. "Resistance of Kochia (*Kochia scoparia*) to Sulfonylurea and Imidazolinone Herbicides," *Weed Technol.* 4:169-172 (1990).
42. Eoyang, L., and P. M. Silverman. "Purification and Subunit Composition of Acetohydroxyacid Synthase I from *Escherichia coli* K-12," *J. Bacteriol.* 157:184-189 (1984).
43. Grimminger, H., and H. Umbarger. "Acetohydroxyacid Synthase I of *Escherichia coli*: Purification and Properties," *J. Bacteriol.* 137:846-853 (1979).
44. Schloss, J. V., D. E. Van Dyk, J. F. Vasta, and R. M. Kutny. "Purification and Properties of *Salmonella typhimurium* Acetolactate Synthase Isozyme II from *Escherichia coli* HB101/pDU9," *Biochemistry* 24:4952-4959 (1985).
45. Eoyang, L., and P. M. Silverman. "Role of Small Subunit (ilvN Polypeptide) of Acetohydroxyacid Synthase I from *Escherichia coli* K-12 in Sensitivity of the Enzyme to Valine Inhibition," *J. Bacteriol.* 166:901-904 (1986).
46. Weinstock, O., C. Sella, D. M. Chipman, and Z. Barak. "Properties of Subcloned Subunits of Bacterial Acetohydroxy Acid Synthases," *J. Bacteriol.* 174(17):5560-5566 (1992).

47. Singh, B., I. Szamosi, J. M. Hand, and R. Misra. "*Arabidopsis* Acetohydroxyacid Synthase Expressed in *Escherichia coli* Is Insensitive to the Feedback Inhibitors," *Plant Physiol.* 99:812-816 (1992).
48. Trebst, A. "The Molecular Basis of Resistance of Photosystem II Herbicides," in *Herbicide Resistance in Weeds and Crops*, J. C. Caseley, G. W. Cussans, and R. K. Atkin, Eds. (Oxford, England: Butterworth-Heinemann Ltd., 1991), pp. 145-164.
49. Mallory-Smith, C. A., D. C. Thill, and M. J. Dial. "Identification of Sulfonylurea Herbicide-Resistant Prickly Lettuce (*Lactuca serriola*)," *Weed Technol.* 4:163-168 (1990).
50. LeBaron, H. M. "Distribution and Seriousness of Herbicide-Resistant Weed Infestations Worldwide," in *Herbicide Resistance in Weeds and Crops*, J. C. Caseley, G. W. Cussans, and R. K. Atkin, Eds. (Oxford, England: Butterworth-Heinemann Ltd., 1991), pp. 27-43.
51. LeBaron, H. M. "Herbicide Resistant Weeds Continue to Spread," in *Resistant Pest Management Newsletter*, Vol. 3, No. 1, M. Whalon and R. Hollingworth, Eds. (East Lansing, MI: Michigan State University, 1991), pp. 36-37.
52. Rubin, B., M. Sibony, Y. Benyamini, and Y. Danino. "Resistance to Sulfonylurea Herbicides in Redroot Pigweed (*Amaranthus retroflexus* L.)," in *WSSA Abstracts, Vol. 32, Proceedings of the 1992 Meeting of the Weed Science Society of America* (Champaign, IL: WSSA, 1992), p. 66.
53. Pappas-Fader, T., J. F. Cook, T. Butler, P. J. Lana, and J. Hare. "Resistance Of California Arrowhead And Smallflower Umbrella plant To Sulfonylurea Herbicides," *Proceedings of the Western Weed Science Society* (Newark, CA: Western Society of Weed Science, 1993), p. 76.
54. Christopher, J. T., S. B. Powles, and J. A. M. Holtum. "Resistance to Acetolactate Synthase-Inhibiting Herbicides in Annual Ryegrass (*Lolium rigidum*) Involves at Least Two Mechanisms," *Plant Physiol.* 100:1909-1913 (1992).
55. Saari, L. L., J. C. Cotterman, W. F. Smith, and M. M. Primiani. "Sulfonylurea Herbicide Resistance in Common Chickweed, Perennial Ryegrass and Russian Thistle," *Pestic. Biochem. Physiol.* 42:110-118 (1992).
56. Hall, L. M., and M. D. Devine. "Cross-Resistance of a Chlorsulfuron-Resistant Biotype of *Stellaria media* to a Triazolopyrimidine Herbicide," *Plant Physiol.* 93:962-966 (1990).
57. Shaner, D. L., American Cyanamid Co., Princeton, NJ. Unpublished results (1992).
58. LeClair, J. J., and J. C. Cotterman. "Resistant Ryegrass in Texas Winter Wheat," in *4th Annual Texas Plant Protection Conference Abstracts*, (College Station, TX: 1992), p. 29.
59. DuPont Agricultural Products. Newark, DE, Unpublished results (1988-1993).
60. Mallory-Smith, C. A., D. C. Thill, and L. L. Saari. "Survey of Sulfonylurea Herbicide Resistant Agricultural Sites in North America," in *WSSA Abstracts, Vol. 31, Proceedings of the 1991 Meeting of the Weed Science Society of America* (Champaign, IL: WSSA, 1991), p. 83.
61. Gressel, J., and L. A. Segel. "Interrelating Factors Controlling the Rate of Appearance of Resistance: The Outlook for the Future," in *Herbicide Resistance in Plants*, H. M. LeBaron, and J. Gressel, Eds. (New York: John Wiley & Sons, Inc., 1982), pp. 325-347.

62. Fay, P. K., D. M. Mulugeta, and W. E. Dyer. "The Role of Seed Dispersal in the Spread of Sulfonylurea Resistant *Kochia scoparia*," in *WSSA Abstracts, Vol. 32, Proceedings of the 1992 Meeting of the Weed Science Society of America* (Champaign, IL: WSSA, 1992), p. 17.
63. Stallings, G. P., and D. C. Thill. University of Idaho, Moscow, ID. Unpublished results (1992).
64. Mulugeta, D., P. K. Fay, and W. E. Dyer. "The Role of Pollen in the Spread of Sulfonylurea Resistant *Kochia scoparia* L. (Schrad.)," in *WSSA Abstracts, Vol. 32, Proceedings of the 1992 Meeting of the Weed Science Society of America* (Champaign, IL: WSSA, 1992), p. 16.
65. Mallory-Smith, C. A., D. C. Thill, and G. P. Stallings. University of Idaho, Moscow, ID, Unpublished results (1992).
66. Stallings, G. P., D. C. Thill, and C. A. Mallory-Smith. "Sulfonylurea-resistant RussianThistle Survey in Washington State," in *Proceedings of the Western Weed Science Society* (Newark, CA: Western Society of Weed Science, 1992), pp. 38-39.
67. Fay, P. K., Department of Plant and Soil Science, Montana State University, Bozeman, MT. Unpublished results (1992).
68. Westra, P., Colorado State University, Fort Collins, CO. Unpublished results (1992).
69. Barrentine, W., Delta Research and Extension Center, Stoneville, MS. Unpublished results (1992).
70. Gressel, J., and L. A. Segel. "Herbicide Rotations and Mixtures: Effective Strategies to Delay Resistance," in *Managing Resistance to Agrochemicals. From Fundamental Research to Practical Strategies*, M. B. Green, H. M. LeBaron, and W. K. Moberg, Eds. (Washington, D.C.: American Chemical Society, 1990), pp. 430-458.
71. Maxwell, B. D., M. L. Roush, and S. R. Radosevich. "Predicting the Evolution and Dynamics of Herbicide Resistance in Weed Populations," *Weed Technol.* 4:2-13 (1990).
72. Mortimer, A. M., P. F. Ulf-Hansen, and P. D. Putwain. "Modelling Herbicide Resistance — A Study of Ecological Fitness," in *Resistance '91: Achievements and Developments in Combatting Pesticide Resistance*, I. Denholm, A. L. Devonshire, and D. W. Hollomon, Eds. (London: Elsevier Applied Science, 1992) pp. 148-164.
73. Haughn, G., and C. R. Somerville. "Selection for Herbicide Resistance at the Whole Plant Level," in *Biotechnology in Agricultural Chemistry*, Symposium Series No. 334, H. M. LeBaron, R. O. Mumma, R. C. Honeycutt, and J. H. Duesing, Eds. (Washington, D.C.: American Chemical Society, 1987), pp. 98-107.
74. Harms, C. T., and J. J. DiMaio. "Primisulfuron Herbicide-Resistant Tobacco Cell Lines. Application of Fluctuation Test Design to *in vitro* Mutant Selection with Plant Cells," *J. Plant Physiol.* 137:513-519 (1991).
75. Stannard, M. E. "Weed Control in Alfalfa (*Medicago sativa* L.) Grown for Seed," M. S. Thesis, Montana State University, Bozeman, MT (1987).
76. Mauvais, C., DuPont Agricultural Products, Wilmington, DE. Unpublished results (1989).
77. Saari, L. L., J. C. Cotterman, and M. M. Primiani. "Mechanism of Sulfonylurea Herbicide Resistance in the Broadleaf Weed, *Kochia scoparia*," *Plant Physiol.* 93:55-61 (1990).

78. Holtum, J. A. M., and S. B. Powles. "Annual Ryegrass: An Abundance of Resistance, A Plethora of Mechanisms," in *Brighton Crop Protection Conference — Weeds,* (Farnham, U.K.: The British Crop Protection Council, 1991), pp. 1071-1078.
79. Schmitzer, P. R., R. J. Eilers, and C. Cséke. "Lack of Cross-Resistance of Imazaquin-Resistant *Xanthium strumarium* Acetolactate Synthase to Flumetsulam and Chlorimuron," *Plant Physiol.* 103:281-283 (1993).
80. Mourad, G., and J. King. "Effect of Four Classes of Herbicides on Growth and Acetolactate-Synthase Activity in Several Variants of *Arabidopsis thaliana,*" *Planta* 188:491-497 (1992).
81. Chaleff, R. S., and T. B. Ray. "Herbicide-Resistant Mutants from Tobacco Cell Cultures," *Science* 223:1148-1151 (1984).
82. Harms, C. T., J. J. DiMaio, S. M. Jayne, L. A. Middlesteadt, D. V. Negrotto, H. Thompson-Taylor, and A. L. Montoya. "Primisulfuron Herbicide-Resistant Tobacco Plants: Mutant Selection In Vitro by Adventitious Shoot Formation from Cultured Leaf Disks," *Plant Sci.* 79:77-85 (1991).
83. Subramanian, M. V., V. Loney-Gallant, J. M. Dias, and L. C. Mireles, "Acetolactate Synthase Inhibiting Herbicides Bind to the Regulatory Site," *Plant Physiol.* 96:310-313 (1991).
84. Chaleff, R. S., and N. F. Bascomb. "Genetic and Biochemical Evidence for Multiple Forms of Acetolactate Synthase in *Nicotiana tabacum,*" *Mol. Gen. Genet.* 210:33-38 (1987).
85. Creason, G. L., and R. S. Chaleff. "A Second Mutation Enhances Resistance of a Tobacco Mutant to Sulfonylurea Herbicides," *Theor. Appl. Genet.* 76:177-182 (1988).
86. Yadav, N., R. E. McDevitt, S. Benard, and S. C. Falco. "Single Amino Acid Substitutions in the Enzyme Acetolactate Synthase Confer Resistance to the Herbicide Sulfometuron Methyl," *Proc. Natl. Acad. Sci. U.S.A.* 83:4418-4422 (1986).
87. Friedberg, D., and J. Seijffers. "Sulfonylurea-Resistant Mutants and Natural Tolerance of *Cyanobacteria,*" *Arch. Microbiol.* 150:278-281 (1988).
88. Friedberg, D., and J. Seijffers. "Molecular Characterization of Genes Coding for Wild-Type and Sulfonylurea-Resistant Acetolactate Synthase in the *Cyanobacterium Synechococcus* PCC7942," *Z. Naturforsch.* 45c:538-543 (1990).
89. Sebastian, S. A., G. M. Fader, J. F. Ulrich, D. R. Forney, and R. S. Chaleff. "Semidominant Soybean Mutation for Resistance to Sulfonylurea Herbicides," *Crop Sci.* 29:1403-1408 (1989).
90. Newhouse, K., B. Singh, D. Shaner, and M. Stidham. " Mutations in Corn (*Zea mays* L.) Conferring Resistance to Imidazolinone Herbicides," *Theor. Appl. Genet.* 83:65-70 (1991).
91. Swanson, E. B., M. J. Herrgesell, M. Arnoldo, D. W. Sippell, and R. S. C. Wong. "Microspore Mutagenesis and Selection: Canola Plants with Field Tolerance to the Imidazolinones," *Theor. Appl. Genet.* 78:525-530 (1989).
92. Hartnett, M. E., J. R. Newcomb, and R. C. Hodson. "Mutations in *Chlamydomonas reinhardtii* Conferring Resistance to the Herbicide Sulfometuron Methyl," *Plant Physiol.* 85:898-901 (1987).
93. Winder, T., and M. H. Spalding. "Imazaquin and Chlorsulfuron Resistance and Cross Resistance in Mutants of *Chlamydomonas reinhardtii,*" *Mol. Gen. Genet.* 213:394-399 (1988).

94. Saxena, P. K., and J. King. "Herbicide Resistance in *Datura innoxia*. Cross-Resistance of Sulfonylurea-Resistant Cell Lines to Imidazolinones," *Plant Physiol.* 86:863-867 (1988).
95. Rathinasabapathi, B., D. Williams, and J. King. "Altered Feedback Sensitivity to Valine, Leucine, and Isoleucine of Acetolactate Synthase from Herbicide-Resistant Variants of *Datura innoxia*," *Plant Sci.* 67:1-6 (1990).
96. Rathinasabapathi, B., and J. King. "Herbicide Resistance in *Datura innoxia*. Kinetic Characterization of Acetolactate Synthase from Wild-Type and Sulfonylurea-Resistant Cell Variants," *Plant Physiol.* 96:255-261 (1991).
97. Pofelis, S., H. Le, and W. F. Grant. "The Development of Sulfonylurea Herbicide-Resistant Birdsfoot Trefoil (*Lotus corniculatus*) Plants from *in vitro* Selection," *Theor. Appl. Genet.* 83:480-488 (1992).
98. Newhouse, K. E., W. A. Smith, M. A. Starrett, T. J. Schaefer, and B. K. Singh. "Tolerance to Imidazolinone Herbicides in Wheat," *Plant Physiol.* 100:882-886 (1992).
99. Hart, S. E., J. W. Saunders, and D. Penner. "Chlorsulfuron-Resistant Sugarbeet: Cross-Resistance and Physiological Basis of Resistance," *Weed Sci.* 40:378-383 (1992).
100. Saunders, J. W., G. Acquaah, K. A. Renner, and W. P. Doley. "Monogenic Dominant Sulfonylurea Resistance in Sugarbeet from Somatic Cell Selection," *Crop Sci.* 32:1357-1360 (1992).
101. Gressel, J., and L. A. Segel. "Negative Cross Resistance; a Possible Key to Atrazine Resistance Management: A Call for Whole Plant Data," *Z. Naturforsch.* 45c:470-473 (1990).
102. Hattori, J., R. Rutledge, H. Labbe, D. Brown, G. Sunohara, and B. Miki. "Multiple Resistance to Sulfonylureas and Imidazolinones Conferred by an Acetohydroxyacid Synthase Gene with Separate Mutations for Selective Resistance," *Mol. Gen. Genet.* 232:167-173 (1992).
103. Guttieri, M. J., C. V. Eberlein, C. A. Mallory-Smith, D. C. Thill, and D. L. Hoffman. "DNA Sequence Variation in Domain A of the Acetolactate Synthase Genes of Herbicide-Resistant and -Susceptible Weed Biotypes," *Weed Sci.* 40:670-677 (1992).
104. Guttieri, M. J., and C. V. Eberlein. University of Idaho, Aberdeen, ID. Unpublished results (1993).
105. Mauvais, C., and Saari, L. L., DuPont Agricultural Products, Wilmington, DE. Unpublished results (1989).
106. Gabard, J. M., P. J. Charest, V. N. Iyer, and B. L. Miki. "Cross-Resistance to Short Residual Sulfonylurea Herbicides in Transgenic Tobacco Plants," *Plant Physiol.* 91:574-580 (1989).
107. Charest, P. J., J. Hattori, J. DeMoor, V. N. Iyer, and B. L. Miki. "In vitro Study of Transgenic Tobacco Expressing *Arabidopsis* Wild Type and Mutant Acetohydroxyacid Synthase Genes," *Plant Cell Rep.* 8:643-646 (1990).
108. Smith, J. K., J. V. Schloss, and B. J. Mazur. "Functional Expression of Plant Acetolactate Synthase Genes in *Escherichia coli*," *Proc. Natl. Acad. Sci. U.S.A.* 86:4179-4183 (1989).
109. Miki, B. L., H. Labbe, J. Hattori, T. Ouellet, J. Gabbard, G. Sunohara, P. J. Charest, and V. N. Iyer. "Transformation of *Brassica napus* Canola Cultivars with *Arabidopsis thaliana* Acetohydroxyacid Synthase Genes and Analysis of Herbicide Resistance," *Theor. Appl. Genet.* 80:449-458 (1990).

110. Wiersma, P. A., J. E. Hachey, W. L. Crosby, and M. M. Moloney. "Specific Truncations of an Acetolactate Synthase Gene from *Brassica napus* Efficiently Complement ilvB/ilvG Mutants of *Salmonella typhimurium*," *Mol. Gen. Genet.* 224:155-159 (1990).
111. Harms, C. T., S. L. Armour, J. J. DiMaio, L. A. Middlesteadt, D. Murray, D. V. Negrotto, H. Thompson-Taylor, K. Weymann, A. L. Montoya, R. D. Shillito, and G. C. Jen. "Herbicide Resistance Due to Amplification of a Mutant Acetohydroxyacid Synthase Gene," *Mol. Gen. Genet.* 233:427-435 (1992).
112. Matthews, J. M., J. A. M. Holtum, D. R. Liljegren, B. Furness, and S. B. Powles. "Cross-Resistance to Herbicides in Annual Ryegrass (*Lolium rigidum*). I. Properties of the Herbicide Target Enzymes Acetyl-Coenzyme A Carboxylase and Acetolactate Synthase," *Plant Physiol.* 94:1180-1186 (1990).
113. Odell, J. T., P. G. Caimi, N. S. Yadav, and C. J. Mauvais. "Comparison of Increased Expression of Wild-Type and Herbicide-Resistant Acetolactate Synthase Genes in Transgenic Plants, and Indication of Posttranscriptional Limitation on Enzyme Activity," *Plant Physiol.* 94:1647-1654 (1990).
114. Forlani, G., E. Nielsen, P. Landi, and R. Tuberosa. "Chlorsulfuron Tolerance and Acetolactate Synthase Activity in Corn (*Zea mays* L.) Inbred Lines," *Weed Sci.* 39:553-557 (1991).
115. Meredith, C. P. "Selecting Better Crops from Cultured Cells," in *Gene Manipulation in Plant Improvement*, J. P. Gustafson, Ed. (New York: Plenum Press, 1984), pp. 503-528.
116. Schloss, J. V., L. M. Ciskanik, and D. E. Van Dyk. "Origin of the Herbicide Binding Site of Acetolactate Synthase," *Nature* 331:360-362 (1988).
117. Grabau, C., and J. E. Cronan, Jr. "Nucleotide Sequence and Deduced Amino Acid Sequence of *Escherichia coli* Pyruvate Oxidase, A Lipid-Activated Flavoprotein," *Nucleic Acids Res.* 14:5449-5460 (1986).
118. Schloss, J. V. "Interaction of the Herbicide Sulfometuron Methyl with Acetolactate Synthase: A Slow-Binding Inhibitor," in *Flavins and Flavoproteins*, R. C. Bray, P. C. Engel, and S. G. Mayhew, Eds. (New York: Walter de Gruyter and Co., 1984), pp. 737-740.
119. Hawkes, T. R., J. L. Howard, and S. E. Pontin. "Herbicides that Inhibit the Biosynthesis of Branched Chain Amino Acids," in *Herbicides and Plant Metabolism*, A. D. Dodge, Ed. (Cambridge, U.K.: Cambridge University Press, 1989), pp. 113-136.
120. Xing, R., and W. B. Whitman. "Sulfometuron Methyl-Sensitive and -Resistant Acetolactate Synthases of the Archaebacteria *Methanococcus* spp.," *J. Bacteriol.* 169:4486-4492 (1987).
121. Durner, J., V. Gailus, and P. Böger. "New Aspects on Inhibition of Plant Acetolactate Synthase by Chlorsulfuron and Imazaquin," *Plant Physiol.* 95:1144-1149 (1991).
122. Subramanian, M. V., and B. C. Gerwick. "Inhibition of Acetolactate Synthase by Triazolopyrimidines: A Review of Recent Developments," in *Biocatalysis in Agricultural Biotechnology*, ACS Symposium Series No. 389, J. R. Whitaker, and P. E. Sonnet, Eds. (Washington, D.C.: American Chemical Society, 1989), pp. 277-288.
123. Shaner, D. L., and B. K. Singh. "Imidazolinone-Induced Loss of Acetolactate Synthase Activity in Maize Is Not Due to the Enzyme Degradation," *Plant Physiol.* 97:1339-1341 (1991).

124. Schloss, J. V., and A. Aulabaugh. "Acetolactate Synthase and Ketol-Acid Reductoisomerase: Targets for Herbicides Obtained by Screening and de novo Design," *Z. Naturforsch.* 45c:544-551 (1990).
125. Meyer, A. M., and F. Muller. "Triasulfuron and Its Selective Behaviour in Wheat and *Lolium perenne*," in *Brighton Crop Protection Conference — Weeds*, (Farnham, U.K.: The British Crop Protection Council, 1989), pp. 441-448.
126. Frear, D. S., H. R. Swanson, and F. W. Thalacker. "Induced Microsomal Oxidation of Diclofop, Triasulfuron, Chlorsulfuron, and Linuron in Wheat," *Pestic. Biochem. Physiol.* 41:274-287 (1991).
127. Eberlein, C. V., K. M. Rosow, J. L. Geadelmann, and S. J. Openshaw. "Differential Tolerance of Corn Genotypes to DPX-M6316," *Weed Sci.* 37:651-657 (1989).
128. Brown, H. M., V. A. Wittenbach, D. R. Forney, and S. D. Strachan. "Basis for Soybean Tolerance to Thifensulfuron Methyl," *Pestic. Biochem. Physiol.* 37:303-313 (1990).
129. Hall, J. C., C. J. Swanton, and M. D. Devine. "Physiological and Biochemical Investigation of the Selectivity of Ethametsulfuron in Commercial Brown Mustard and Wild Mustard," *Pestic. Biochem. Physiol.* 42:188-195 (1992).
130. Sweetser, P. B., G. S. Schow, and J. M. Hutchison. "Metabolism of Chlorsulfuron by Plants: Biological Basis for Selectivity of a New Herbicide for Cereals," *Pestic. Biochem. Physiol.* 17:18-23 (1982).
131. Cotterman, J. C., and L. L. Saari. "Rapid Metabolic Inactivation is the Basis for Cross-Resistance to Chlorsulfuron in Diclofop-Methyl-Resistant Rigid Ryegrass (*Lolium rigidum*) Biotype SR4/84," *Pestic. Biochem. Physiol.* 43:182-192 (1992).
132. Jensen, K. I. N. "The Roles of Uptake, Translocation, and Metabolism in the Differential Intraspecific Responses to Herbicides," in *Herbicide Resistance in Plants*, H. M. LeBaron, and J. Gressel, Eds. (New York: John Wiley & Sons, Inc., 1982), pp. 133-162.
133. Schafer, D. E., and D. O. Chilcote. "Translocation and Degradation of Bromoxynil in a Resistant and a Susceptible Species," *Weed Sci.* 18:729-732 (1970).
134. Brown, H. M., R. F. Dietrich, W. H. Kenyon, and F. T. Lichtner. "Prospects for the Biorational Design of Crop Selective Herbicides," in *Brighton Crop Protection Conference — Weeds*, (Farnham, U.K.: The British Crop Protection Council, 1991), pp. 847-856.
135. Brown, H. M. "Mode of Action, Crop Selectivity, and Soil Relations of the Sulfonylurea Herbicides," *Pestic. Sci.* 29:263-281 (1990).
136. Fonne-Pfister, R., J. Gaudin, K. Kreuz, K. Ramsteiner, and E. Ebert. "Hydroxylation of Primisulfuron by an Inducible Cytochrome P450-Dependent Monooxygenase System from Maize," *Pestic. Biochem. Physiol.* 37:165-173 (1990).
137. Omer, C. A., R. Lenstra, P. J. Litle, C. Dean, J. M. Tepperman, K. J. Leto, J. A. Romesser, and D. P. O'Keefe. "Genes for Two Herbicide-Inducible Cytochromes P-450 from *Streptomyces griseolus*," *J. Bacteriol.* 172:3335-3345 (1990).
138. Zimmerlin, A., and F. Durst. "Aryl Hydroxylation of the Herbicide Diclofop by a Wheat Cytochrome P-450 Monooxygenase," *Plant Physiol.* 100:874-881 (1992).
139. Brown, H. M., and S. M. Neighbors. "Soybean Metabolism of Chlorimuron Ethyl: Physiological Basis for Soybean Selectivity," *Pestic. Biochem. Physiol.* 29:112-120 (1987).

140. Neighbors, S., and L. S. Privalle. "Metabolism of Primisulfuron by Barnyard Grass," *Pestic. Biochem.Physiol.* 37:145-153 (1990).
141. Swisher, B. A., B. C. Gerwick, III, M. Chang, V. W. Miner, and G. J. deBoer. "Metabolism of the Triazolopyrimidine Sulfonanilide DE-498 in Plants," in *WSSA Abstracts,* Vol. 31, *Proceedings of the 1991 Meeting of the Weed Science Society of America,* (Champaign, IL: WSSA, 1991), p. 50.
142. Shaner, D. L., and N. M. Mallipudi. "Mechanisms of Selectivity of the Imidazolinone Herbicides," in *The Imidazolinone Herbicides,* D. L. Shaner and S. L. O'Connor, Eds. (Boca Raton, FL: CRC Press, Inc., 1991), pp. 91-102.
143. Brown, M. A., T. Y. Chiu, and P. Miller. "Hydrolytic Activation Versus Oxidative Degradation of Assert Herbicide, an Imidazolinone Aryl-Carboxylate, in Susceptible Wild Oat versus Tolerant Corn and Wheat," *Pestic. Biochem. Physiol.* 27:24-29 (1987).
144. Brown, H. M., T. P. Fuesler, T. B. Ray, and S. D. Strachan. "Role of Plant Metabolism in Crop Selectivity of Herbicides," in *Pesticide Chemistry: Advances in International Research, Development, and Legislation,* H. Frehse, Ed. (Weinheim: VCH, 1991), pp. 257-266.
145. Ray, T. B. "Sulfonylurea Herbicides as Inhibitors of Amino-Acid Biosynthesis in Plants," *Trends Biochem. Sci.* 11:180-183 (1986).
146. Christopher, J. T., S. B. Powles, D. R. Liljegren, and J. A. M. Holtum. "Cross-Resistance to Herbicides in Annual Ryegrass (*Lolium rigidum*). II. Chlorsulfuron Resistance Involves a Wheat-Like Detoxification System," *Plant Physiol.* 100:1036-1043 (1991).
147. Kemp, M. S., S. R. Moss, and T. H. Thomas. "Herbicide Resistance in *Alopecurus myosuroides*," in *Managing Resistance to Agrochemicals. From Fundamental Research to Practical Strategies,* M. B. Green, H. M. LeBaron, and W. K. Moberg, Eds. (Washington, D.C.: American Chemical Society, 1990), pp. 376-393.
148. Sweetser, P. B., DuPont Agricultural Products, Wilmington, DE. Unpublished results (1984).
149. Anderson, M. P., and J. W. Gronwald. "Atrazine Resistance in a Velvetleaf (*Abutilon theophrasti*) Biotype Due to Enhanced Glutathione S-Transferase Activity," *Plant Physiol.* 96:104-109 (1991).
150. Powles, S. B., and P. D. Howat. "Herbicide-Resistant Weeds in Australia," *Weed Technol.* 4:178-185 (1990).
151. Gressel, J. *Wheat Herbicides. The Challenge of Emerging Resistance* (Reading, U.K.: Biotechnology Affiliates, 1988), 247 pp.
152. Shimabukuro, R. H., and B. L. Hoffer. "Metabolism of Diclofop-Methyl in Susceptible and Resistant Biotypes of *Lolium rigidum*," *Pestic. Biochem. Physiol.* 39:251-260 (1991).
153. Holtum, J. A. M., J. M. Matthews, R. E. Hausler, D. R. Liljegren, and S. B. Powles. "Cross-Resistance to Herbicides in Annual Ryegrass (*Lolium rigidum*). III. On the Mechanism of Resistance to Diclofop-Methyl," *Plant Physiol.* 97:1026-1034 (1991).
154. Powles, S. B., and J. M. Matthews. "Multiple Herbicide Resistance in Annual Ryegrass (*Lolium rigidum*): A Driving Force for the Adoption of Integrated Weed Management," in *Resistance '91: Achievements and Developments in Combatting Pesticide Resistance,* I. Denholm, A. L. Devonshire, and D. W. Hollomon, Eds. (London: Elsevier Applied Science, 1992) pp. 75-87.

155. Clarke, J. H., and S. R. Moss. "The Occurrence of Herbicide Resistant *Alopecurus myosuroides* (Black-Grass) in the United Kingdom and Strategies for Its Control," in *Brighton Crop Protection Conference — Weeds*, (Farnham, U.K.: The British Crop Protection Council, 1991), pp. 1041-1048.
156. McHughen, A. "*Agrobacterium* Mediated Transfer of Chlorsulfuron Resistance to Commercial Flax Cultivars," *Plant Cell Rep.* 8:445-449 (1989).
157. McHughen, A., and F. Holm. "Herbicide Resistant Transgenic Flax Field Test: Agronomic Performance in Normal and Sulfonylurea-Containing Soils," *Euphytica* 55:49-56 (1991).
158. McHughen, A., and G. G. Rowland. "The Effect of T-DNA on the Agronomic Performance of Transgenic Flax Plants," *Euphytica* 55:269-275 (1991).
159. McSheffrey, S. A., A. McHughen, and M. D. Devine. "Characterization of Transgenic Sulfonylurea-Resistant Flax (*Linum usitatissimum*)," *Theor. Appl. Genet.* 84:480-486 (1992).
160. Swanson, E. B., M. P. Coumans, G. L. Brown, J. D. Patel, and W. D. Beversdorf. "The Characterization of Herbicide Tolerant Plants in *Brassica napus* L. after In Vitro Selection of Microspores and Protoplasts," *Plant Cell Rep.* 7:83-87 (1988).
161. Bornman, C. H. "Molecular Biology in Sugar Beet," in *Proceedings of the International. Institute for Sugar Beet Research* (Brussels: International Institute for Sugar Beet Research, 1990), pp. 31-38.
162. Stougaard, P., K. Bojsen, and T. Christensen. "Herbicide Resistant Mutants of Sugar Beet *Beta vulgaris*," in *Proceedings of the Symposium on Molecular Strategies for Crop Improvement*, J. Cell Biochem. Suppl. 0 (14 Part E):310 (1990).
163. Vermeulen, A., H. Vaucheret, V. Pautot, and Y. Chupeau. "*Agrobacterium* mediated Transfer of a Mutant *Arabidopsis* Acetolactate Synthase Gene Confers Resistance to Chlorsulfuron in Chicory (*Cichorium intybus* L.)," *Plant Cell Rep.* 11:243-247 (1992).
164. Terakawa, T., and K. Wakasa. "Rice Mutant Resistant to the Herbicide Bensulfuron Methyl BSM by In-Vitro Selection," *Jpn. J. Breed.* 42:267-275 (1992).
165. Li, Z., A. Hayashimoto, and N. Murai. "A Sulfonylurea Herbicide Resistance Gene from *Arabidopsis thaliana* as a New Selectable Marker for Production of Fertile Transgenic Rice Plants," *Plant Physiol.* 100:662-668 (1992).
166. O'Keefe, D. P., K. R. Bozak, R. E. Christoffersen, J. A. Tepperman, C. Dean, and P. A. Harder. "Endogenous and Engineered Cytochrome P-450 Mono-Oxygenases in Plants," *Biochem. Soc. Trans.* 20:357-361 (1992).
167. Stalker, D. M., K. E. McBride, and L. D. Malyj. "Herbicide Resistance in Transgenic Plants Expressing a Bacterial Detoxification Gene," *Science* 242:419-422 (1988).
168. Freyssinet, G., B. Leroux, M. Lebrun, B. Pelissier, and A. Sailland. "Transfer of Bromoxynil Resistance into Crops," in *Brighton Crop Protection Conference — Weeds*, (Farnham, U.K.: The British Crop Protection Council, 1989), pp. 1225-1234.
169. DeBlock, M., J. Botterman, M. Vandewiele, J. Dockx, C. Thoen, V. Gossele, N. Movva, C. Thompson, M. Van Montagu, and J. Leemans. "Engineering Herbicide Resistance in Plants by Expression of a Detoxifying Enzyme," *EMBO J.* 6:2513-2518 (1987).
170. Wohlleben, W., W. Arnold, I. Broer, D. Hillemann, E. Strauch, and A. Puhler. "Nucleotide Sequence of the Phosphinothricin-N-Acetyl Transferase Gene from *Streptomyces viridochromogenes* Tu 494 and its Expression in *Nicotiana tabacum*," *Gene* 70:25-37 (1988).

171. Gressel, J. "The Needs for New Herbicide-Resistant Crops," in *Resistance '91: Achievements and Developments in Combating Pesticide Resistance*, I. Denholm, A. L. Devonshire, and D. W. Hollomon, Eds. (London: Elsevier Applied Science, 1992) pp. 283-294.
172. LaRossa, R. A., T. K. Van Dyk, and D. R. Smulski. "Toxic Accumulation of α-Ketobutyrate Caused by Inhibition of the Branched-Chain Amino Acid Biosynthetic Enzyme Acetolactate Synthase in *Salmonella typhimurium*," *J. Bacteriol.* 169:1372-1378 (1987).
173. Sebastian, S. A., and R. S. Chaleff. "Soybean Mutants with Increased Tolerance for Sulfonylurea Herbicides," *Crop Sci.* 27:948-952 (1987).
174. Harms, C. T., A. L. Montoya, L. S. Privalle, and R. W. Briggs. "Genetic and Biochemical Characterization of Corn Inbred Lines Tolerant to the Sulfonylurea Herbicide Primisulfuron," *Theor. Appl. Genet.* 80:353-358 (1990).
175. Anderson, P. C., and M. Georgeson. "Herbicide-Tolerant Mutants of Corn," *Genome* 31:994-999 (1989).
176. Anderson, P. C., and M. Georgeson. "Selection and Characterization of Imidazolinone Tolerant Mutants of Maize," in *Biochemical Basis of Herbicide Action*, (Ashford, U.K.: 27th Harden Conf. Prog. Abstr. Wye. College, 1986).
177. Mallory-Smith, C. A., D. C. Thill, M. J. Dial, and R. S. Zemetra. "Inheritance of Sulfonylurea Herbicide Resistance in *Lactuca* spp.," *Weed Technol.* 4:787-790 (1990).
178. Thompson, C. R., and D. C. Thill. University of Idaho, Moscow, ID. Unpublished results. (1992).
179. Mulugeta, D., P. K. Fay, W. E. Dyer, and L. E. Talbert. "Inheritance of Resistance to the Sulfonylurea Herbicides in *Kochia scoparia* L. (Schrad.)," in *Proceedings of the Western Weed Science Society* (Newark, CA: Western Society of Weed Science, 1991), p. 81-82.
180. Silvertown, J. W. *Introduction to Plant Population Ecology*, 2nd Ed. (New York: Longman Scientific and Technical, 1987), Chapters 1, 7.
181. Holt, J. S., and S. R. Radosevich. "Differential Growth of Two Common Groundsel *(Senecio vulgaris)* Biotypes," *Weed Sci.* 31:112-119 (1983).
182. Radosevich, S. R. "Mechanism of Atrazine Resistance in Lambsquarters and Pigweed," *Weed Sci.* 25:316-318 (1977).
183. Conard, S. G., and S. R. Radosevich. "Ecological Fitness of *Senecio vulgaris* and *Amaranthus retroflexus* Biotypes Susceptible or Resistant to Atrazine," *J. Appl. Ecol.* 16:171-177 (1979).
184. Jacobs, B. F., J. H. Duesing, J. Antonovics, and D. T. Patterson. "Growth Performance of Triazine-Resistant and -Susceptible Biotypes of *Solanum nigrum* Over a Range of Temperatures," *Can. J. Bot.* 66:847-850 (1988).
185. Warwick, S. I., and L. Black. "The Relative Competitiveness of Atrazine Susceptible and Resistant Populations of *Chenopodium album* and *C. strictum*," *Can. J. Bot.* 59:689-693 (1981).
186. Alcocer-Ruthling, M., D. C. Thill, and B. Shafii. "Seed Biology of Sulfonylurea-Resistant and -Susceptible Biotypes of Prickly Lettuce (*Lactuca serriola*)," *Weed Technol.* 6:858-864 (1992).
187. Dyer, W. E., P. W. Chee, and P. K. Fay. "Low Temperature Seed Germination Characteristics of Sulfonylurea Herbicide-Resistant *Kochia scoparia* L. Accessions," in *Proceedings of the Western Weed Science Society* (Newark, CA: Western Society of Weed Science, 1992), pp. 117-118.

188. Alcocer-Ruthling, M., D. C. Thill, and B. Shafii. "Differential Competitiveness of Sulfonylurea-Resistant and -Susceptible Prickly Lettuce (*Lactuca serriola*)," *Weed Technol.* 6:303-309 (1992).
189. Thompson, C. R., and D. C. Thill. "Sulfonylurea Herbicide-Resistant and -Susceptible Kochia (*Kochia scoparia* L. Schrad.) Growth Rate and Seed Production," in *WSSA Abstracts,* Vol. 32, *Proceedings of the 1992 Meeting of the Weed Science Society America*, (Champaign, IL: WSSA, 1992), p. 44.
190. Mallory-Smith, C. A., D. C. Thill, M. Alcocer-Ruthling, and C. Thompson. "Growth Comparisons of Sulfonylurea Resistant and Susceptible Biotypes," *Proceedings of the First International Weed Control Congress,* Vol. 2 (Melbourne, Australia: Weed Science Society of Victoria Inc., 1992), pp. 301-303.
191. Christoffoletti, P. J., and P. Westra. "Competition and Coexistence of Sulfonylurea Resistant and Susceptible Kochia (*Kochia scoparia*) Biotypes in Unstable Environments," in *WSSA Abstracts,* Vol. 32, *Proceedings of the 1992 Meeting of the Weed Science Society of America* (Champaign, IL: WSSA, 1992), p. 17.
192. Spitters, C. J. T. "An Alternative Approach to the Analysis of Mixed Cropping Experiments. I. Estimation of Competition Effects," *Neth. J. Agric. Sci.* 31:1-11 (1983).
193. Rejmanek, M., G. R. Robinson, and E. Rejmankova. "Weed-Crop Competition: Experimental Designs and Models for Data Analysis," *Weed Sci.* 37:276-284 (1989).
194. Singh, B. K., M. A. Stidham, and D. L. Shaner. "Separation and Characterization of Two Forms of Acetohydroxy Acid Synthase from Black Mexican Sweet Corn Cells," *J. Chromatogr.* 444:251-261 (1988).
195. Alcocer-Ruthling, M., D. C. Thill, and C. Mallory-Smith. "Monitoring the Occurrence of Sulfonylurea-Resistant Prickly Lettuce (*Lactuca serriola*)," *Weed Technol.* 6: 437-440 (1992).
196. Richter, J., and S. B. Powles. "Pollen Expression of Herbicide Target Site Resistance Genes in Annual Ryegrass (*Lolium rigidum*)," *Plant Physiol.* (in press) (1993).
197. Mulugeta, D. M. "Management, Inheritance, and Gene Flow of Resistance to Chlorsulfuron in *Kochia scoparia* (L.) Schrad.," M. S. Thesis, Montana State University, Bozeman, MT (1991).
198. Rubin, B., Hebrew University of Jerusalem, Rehovot, Israel. Unpublished results (1992).
199. Mallory-Smith, C. A. "Identification and Inheritance of Sulfonylurea Herbicide Resistance in Prickly Lettuce (*Lactuca serriola* L.)," Ph. D. Dissertation, University of Idaho, Moscow, ID (1990).
200. Boutsalis, P., and S. B. Powles, University of Adelaide, Waite Institute, Glen Osmond, South Australia, Personal communication (1992).
201. Gill, G. S. "Herbicide Resistance in Annual Ryegrass in Western Australia," in *National Herbicide Resistance Extension Workshop*, Glen Osmond, S. A., Australia (1992), pp. 8-11.
202. Friesen, L. F. "Herbicide Resistance Summary," *Research Report, Expert Committee on Weeds (Western Section)*, Agriculture Canada (1992), pp. 408-413.
203. Devine, M. D., M. A. S. Marles, and L. M. Hall. "Inhibition of Acetolactate Synthase in Susceptible and Resistant Biotypes of *Stellaria media*," *Pestic. Sci.* 31:273-280 (1991).

204. Kudsk, P., S. K. Mathiassen, and E. F. Petersen. "Resistance to Sulfonylurea Herbicides in *Stellaria media*," *Tidsskr. Planteavl.* 86:147-156 (1992).
205. Moss, S. R. "Herbicide Resistance in Black-Grass (*Alopecurus myosuroides*)," in *British Crop Protection Conference — Weeds*, (Farnham, U.K.: The British Crop Protection Council, 1987), pp. 879-886.
206. Friesen, L. F., I. N. Morrison, A. Rashid, and M. D. Devine. "Response of a Chlorsulfuron-Resistant Biotype of *Kochia scoparia* to Sulfonylurea and Alternative Herbicides," *Weed Sci.* 41:100-106 (1993).
207. Takeda, S., D. L. Erbes, P. B. Sweetser, J. V. Hay, and T. Yuyama. "Mode of Herbicidal and Selective Action of DPX-F5384 Between Rice and Weeds," *Weed Res. (Tokyo)* 31:157-163 (1986).
208. Hutchison, J. M., R. Shapiro, and P. B. Sweetser. "Metabolism of Chlorsulfuron by Tolerant Broadleaves," *Pestic. Biochem. Physiol.* 22:243-247 (1984).
209. Anderson, J. J., T. M. Priester, and L. M. Shalaby. "Metabolism of Metsulfuron Methyl in Wheat and Barley," *J. Agric. Food Chem.* 37:1429-1434 (1989).
210. Suzuki, K., T. Nawamaki, S. Watanabe, S. Yamamoto, T. Sato, K. Morimoto, and B. H. Wells. "NC-319 — A New Herbicide for Control of Broad-Leaved Weeds and *Cyperus* Spp. in Corn," in *Brighton Crop Protection Conference — Weeds*, (Farnham, U.K.: The British Crop Protection Council, 1991), pp. 31-37.
211. Molin, W., K. Kretzmer, T. Lee, R. White, and D. Loussaert. "The Mechanism of Action of and Corn Tolerance to MON 12000," in *WSSA Abstracts,* Vol. 32, *Proceedings of the 1992 Meeting of the Weed Science Society of America* (Champaign, IL: WSSA, 1992), p. 63.
212. Koeppe, M. K., W. H. Kenyon, R. C. Benson, and S. W. Maciag. "Mechanism of Selectivity of DPX-E9636 Herbicide in Corn: Uptake, Translocation and Metabolism," in *Proceedings 202nd National ACS Meeting* (New York: American Chemical Society, August, 1991), AGRO no. 27.
213. Reinke, H., A. Rosenzweig, J. Claus, M. Kreidi, C. Chisholm, and P. Jensen. "DPX-E9636, Experimental Sulfonylurea Herbicide for Potatoes," in *Brighton Crop Protection Conference — Weeds*, (Farnham, U.K.: The British Crop Protection Council, 1991), pp. 445-451.
214. Cotterman, J. C., and L. L. Saari. "Selectivity of DPX-M6316 in Wheat: Differential Metabolism in Wheat and Sensitive Weeds," in *WSSA Abstracts,* Vol. 29, *Proceedings of the 1989 Meeting of the Weed Science Society of America* (Champaign, IL: WSSA, 1989), pp 73-74.
215 Peeples, K. A., M. P. Moon, F. T. Lichtner, V. A. Wittenbach, T. H. Carski, and M. D. Woodward. "DPX-66037 — A New Low-Rate Sulfonylurea for Post-Emergence Weed Control in Sugar Beet and Fodder Beet," in *Brighton Crop Protection Conference — Weeds*, (Farnham, U.K.: The British Crop Protection Council, 1991), pp. 25-30.
216. Swisher, B. A., and M. R. Weimer. "Comparative Detoxification of Chlorsulfuron in Leaf Disks and Cell Cultures of Two Perennial Weeds," *Weed Sci.* 34:507-512 (1986).
217. Wilcut, J. W., G. R. Wehtje, M. G. Patterson, and T. A. Cole. "Absorption, Translocation, and Metabolism of Foliar-Applied Imazaquin in Soybeans (*Glycine max*), Peanuts (*Arachis hypogaea*), and Associated Weeds," *Weed Sci.* 36:5-8 (1988).
218. Cotterman, J. C., DuPont Agricultural Products, Newark, DE. Unpublished results (1988).

CHAPTER 5

Resistance to Acetyl Coenzyme A Carboxylase Inhibiting Herbicides

M. D. Devine and R. H. Shimabukuro

INTRODUCTION

Chemistry of the Herbicide Groups

The aryloxyphenoxypropanoate and cyclohexanedione herbicides are two important groups of postemergence herbicides used to control grass weeds in grass and dicot crops. The chemistry of these two groups differs substantially, but similarities in their activity at the whole plant and physiological/biochemical levels suggest that they share some common structural element(s) responsible for their herbicidal activity.

The aryloxyphenoxypropanoate herbicides have been referred to by various group names in the literature. These include phenoxy phenoxypropionates, aryl propanoates, and polycyclic alkanoic acids. The most commonly used name, aryloxyphenoxypropanoate will be used in this review (abbreviated here as APP). Examples of APP herbicides are shown in Figure 1. APP herbicides are characterized by the two aryl groups, one of which is invariably substituted with an isopropyl group on the 1 position. This group contains a chiral center, with the two possible enantiomers designated R(+) or S(−). Most studies indicate that only the R(+) enantiomer is active as a herbicide,[1,2] although the S(−) enantiomer does have some biological activity.[3] The biochemical and physiological activities of the

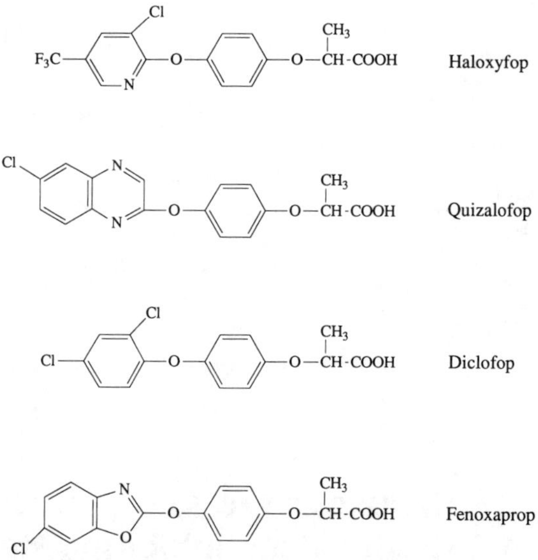

Figure 1. Structures of some agronomically important aryloxyphenoxypropanoate herbicides.

R and S enantiomers are discussed in subsequent sections of this review. Considerable variation is possible in the second aryl group, giving rise to herbicides with differing biological activity and selectivity. Possible structures here can include substituted phenoxy, benzoxazolyloxy, quinoxalinyloxy, or pyridinyloxy groups, with chloro- or trifluoromethyl substituents as the most common. Some examples of these possible substituents are included in Figure 1.

The APP herbicides are usually formulated as esters of the parent acid (e.g., diclofop-methyl, fenoxaprop-ethyl, fluazifop-butyl, etc.). Formulation as an ester facilitates penetration into leaf tissue. Shortly after entry into the leaf, the ester is hydrolyzed to release the parent acid. Thus, APP herbicides form weak acids in plant tissue. However, APP herbicides are generally lipophilic, and do not show the high phloem mobility typically seen with weak acid herbicides.

Cyclohexanedione (CHD) herbicides vary considerably in the substitution at the 2 and 5 positions of the cyclohexeneone ring (Figure 2). Again, different substituents give rise to variation in biological activity and selectivity. The 3-hydroxy group in CHD herbicides can deprotonate, conferring weak acid characteristics on these herbicides. The CHD herbicides are generally very labile, and are subject to rapid photolysis.

Use in Agriculture

APP and CHD herbicides are used for the selective control of grass weeds in cereal and dicot crops. A selection of the registered uses of some of these herbicides is given in Table 1. Sethoxydim, the first CHD herbicide to be

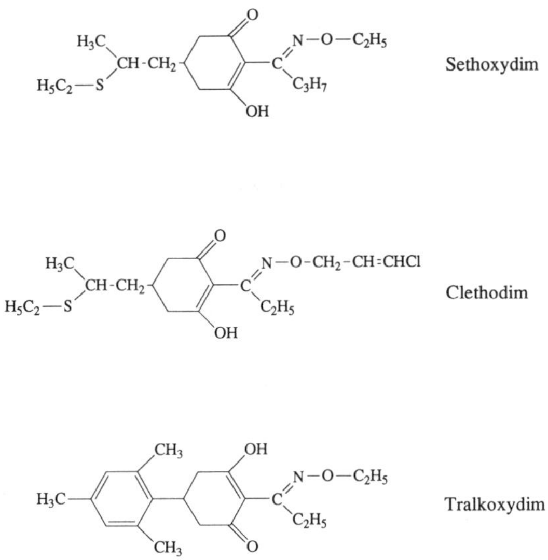

Figure 2. Structures of some agronomically important cyclohexanedione herbicides.

Table 1. A selection of common uses of aryloxyphenoxypropanoate and cyclohexanedione herbicides in crop protection

Herbicide	Crops	Representative weeds controlled
Diclofop-methyl	Wheat, barley, alfalfa, pea, canola, flax	*Avena fatua, Setaria viridis, Setaria glauca, Echinochloa crus-galli, Lolium rigidum*
Fluazifop-butyl	Flax, canola, alfalfa	*A. fatua, S. viridis, S. glauca, E. crus-galli,* volunteer barley
Fenoxaprop-ethyl (with safener)	Canola, pea Wheat	*A. fatua, S. viridis, S. glauca, E. crus-galli*
Sethoxydim	Canola, flax, pea, lentil, alfalfa	*A. fatua, S. viridis, S. glauca E. crus-galli,* volunteer cereals
Tralkoxydim	Wheat, barley, some dicot crops	*A. fatua, S. viridis, S. glauca, E. crus-galli*
Clethodim	Canola, flax, cotton	*A. fatua, S. viridis, S. glauca, E. crus-galli,* volunteer cereals

widely registered for use, is selective only in dicot crops, and can provide some control of volunteer cereal crops. Some members of these groups, including diclofop-methyl, tralkoxydim, and fenoxaprop-ethyl (the latter with the safener, fenchlorazole-ethyl), can also be used in some cereal crops (see Table 1).

Most APP and CHD herbicides are particularly active on annual grass weeds, with little activity on perennials. However, some do have high activity on perennial weeds such as *Elytrigia/Elymus repens*.[4,5] Differences in phloem mobility, which give rise to differences in the amount of herbicide translocated from the foliage to the rhizomes, and in the inherent biological activity of the individual compounds, may account for the varying levels of activity in perennial weeds.[6]

Mode of Action of Aryloxyphenoxpropanoate and Cyclohexanedione Herbicides

The overall phytotoxic action of a herbicide may be separated into two distinct processes.[7] The "mode of action" may be defined as the sum total of anatomical, physiological, and biochemical responses that constitute the total phytotoxic action of a herbicide. Therefore, this process will account for both primary and secondary effects in response to the herbicide. The "mechanism of action" is the primary biochemical or biophysical process or lesion (enzyme or other target site) that is affected by the herbicide. An effect at the primary site may or may not be sufficient to cause complete phytotoxicity. In this review the terms will be used as defined above.

Published reviews on APP and CHD herbicides discuss much of the literature on the mechanism and mode of action of these herbicides.[3,8-12] Readers are encouraged to refer to these reviews for discussions related to references not cited in this review. Generally, only more recent publications on the mechanism and mode of action of these herbicides are referred to in this chapter.

The mode of action of APP and CHD herbicides must explain the observed morphological and physiological plant responses induced by the herbicides in their phytotoxic action. The meristems (apical and intercalary) and not the mature leaves or roots are the primary sensitive sites responsible for ultimate plant death in susceptible species.[3,10,13] Growth of meristems is inhibited shortly after contact with diclofop-methyl and other APP and CHD herbicides, and chlorosis of emerged leaves is observed 3 to 4 days after herbicide application.[3,10] Auxinic compounds such as 2,4-D, MCPA, and dicamba reverse or antagonize the phytotoxic action of some APP herbicides by preventing the inhibition of meristematic growth while having little or no effect on the development of chlorosis in mature leaves. Therefore, any proposed mechanism of action for APP and CHD herbicides must be evaluated in terms of their relationship to the observed responses in whole plants.

Two mechanisms of action have been proposed for APP and CHD herbicides: (1) a biochemical mechanism involving the inhibition of acetyl-CoA carboxylase (ACCase) and subsequent fatty acid biosynthesis in plastids, and (2) a biophysical mechanism involving the perturbation of the transmembrane proton gradient across the plasma membrane.

Inhibition of Fatty Acid Biosynthesis

Biosynthesis of fatty acids and lipids is essential to the normal growth and development of plants. Lipids are involved in the biogenesis and function of various membranes, cellular signal transduction, and other physiological functions.[14] Several reviews on this topic have appeared and readers are directed to these articles for details.[11,12,14-17]

General information on lipid biosynthesis is helpful in evaluating the significance of the effects of APP and CHD herbicides on plant lipids. An abbreviated scheme for lipid biosynthesis based on information in the reviews is presented in Figure 3. This scheme represents the eukaryotic pathway.[14] Fatty acid biosynthesis is localized in the chloroplasts and plastids of non-green tissues. The synthesis of malonyl-CoA catalyzed by acetyl-coenzyme A carboxylase (ACCase; EC 6.4.1.2) is the first committed step in fatty acid biosynthesis. The condensation of acetyl-CoA and malonyl-ACP catalyzed by 3-ketoacyl-ACP synthase is the first step in a series of reactions in the fatty acid synthase complex (FAS) that results in 16:0-ACP and 18:1-ACP as major products of fatty acid synthesis in plastids.[14] The antibiotic cerulenin inhibits this condensation reaction to give similar end results to APP and CHD herbicides.[17] Phosplːolipids are synthesized at the endoplasmic reticulum through initial acylation of glycerol phosphate by acyl-CoA-specific acyltransferases. Desaturation of fatty acids occurs by the action of desaturases on phosphatidyl glycerol and galactolipids at the endoplasmic reticulum and chloroplasts, respectively.[14,16]

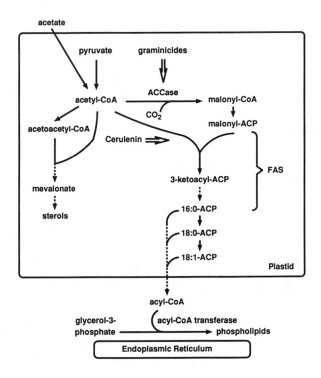

Figure 3. Acyl lipid and glycerolipid biosynthesis in plants. Fatty acid biosynthesis occurs in the chloroplasts and plastids of non-green plant tissues. Long-chain fatty acids and complex glycerolipids are synthesized at the endoplasmic reticulum. Sites of inhibition by postemergence graminicides (APP and CHD herbicides) and the antibiotic cerulenin are indicated by the wide arrows.

Many aspects of fatty acid and lipid biosynthesis are still not well understood. Nevertheless, it is clear that ACCase is the key enzyme that regulates the biosynthesis of fatty acids.[14,16] However, the synthesis of longer chain fatty acids other than palmitic (C16) and stearic (C18) acids, desaturation of fatty acids, and synthesis of complex lipids (phospholipids, glycolipids, etc.) are not regulated by ACCase.[14] Therefore, total lipid biosynthesis may be reduced by the inhibition of ACCase, but compositional changes in acyl lipids and glycerolipids are unaffected by ACCase activity.[14] Such compositional changes are under different regulatory mechanisms.[14,16]

APP and CHD herbicides are effective and potent inhibitors of ACCase from susceptible grass species but not from resistant dicot species.[11,12,18] An example of typical ACCase inhibition data is shown in Figure 4. ACCase from susceptible grasses is more sensitive to inhibition by the R(+) than the S(–) enantiomer of APP herbicides.[11] Readers are referred to a recent review for a detailed description of ACCase isolation, characterization, and reaction mechanisms and kinetics.[18]

Diclofop-methyl and other APP and CHD compounds were initially shown to inhibit lipid biosynthesis. Subsequently, ACCase was demonstrated to be the site of inhibition.[2,19-24] Inhibition of ACCase was assayed *in vitro* by determining the incorporation of [^{14}C]acetate into lipids in whole or disrupted chloroplasts[19-21,24] and the carboxylation of acetyl-CoA using NaH[^{14}C]O$_3$ in cell-free preparations.[2,19,22-24] Extremely low I_{50} concentrations for *in vitro* inhibition of ACCase (0.01 to 0.5 μM)[12,18] have been reported in sensitive species, suggesting that this is the primary target site in higher plants.

Detailed kinetic analyses strongly suggest that both APP and CHD herbicides are reversible, linear, noncompetitive inhibitors of ACCase.[25,26] However, inhibition was

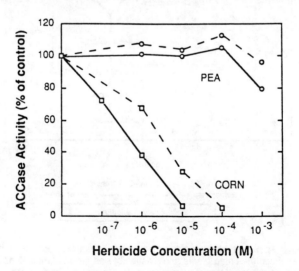

Figure 4. Inhibition of maize and pea ACCase by haloxyfop (- - -) and sethoxydim (- - -). Taken from Reference 18.

"nearly competitive" versus acetyl-CoA, indicating that the herbicides may interact with an acetyl-CoA binding site and perhaps with the release of malonyl-CoA.[26] In addition, double-inhibition experiments have shown that APP and CHD herbicides may act at the same site on ACCase.[25,26] Of the two possible steps in the normal catalytic sequence of ACCase action, biotin carboxylation and transcarboxylation (acetate -> malonate), the latter reaction is more sensitive to inhibition by herbicides.[25,26] Collectively, the enzyme inhibition data clearly indicate that APP and CHD herbicides are potent inhibitors of ACCase from grass species, and that selectivity between grass and dicot species is due to the relative insensitivity of dicot ACCase to these herbicides.

ACCase sensitivity to herbicides is best assayed by measuring [^{14}C]acetate incorporation into lipids by isolated plastids or H[^{14}C]O$_3$ incorporation by purified or partially purified enzyme preparations. In intact cells the incorporation of [^{14}C] from [^{14}C]acetate into cellular lipids is a function of substrate uptake and *de novo* inhibition of ACCase by APP and CHD herbicides. The uptake of [^{14}C]acetate is indirectly inhibited by the perturbation of the transmembrane proton gradient, as discussed in the following section. Therefore, labeling of cellular lipids in whole cells with pulsed treatments of [^{14}C]acetate reflects the net effect of both mechanisms. To determine the effect of APP and CHD herbicides on cellular lipid biosynthesis, meristematic and actively growing tissues such as root tips, coleoptiles, and stem tissue near coleoptilar nodes in grasses should be used.[13] Since growth inhibition and irreversible injury to meristematic tissues occur within 3 to 8 h,[3,27] the effect of APP herbicides on *in vivo* lipid biosynthesis should be determined within this period before senescence and death occur.

Perturbation of the Transmembrane Proton Gradient

All plant cells establish and maintain a transmembrane proton gradient that is vital to their growth and development (Figure 5).[28-30] The proton gradient is established (inside negative and outside positive at the plasma membrane) by an electrogenic (energy requiring) proton pump (H$^+$-ATPase) that is driven by the hydrolysis of ATP.[15,28,31] The free energy or proton motive force (pmf) ($\Delta\mu_{H^+}$) available at steady state to do useful work has the following relationship at 30°C:[1,31]

$$\Delta\mu_{H^+} = \Delta Em - 59\, \Delta pH$$

The free energy available is the difference between the membrane potential (Em) measured in mV and the pH difference that exists between the inside and outside of the membrane.

Changes in Em often reflect the activity of the electrogenic proton pump.[32] A more negative Em (hyperpolarization) reflects the increased outward transport of protons that results in a steeper proton gradient that can be responsible for a decrease in pH (acidification) of a bathing solution containing plant

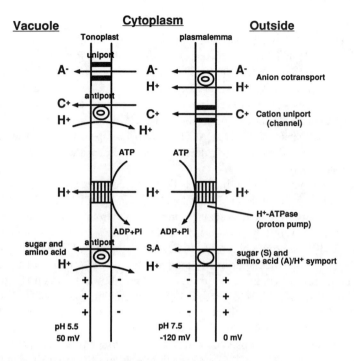

Figure 5. The energy transduction mechanism in the plasma membrane and tonoplast of plant cells. The electrogenic proton gradient with its corresponding membrane potential is established by the activity of the H+-ATPase proton pump. The figure illustrates the active transport of different inorganic and organic solutes across the membrane powered by the proton motive force (pmf).

tissues.[32,33] Not only does the H+-ATPase energy transduction mechanism regulate the active transport of organic and inorganic solutes across the membrane,[29,32-36] but it is also involved in the regulation and control of intracellular pH and its many ramifications.[36-40] This includes hormonal control of plant growth,[15,29,32,33] regulation of sieve tube loading and unloading in phloem transport,[15] cell division,[37,38] etc. Therefore, any chemical or xenobiotic such as diclofop-methyl that perturbs the transmembrane proton gradient may have significant effects on many aspects of the physiology of a plant.[3]

The free acids diclofop and haloxyfop depolarize the membrane potential or dissipate the transmembrane proton gradient in *Chara* spp.[39] and parenchyma cells of oat, wheat, *Avena fatua,* and *Lolium rigidum* coleoptiles and root tips.[40-43] Similar effects have also been observed in isolated vesicles of oat plasma membrane and tonoplast.[44,45] The depolarization is due to the specific influx of protons caused by diclofop.[39,40] The mechanism for the increased influx of protons by diclofop is unknown, but the herbicide action differs from that of a nonspecific, mobile protonophore such as CCCP (carbonyl cyanide-*m*-chlorophenyl hydrazone).[43] The impermeant SH-group inhibitor PCMBS (*p*-chloromercuribenzenesulfonic acid) inhibits the depolarizing action of diclofop but not that of CCCP, indicating the possibility of a specific diclofop-receptor interaction at the plasma membrane that causes the increased permeability to protons.[43]

A characteristic response of susceptible plants to the action of diclofop-methyl and other APP herbicides is the rapid inhibition of meristematic growth.[3,10] Auxin initiates rapid cell elongation and growth of meristematic cells by stimulating proton excretion and acidification of the apoplast (the acid growth theory).[46] Diclofop does not dissipate the proton gradient by inhibiting the plasma membrane H^+-ATPase.[45,47] This is indicated further by the apparent *in vivo* stimulation of H^+-ATPase by diclofop as detected by a transient decrease in cellular ATP.[39] The stimulation of the proton pump is consistent with a decrease in cytoplasmic pH[37] that probably occurs with the influx of protons upon treatment with diclofop. Antagonism of auxin-induced growth by diclofop appears to occur through the dissipation of the proton gradient which eliminates net acidification of the apoplast without affecting the proton pump. Indeed, antagonism of auxin-induced oat coleoptile growth occurs almost immediately and is correlated with a rapid depolarization of Em and a concomitant alkalinization of the external solution.[48,49] Since the activity of the proton pump is unaffected by diclofop, growth inhibition may be reversed by either increasing the concentration of auxin or auxinic compounds such as 2,4-D and dicamba, or by removing the herbicide from the treatment solution within 3 h.[3,48,49] Either process will result in the resumption of growth correlated with the repolarization of Em and reacidification of external solution.[3,48,49] In whole plant and field applications, antagonism of phytotoxicity in susceptible plants is most effective when auxinic compounds (2,4-D, MCPA, dicamba) are applied simultaneously with the herbicide or within 8 h following herbicide application.[3,10,50]

There are many possible functional consequences for cell survival following dissipation of the proton gradient. A decrease in pmf will reduce the uptake of leucine and sucrose, although contradictory results have been reported.[51-54] Sucrose uptake is significantly reduced by haloxyfop in susceptible maize cell culture[53] while nucleotide uptake is reduced by fluazifop[51] and diclofop[52,54] in susceptible tissues. The reduction in protein and nucleic acid synthesis is primarily due to the inhibition of substrate uptake by fluazifop.[51] 2,4-D reduced the inhibitory effect of fluazifop on leucine uptake,[51] probably by stimulating the proton pump (hyperpolarization) with a concomitant increase in pmf,[49] but 2,4-D did not affect *de novo* protein synthesis.[51] The significant inhibition of protein, RNA and lipid biosynthesis in oat coleoptiles by quizalofop-ethyl[55] may be due primarily to inhibition of substrate uptake, especially since protein and RNA synthesis are not target sites for APP herbicides.

The uptake of [^{14}C]acetate by plant tissues may have a significant effect on *de novo* biosynthesis of fatty acids. Acetate uptake is a passive, nonsaturable process that is dependent on the pH at the outer surface of the plasma membrane.[38,56] The rapid proton influx caused by diclofop results in the alkalinization of the outer membrane surface,[48] which decreases the passive uptake of acetate by susceptible plant cells.[57] Inhibition of acetate uptake ranging from

30% to 70% has been reported in cell cultures, root tips, and coleoptiles of susceptible plants following treatment with APP herbicides.[52-54,57] However, in older leaf tissues, differences in acetate uptake between treated and control tissues were not detected.[58-60] The biophysical effect will invariably affect the results of *in vivo* assays of the biochemical mechanism based on [^{14}C]acetate incorporation into lipids.

The biophysical mechanism alone does not account for the phytotoxic action of APP herbicides.[3] The R(+) enantiomer is more phytotoxic than the S(−) enantiomer on whole plants.[3] However, the two enantiomers are almost equal in their depolarizing action.[3,45] Resistant dicots such as mung bean[57] and sunflower[61] are insensitive to membrane depolarization by APP herbicides, although pea has been reported to be sensitive.[62] It has been shown recently that diclofop inhibits acetate uptake in pea stem segments, but has no effect on the incorporation of absorbed acetate into lipid components.[63] Therefore, although the plasma membrane in pea is sensitive to high diclofop concentrations, the resistance exhibited at the whole plant level when pea is treated with normal doses of diclofop suggests that at these doses the effect on the membrane is not phytotoxic.

Less information is available on the effects of CHD herbicides on membrane function. Sethoxydim depolarizes Em of leaf cells from sensitive grasses[64,65] and coleoptiles of resistant and susceptible *L. rigidum* biotypes.[42] This effect on Em inhibits the H$^+$-symport of alanine into cells.[64] Sethoxydim does not affect H$^+$-ATPase activity, but it inhibits the plasma membrane-bound transmembrane redox system,[65] an alternative mechanism for the establishment of the transmembrane proton gradient.[66] However, the very high concentrations of sethoxydim required to affect the redox system (0.3 to 1 mM) raise some doubt as to the physiological significance of this result.

Summary of Modes of Action

Of the target sites discussed above, ACCase is better characterized as a herbicide target site, based on detailed *in vitro* assays and correlation between whole plant and enzyme sensitivity. The role of the membrane site is more speculative. The discussion on mechanisms of action of APP and CHD herbicides indicates that in intact tissues the biochemical and biophysical effects probably occur simultaneously; at least during the first 12 h following exposure to the herbicide.

Observations that both the uptake of acetate, which is affected by Em, and the incorporation of acetate into lipids, which is affected by ACCase activity, can be inhibited simultaneously in the presence of APP herbicides[12,54,58,70,71] have been used to argue that the biophysical and biochemical mechanisms operate simultaneously. The effect of diclofop on acetate incorporation in lipids is more pronounced in meristem-containing root tips than in entire roots,[13] indicating that caution must be exercised when comparing results from different sources.

Interpretation of data on the ACCase-based inhibition by APPs of acetate incorporation into fatty acids is complicated because, although inhibition of acetyl-CoA conversion to malonyl-CoA by ACCase might be expected to inhibit subsequent net fatty acid synthesis, it appears that some plants have an ability to synthesize fatty acids when ACCase activity is inhibited (at least in the short term). In contrast, plants cannot synthesize fatty acids in the absence of activity of the fatty acid synthetase (FAS) complex (Figure 3). In short-term whole tissue experiments with both APP-sensitive and -resistant species, it has been observed that while growth may be inhibited and there are changes in lipid unsaturation and phospholipid content, the conversion of acetate to fatty acids may be minimally affected.[27,67,68] In contrast, in oat coleoptiles treated with cerulenin, an inhibitor of FAS, lipid biosynthesis was reduced significantly but growth, Em, and lipid composition were unaffected.[17,57] Inhibition of ACCase-catalyzed *de novo* fatty acid synthesis may reduce long-term glycerolipid synthesis, affecting plasma membrane and chloroplast membrane biogenesis and maintenance. It has been recognized that composition changes in acyl lipids and polar lipids are not regulated by ACCase,[14,16] but can be induced by environmental and physiological stresses.[72,73] Therefore, an unresolved question is whether herbicide-induced stresses that cause the inhibition of growth result in compositional changes to cellular lipids, or whether the changes in lipids precede growth inhibition and cell death.

The antagonism of APP herbicide phytotoxicity by auxinic compounds[3,10,50] does not appear to involve any interaction with ACCase,[22,69] but may be associated with stimulated H^+-ATPase activity.[49] Such activity might affect the transmembrane distribution of herbicide.

There is evidence that, in whole plants and intact tissues, the rapid inhibition of growth and the effect on fatty acid synthesis are distinctly separate events.[3,27,57,67,68] An abbreviated scheme involving both mechanisms of action is proposed in Figure 6. The plasma membrane is the initial barrier or target site that diclofop and other APP herbicides encounter in their action. The lipophilic ester penetrates the plasma membrane (without interacting with the membrane target site) at a higher rate than the amphophilic free acid.[3,48] However, the ester is rapidly hydrolyzed to the active acid, resulting in a higher cytoplasmic concentration of the acid than in cells initially treated with diclofop.[48] The intracellular location of diclofop hydrolysis is unknown. Diclofop appears to interact more directly than the ester with diclofop specific sites located in the plasma membrane,[43] as indicated by the more rapid depolarization of Em[40] and alkalinization of external solution caused by the acid than its ester.[48] This rapid proton influx (mechanism unknown) lowers intracellular pH to cause other biochemical and physiological effects.[37,39] The collapse of the proton gradient and elimination of net acidification inhibits growth in meristematic tissues[46] and impairs active transport functions (see previous discussion).

The initial growth inhibition of meristematic tissues in susceptible plants may be due to the rapid effects of the biophysical mechanism which acts more rapidly than the biochemical mechanism in intact cells. However, the fact that

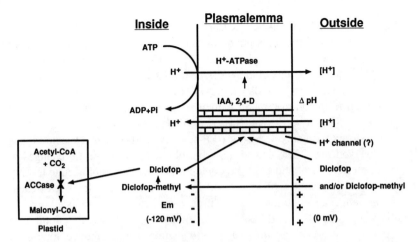

Figure 6. A diagrammatic presentation of the biophysical and biochemical mechanisms of action of aryloxyphenoxypropanoate herbicides. The plasma membrane (biophysical) and plastids (biochemical) are the two sensitive sites of action. The mechanism for the increased influx of protons caused by APP herbicides is unknown, and the transport of protons via an H+ channel regulated by the action of diclofop is speculative. Diclofop interacts more directly with the plasma membrane site than diclofop-methyl, which must be deesterified. The penetration of diclofop into the plastids will be influenced by cytoplasmic pH changes caused by the action of diclofop on the plasma membrane.

in some species (e.g., pea) the membrane appears to be sensitive,[63] but the plants (and their component ACCase) are resistant to these herbicides at the whole plant level, indicates that there is no simple correlation between membrane effects and phytotoxicity. The long-term effects resulting in the loss of membrane integrity (electrolyte leakage, chlorosis due to chloroplast membrane damage),[3] and necrosis of meristem cells[75] may be caused by the effects of the biochemical mechanism or a combination of both mechanisms. The biochemical and physiological interactions between the two mechanisms in the cause of plant death are not clearly understood. Until the proposed target site associated with the membrane has been identified and its herbicide sensitivity characterized in detail, no definitive conclusion can be reached on its role in herbicide phytotoxicity.

Mechanisms Endowing Selectivity in Resistant Crop Species

Two distinct mechanisms confer selectivity on grass and dicot crops that are resistant to APP and CHD herbicides. ACCase from dicot species is much less sensitive to *in vitro* inhibition by APP and CHD herbicides than is ACCase from grasses (Figure 4). Many dicot crops do exhibit a substantial capacity for APP detoxification[3] and may be insensitive to membrane depolarization. The role of ACCase in conferring selectivity between resistant and susceptible weed species is discussed in detail in a subsequent section of this chapter. Selectivity in different grass species is based upon at least one other mechanism. Grass crops such as wheat rapidly metabolize some APP herbicides to inactive products.[76] Since the ester forms of these herbicides are nonphytotoxic, the first step in the metabolism of APP herbicides in both R and S species is usually deesterification of the nonphytotoxic ester to the parent acid. Although this bioactivation reaction has not been studied extensively, there is evidence that it is catalyzed by a carboxylesterase.

The metabolism pathway for diclofop-methyl in wheat and *A. fatua* is shown in Figure 7. Following deesterification, possibly outside the plasma membrane, diclofop acid is subject to aryl hydroxylation;[76,77] recent evidence suggests that the hydroxylation is catalyzed by a cytochrome P450 monooxygenase (Cyt P450).[78] The hydroxylated product is usually found in only very small quantities, due to rapid glycosylation of the hydroxyl group.[77,79] Thus, the parent acid is depleted and the aryl glycoside accumulates rapidly in resistant grasses. The same process can occur in susceptible grasses, such as *A. fatua*; however, less of the aryl glycoside is formed, and more of the glucose ester of diclofop. The latter product can be hydrolyzed in the tissue to release the parent acid again, thus prolonging the presence of the herbicidally active compound.[76,77]

In wheat, the plasma membrane and plastid target sites are both sensitive to the action of diclofop. Nevertheless, inhibition of lipid biosynthesis and membrane depolarization are overcome by the rapid metabolism and detoxification of diclofop in intact tissues.[3] However, wheat is susceptible to haloxyfop, fluazifop, and some other APP herbicides because of its inability to rapidly detoxify these herbicides. Both target sites are sensitive in those grasses that are susceptible to diclofop-methyl and other APP herbicides, and detoxification of these compounds appears to be very limited in these species. Haloxyfop-methyl was rapidly converted to haloxyfop in soybean, *Setaria glauca*, and

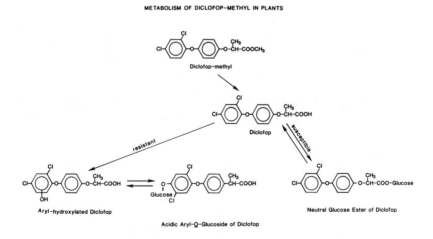

Figure 7. Metabolism of diclofop-methyl in plants. This scheme is typical for grasses such as wheat and *Avena fatua*, although the final proportions of the end products may vary between species.

Sorghum bicolor, with no apparent differences between species in intact plants.[80] However, in cell culture *S. glauca* converted the methyl ester to the free acid more extensively than did soybean. Similarly, pyridinyloxyphenoxypropionate derivatives are readily deesterified in *Elymus repens*, and then further conjugated to polar compounds.[4] Since the free acid could be regenerated upon hydrolysis of these polar metabolites, they may represent glucose esters of the type formed with diclofop in *A. fatua*. However, these compounds were not identified in this study. With both haloxyfop-methyl and fluazifop-butyl there is some evidence of formation of lipophilic metabolites.[80,81] Again, these products have not been identified, or their phytotoxicities determined.

In some cases the margin of selectivity with the herbicide alone is inadequate, and a safener is added to selectively enhance detoxification of the herbicide in the crop plant. For example, metabolism of fenoxaprop-ethyl in wheat is enhanced by the safener fenchlorazole-ethyl.[82]

Little information has been published on the metabolism of CHD herbicides in plants. There is little evidence to suggest that selectivity between R and S grasses is based on differential rates of herbicide metabolism between the two groups. In general, CHD herbicides such as sethoxydim are very unstable, and are subject to rapid degradation both in the plant and under UV radiation.[83]

EVOLUTION AND DEVELOPMENT OF RESISTANCE

Grass Weeds

APP herbicides were introduced in the late 1970s, and CHD herbicides several years later. Since then they have gained widespread popularity for control of grass weeds in both cereal and dicot crops. In particular, diclofop-methyl has become one of the most widely used postemergence herbicides for control of annual grass weeds in cereal crops. The practice of growing wheat in a monoculture, as is common in parts of Australia and Canada, has led to the continuous use of this herbicide by many farmers. More recently, other APP and CHD herbicides have been introduced that allow farmers to use these compounds in a wide variety of different crops (Table 1). Consequently, it is now common for farmers who practice diverse crop rotation to use APP or CHD herbicides every year.

Since the APP and CHD herbicides provide control of grass weeds only, evolution of resistance to these herbicides is restricted to grasses. The weed species in which resistance to APP and/or CHD herbicides has been documented are listed in Table 2. In most instances, resistance has developed after repeated use of herbicides from either or both chemical groups over several years. Examples of two herbicide use histories that led to the selection of resistant populations of *A. fatua* in Manitoba, Canada are shown in Table 3. Although it is difficult to determine accurately the number of resistance sites, it has been estimated recently that there are >3000 resistant *L. rigidum* sites in Australia and

Table 2. Weed species in which resistance to aryloxyphenoxy-propanoate and cyclohexanedione herbicides has developed

Species	Source	Herbicides	Ref.
Lolium multiflorum	Oregon, U.S.	Diclofop	84
Lolium rigidum	Australia	APP, CHD	85–87
Avena fatua	Canada, U.S.	APP, CHD	88–90
	Australia	Diclofop	91
Avena sterilis	Australia	APP, CHD	91
Setaria viridis	Canada	APP, CHD	92
Setaria faberii	Wisconsin, U.S.	APP, Sethoxydim	93
Digitaria sanguinalis	Wisconsin, U.S.		94
Sorghum halepense	Mississippi, U.S.	APP, CHD	95
Eleusine indica	Malaysia	APP, CHD	96

>100 resistant *Avena* spp. sites in both Canada and Australia. In these and most other populations it is likely that a low frequency of resistant individuals was present in the unselected populations, and that repeated selection for resistance (by application of APP or CHD herbicides) shifted the populations such that they are now composed predominantly of resistant plants. Since APP and CHD herbicides are becoming more widely used, it is very likely that additional weed species will be added to the list in Table 2 in the coming years.

It should be noted that patterns of resistance and cross resistance within a species are variable. For example, although most of the resistant *A. fatua* lines from Canada are resistant to both APP and CHD herbicides, the relative degree of resistance to different products can vary among lines. In some cases there is a high level of resistance to APP herbicides but no resistance to CHD herbicides.[89] Similar results have been found with *L. rigidum*,[97] and it is likely that different patterns of cross resistance will be found as more herbicide-resistant weed biotypes are characterized.

Although it is often difficult to determine herbicide use histories, it appears that selection for >4 years is sufficient to give rise to resistant populations. However, in situations where the herbicides may be applied more than once per year, this period may be shortened. Other agronomic practices, such as the use

Table 3. Examples of herbicide use histories that have led to the development of *Avena fatua* biotypes resistant to aryloxyphenoxy-propanoate and cyclohexanedione herbicides in western Canada

	Field 1		Field 2	
Year	Crop	Herbicide	Crop	Herbicide
1981	—	—	Barley	Diclofop-methyl
1982	Wheat	Diclofop-methyl	Wheat	—
1983	Barley	Diclofop-methyl	Wheat	Diclofop-methyl
1984	Wheat	Diclofop-methyl	Flax	Sethoxydim
1985	Barley	Diclofop-methyl	Wheat	Diclofop-methyl
1986	Wheat	Diclofop-methyl	Canola	Sethoxydim
1987	Flax	Sethoxydim	Wheat	—
1988	Barley	Diclofop-methyl	Flax	Sethoxydim
1989	Flax	Sethoxydim	Wheat	Diclofop-methyl
1990	Wheat	Fenoxaprop-ethyl	Wheat	Fenoxaprop-ethyl

of tillage or other mechanical weed control measures, may delay the development of resistance.[98]

One exception to the above is the discovery of low-level diclofop-methyl resistance in *A. fatua* in western Canada in the late 1970s, within 2 or 3 years of the introduction of this herbicide. The resistant biotypes were originally identified as being resistant to triallate,[99] a herbicide believed to act primarily on the fatty acid synthase complex.[100] Resistance to triallate presumably developed in response to many years of selection with this herbicide. However, subsequent testing revealed that the biotypes were cross resistant to diclofop-methyl.[99] Although no further research was reported on these biotypes, it is possible that triallate resistance was conferred by an enhanced detoxification pathway that was also capable of detoxifying diclofop-methyl.

Laboratory-Generated Mutants

The techniques of tissue culture and genetic transformation have introduced the possibility of obtaining herbicide-resistant cultivars in crops that are normally sensitive to those herbicides. The first strategy involves the selection of somaclonal mutants in cell cultures growing on media containing high (normally lethal) herbicide concentrations; resistance may be due to a modified gene encoding a form of the target protein with reduced herbicide senstivity or to enhanced gene copy number or gene amplification. The latter strategy involves the transfer of a gene for resistance from another source (usually plant or bacterial); resistance may be based on the same mechanisms as described for tissue culture.

Two different resistance mechanisms have been reported in maize tissue cultures selected for resistance to APP and CHD herbicides. In one case, cultures were selected that were 40-fold more resistant to sethoxydim and 20-fold more resistant to haloxyfop than susceptible cultures.[101] The I_{50} of ACCase from the resistant cultures was >100- and 16-fold greater than that of the susceptible culture for sethoxydim and haloxyfop, respectively. The mechanism of resistance in these cultures is clearly a mutant ACCase with reduced sensitivity to the herbicides.

Other maize cell lines have been reported that are resistant to APP and CHD herbicides due to overproduction of the normal, sensitive ACCase.[102] ACCase activity in one non-regenerable cell line was 2.6-fold higher than in the parent line. Whether this ACCase overproduction is due to gene amplification or increased gene copy number has not been determined. Since the gene(s) for ACCase has not been characterized or cloned, there has been no opportunity to date for genetic transformation using mutant ACCase genes.

MECHANISMS OF RESISTANCE IN WEEDS

APP and CHD herbicides must reach the sensitive meristematic sites in their toxic forms at sufficient concentrations to disrupt normal growth and development of sensitive cells. In addition, herbicidal injury must be maintained at a threshold level over a critical period, after which irreversible damage results and death of cells becomes inevitable. Three key factors that influence the above processes are herbicide absorption and translocation, herbicide metabolism and detoxification, and sensitivity of the target sites.

Absorption and Translocation

Absorption and translocation of diclofop-methyl and other APP herbicides do not appear to be significant factors in the selective action of the herbicides between resistant and susceptible species.[3,9,10,79] Esters of APP compounds are readily absorbed by leaves and roots of resistant and susceptible plants, but long-distance transport can be limited, particularly in the case of diclofop. Application of the herbicides to mature leaves may cause chlorosis but not death of plants, unless the apical and intercalary meristems are affected; root growth is also unaffected when the herbicides are applied to mature sections of roots and not to the root apex.[3,10] Auxinic compounds appeared to increase slightly the uptake and ester conjugation of diclofop-methyl, but translocation and detoxification are affected very little.[3] Therefore, none of these effects seem to account for the antagonism of diclofop-methyl by 2,4-D in whole plants.[3,10,50]

Metabolism and Detoxification

Metabolism and detoxification of diclofop-methyl (Figure 7) by aryl oxidation and conjugation is a significant mechanism for resistance in certain plants (e.g., wheat).[3,103] Fenoxaprop[104,105] and quizalofop[106] are metabolized and detoxified by resistant grass and dicot species by either cleavage to various phenolic derivatives or aryl oxidation[104] followed by conjugation to polar metabolites. The basis for the loss of biological activity by fenoxaprop, quizalofop, and other APP herbicides appears to be similar to that for diclofop-methyl.[3] The conjugation of the parent acids of APP herbicides to form polar ester conjugates may not be a true detoxification reaction. The herbicides must undergo ether cleavage and/or aryl oxidation followed by conjugation to polar glycosides to detoxify the parent molecules.

There is little evidence in support of enhanced herbicide detoxification as a mechanism of resistance to APP or CHD herbicides in weeds that have developed resistance. In studies on several different *Avena fatua* populations in Canada, no difference has been detected in the rate or pathway of diclofop-methyl and fenoxaprop-ethyl metabolism between the resistant and susceptible biotypes.[79,107] Deesterification of fenoxaprop-ethyl to the acid was slightly faster in susceptible than resistant *Setaria viridis*, but it was concluded that the difference was too small to contribute significantly to the observed whole plant resistance.[108] In this case a much larger difference in ACCase sensitivity between the resistant and susceptible populations was the primary mechanism of resistance (see the following section). A slight increase in the conversion of diclofop-methyl to polar metabolites has been reported in a resistant *L. rigidum* biotype from Australia.[41,117] However, the enhanced detoxification was not considered sufficient to confer the high level of resistance in this biotype. The evidence to date does not indicate that enhanced herbicide metabolism is a significant mechanism of resistance to APP and CHD herbicides in the weed biotypes in which resistance has evolved recently.

Target Site Sensitivity

Reduced target site sensitivity is frequently a significant factor in herbicide resistance. However, target site sensitivity *in vitro* may not be directly proportional to its sensitivity in intact plants (see the previous discussion). Where at least two mechanisms are functioning simultaneously, the differences in resistance or tolerance between species will depend on the degree of sensitivity of each target site within a species.

ACCase from susceptible grasses is sensitive to inhibition by APP and CHD herbicides, whereas the enzyme from resistant dicot species is unaffected (Figure 4).[11,12,18] Mutations in maize that resulted in variably resistant ACCase to haloxyfop and sethoxydim were correlated to their whole plant resistance.[101,109] In addition, several grass species are "naturally" resistant to ACCase inhibitors;[110-112] in some of these, resistance is correlated with reduced ACCase sensitivity.[110,111] However, herbicide metabolism and detoxification and plasma membrane sensitivity to herbicides have not been examined in these resistant grasses.

A more recent development is herbicide-resistant populations of weeds that were normally controlled by diclofop-methyl or other APP or CHD herbicides. Diclofop resistance in a biotype of *L. multiflorum* has been correlated to a diclofop-resistant ACCase.[113] Resistance is controlled by a single, nuclear, partially dominant gene. More information on the ability of this biotype to metabolize and detoxify diclofop, and the sensitivity of its plasma membrane site, should further clarify its resistance mechanism.

Recently research on the mechanism of APP and CHD herbicide resistance in *S. viridis*,[108] *L. rigidum*,[87] *A. sterilis*,[114] and *Sorghum halepense*[115] has revealed altered forms of ACCase that are much less sensitive to these herbicides. The inhibition of ACCase from resistant and susceptible *S. viridis* by diclofop and sethoxydim is shown in Figure 8. At present it is not known if the resistant weed biotypes all contain the same ACCase mutation; however, the variable patterns of resistance to different ACCase inhibitors at the whole plant level described previously suggest that different ACCase mutations may be involved. In some of these resistant biotypes (e.g., the *S. viridis* from Canada and some *L. rigidum* and *A. sterilis* biotypes from Australia) there are no substantial differences in herbicide uptake, translocation, or metabolism between the R and S biotypes, confirming that resistance is due solely to resistant ACCase forms.[87,108,114]

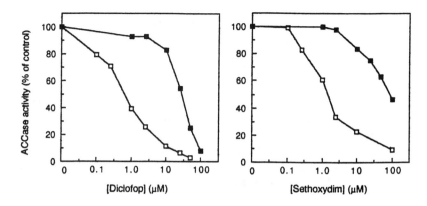

Figure 8. Inhibition of ACCase from a susceptible biotype of *Setaria viridis* and from a resistant biotype carrying a mutant form of ACCase, by diclofop and sethoxydim. Taken from Reference 108.

Other Mechanisms

Diclofop resistance in some biotypes of *L. rigidum* and *A. fatua* is not due to differences in ACCase sensitivity, membrane depolarization, or metabolism and detoxification of diclofop between R and S biotypes. In two reports, ACCase from the R and S biotypes of *L. rigidum*[116] and *A. fatua*[107] examined was equally sensitive to inhibition by diclofop. As described in the previous section, metabolism and detoxification of diclofop in these R and S biotypes of *L. rigidum*[41,117] and *A. fatua*[107] were similar. The R and S biotypes of both species were sensitive to membrane depolarization by diclofop.[41-43,107] However, only the R biotypes of *L. rigidum*[42,43] and *A. fatua*[107] reestablished the electrogenic membrane potential upon the removal of diclofop from the treatment solution. In *L. rigidum* coleoptiles, both R and S biotypes recovered from depolarization by the S(–) enantiomer of diclofop but only the R biotype reestablished the membrane potential following depolarization by the R(+) enantiomer.[118] The differential response to membrane depolarization is the only major difference yet observed between these R and S biotypes of *L. rigidum* and *A. fatua*. The mechanism for depolarization by diclofop and haloxyfop and the recovery of membrane potential is unknown, but neither compound seems to act as a mobile protonophore.[43] Diclofop and haloxyfop appear to interact with specific membrane sites to cause the increased permeability to protons. Therefore, resistance to APP herbicides in the R biotypes may be due to unknown structural or functional membrane changes not found in the S biotypes. However, although a difference has been detected at the membrane level in two different species, there is no clear understanding of how this difference confers resistance on the R biotypes.

Although it is clearly correlated with resistance, the differing ability to reestablish Em after removal of diclofop does not provide a mechanism of resistance. Since ACCase from the resistant *A. fatua* and *L. rigidum* lines are sensitive to APP and CHD herbicides in *in vitro* assays,[107,116] one might speculate that the herbicides are sequestered or compartmentalized in such a way as to prevent them from entering the plastids. Rapid removal of herbicide from the available pool by such a mechanism, for example by binding to the cell wall or accumulation in the vacuole, would effectively limit the herbicide concentration in the plastid, allowing lipid synthesis to proceed normally. In addition, the removal of herbicide in R biotypes may allow the membrane potential in these biotypes to recover to its pretreatment value, as is observed in the R biotypes.[42,107] Recent results indicate reduced herbicide uptake in isolated protoplasts and plasma membrane vesicles from a resistant *A. fatua* biotype compared to a susceptible biotype, providing the first evidence for an exclusion-based herbicide resistance mechanism in this species.[118a]

Summary of Mechanism of Resistance

The discussion of mechanisms of resistance to APP and CHD herbicides is complicated by the fact that the mechanism of action of these herbicide is not yet fully understood. ACCase is well characterized as a herbicide target site, although the link between ACCase inhibition and gross effects on lipid biosynthesis is not clear. ACCase from most grasses is sensitive to both groups of herbicides, whereas resistance in dicot species and in a small group of grasses correlates well with insensitivity of the native ACCase to these herbicides. In addition, in some APP- and CHD-resistant weeds a mutant ACCase with reduced herbicide sensitivity appears to be the primary mechanism of resistance.

The fact that resistance to APP and CHD herbicides is not correlated with reduced ACCase sensitivity in some weeds, or with any other processes often associated with herbicide resistance (e.g., uptake, translocation, metabolism), supports the idea of another mechanism of resistance, a ramification of which is the differential response of the plasma membrane Em in R and S biotypes (i.e., recovery in the resistant biotypes, but not in the susceptible biotypes). At present there is no satisfactory explanation for the finding that these biotypes have a sensitive ACCase, yet are not killed by the herbicides.

At present we can only speculate on the mechanism of resistance in these biotypes. The electrophysiological evidence suggests a subtle difference in membrane function between the R and S biotypes. The antagonism of APP herbicide activity by 2,4-D or other auxinic compounds appears to be associated, in an undetermined way, with effects on membrane function. In addition, APP herbicide activity is reduced under drought stress conditions, and it has been shown that drought stress or abscisic acid pretreatment alters the membrane response to diclofop, such that normally susceptible plants respond in a manner very similar to the resistant *A. fatua* or *L. rigidum* biotypes.[119] Thus, these various protection systems may share the same fundamental mechanism, mediated in different ways, perhaps. A change in the regulation of membrane function, whether constitutive or induced by auxin or abscisic acid, may alter the subcellular distribution of the herbicides such that they do not reach the chloroplast, thus allowing the plants to survive. Clearly, further research is required to determine the precise mechanism of resistance in these plants that may resolve the current controversy over the relevance of the membrane interaction of APP herbicides.

GENETICS AND BIOLOGY OF RESISTANT PLANTS

Genetics

Evidence is accumulating that resistance based on a resistant form of ACCase is carried by a single, partially dominant nuclear gene. This has been demonstrated in *L. multiflorum*,[120] *L. rigidum*,[121] and *A. sterilis*[122] and in some maize cell cultures.[101] It has been suggested recently that ACCase resistance in maize can be conferred by several (at least three) different allelic forms of *Acc*1, the ACCase structural gene.[109] Different mutations may confer different patterns of cross resistance among APP and CHD herbicides.

The second, unexplained, resistance mechanism may also be controlled by a single nuclear gene. Segregation ratios in crosses of resistant and susceptible *A. fatua* biotypes from Canada indicate that resistance is controlled by a single, semidominant nuclear gene.[123] In *Eleusine indica,* resistance to APP and CHD herbicides was not strongly heritable, and there was no indication of the mechanism of inheritance or the number of genes involved.[96]

In diclofop-resistant oat (*A. sativa*) lines, the inheritance of resistance was shown to conform most closely to a two-gene model.[124] It is believed that the genes controlling resistance are recessive in this case. However, the mechanism of resistance in cultivated oat, and the gene(s) conferring resistance, have not been identified.

Biology

Little information is available on the relative fitness of APP- or CHD-susceptible or -resistant weeds. In most instances, field and greenhouse observations of growing plants have not provided any indication of reduced vigor of the resistant plants, although there are some exceptions. Some of the earlier reported resistant *A. fatua* lines from Canada had smaller leaves and produced less total dry weight than susceptible plants.[88] However, such differences are not apparent in the more recently identified, highly resistant lines from Canada.[125] Similarly, no fitness differential has been observed between the R and S *L. rigidum* biotypes studied to date.[126] The resistant *E. indica* biotype is less vigorous than the wild type, and produces less biomass, indicating that it may be at a competitive disadvantage when grown with the wild type.[96] However, no detailed relative fitness comparisons have been reported for any APP- or CHD-resistant weed and its susceptible counterpart. Further research is necessary to determine the reason(s) for the low frequency of resistant plants in unselected populations.

CONCLUSIONS AND LIKELY FUTURE DEVELOPMENTS

Although APP and CHD herbicides have been introduced relatively recently, resistance has developed in several species, and can be expected in others in the near future. The increasing use of these herbicides in world agriculture will result in widespread selection pressure for resistant biotypes. Ironically, although resistance is now seen as an emerging problem that may limit the use of these products, new, potent ACCase inhibitors are still being developed and introduced to the market. This probably reflects the discovery effort in these areas of chemistry over the past 10 years. However, given the widespread increase in ACCase resistance in the recent past, it is likely that many of the companies involved in developing these herbicides are no longer pursuing ACCase inhibitors in their screening programs. Whereas ACCase inhibitors will continue as an important class of herbicides, there is likely to be a decline in the introduction of new products that inhibit ACCase in the coming years.

It is impossible to predict to what extent resistance to ACCase inhibitors will increase or spread. This will be determined by the occurrence and frequency of resistant biotypes, their relative fitness, and the weed control measures used. Given that there are now many ACCase inhibitors on the market, with differing crop selectivities, selection pressure for resistance is likely to be very high. Appropriate herbicide management will be an important determinant of the increase in resistance in the future.

As indicated earlier, much remains to be discovered about the mechanisms of action of these herbicides, and about the mechanisms of resistance. However, the availability of different resistant biotypes, with different mutations conferring resistance, provides a unique set of tools to study these questions. The regulatory role of ACCase in lipid biosynthesis, and the regulation of certain aspects of membrane function, may now be examined using these mutants. Ultimately, this will provide important information on these basic aspects of plant biology.

REFERENCES

1. Gerwick, B. C., L. A. Jackson, J. Handly, N. R. Gray, and J. W. Russell. "Preemergence and Postemergence Activities of the (R) and (S) Enantiomers of Haloxyfop," *Weed Sci.* 36:453-456 (1988).
2. Secor, J., and C. Cséke. "Inhibition of Acetyl-CoA Carboxylase Activity by Haloxyfop and Tralkoxydim," *Plant Physiol.* 86:10-12 (1988).
3. Shimabukuro, R. H. "Selectivity and Mode of Action of the Postemergence Herbicide Diclofop-Methyl," *Plant Growth Regul. Soc. Am. Q.* 18:37-54 (1990).
4. Hendley, P., J. W. Dicks, T. J. Monaco, S. M. Slyfield, O. J. Tummon, and J. C. Barrett. "Translocation and Metabolism of Pyridinyloxyphenoxypropionate Herbicides in Rhizomatous Quackgrass (*Agropyron repens*)," *Weed Sci.* 33:11-24 (1985).
5. Stoltenberg, D. E., and D. L. Wyse. "Regrowth of Quackgrass (*Agropyron repens*) Following Postemergence Applications of Haloxyfop and Sethoxydim," *Weed Sci.* 34:664-668 (1986).

6. Harker, K.N., and J. Dekker. "Effects of Phenology on Translocation Patterns of Several Herbicides in Quackgrass, *Agropyron repens*," *Weed Sci.* 36:463-472 (1988).
7. Ashton, F. M., and A. S. Crafts. *Mode of Action of Herbicides*, (New York: Wiley-Interscience Publications, 1981), p. 2.
8. Köcher, H. "Mode of Action of the Wild Oat Herbicide Diclofop-Methyl," *Canadian Plains Proceedings 12, Wild Oat Symposium: Proceedings* 2:63-77 (1984).
9. Duke, S. O., and W. H. Kenyon. "Polycyclic Alkanoic Acids" in *Herbicides: Chemistry, Degradation, and Mode of Action*, P. C. Kearney, and D. D. Kaufman, Eds. (New York: Marcel Dekker, Inc., 1988), pp. 71-116.
10. O'Sullivan, P. A. "Diclofop" in *Systems of Weed Control in Wheat in North America*, W. W. Donald, Ed. (Champaign, IL: Weed Science Society of America, 1990), pp. 321-345.
11. Hoppe, H. H. "Fatty Acid Biosynthesis — A Target Site of Herbicide Action" in *Target Sites of Herbicide Action*, P. Böger, and G. Sandmann, Eds. (Boca Raton, FL: CRC Press, Inc., 1989), pp. 65-83.
12. Walker, K. A., S. M. Ridley, T. Lewis, and J. L. Harwood. "Action of Aryloxyphenoxy Carboxylic Acids on Lipid Metabolism," *Rev. Weed Sci.* 4:71-84 (1989).
13. DiTomaso, J. M., A. E. Stowe, and P. H. Brown. "Inhibition of Lipid Synthesis by Diclofop-Methyl is Age Dependent in Roots of Oat and Corn," *Pestic. Biochem. Physiol.* 45:210-219 (1993).
14. Browse, J., and C. Somerville. "Glycerolipid Synthesis: Biochemistry and Regulation," *Annu. Rev. Plant Physiol. Plant Mol. Biol.* 42:267-506 (1991).
15. Taiz, L., and E. Zeiger. *Plant Physiology*, (Redwood City,CA: Benjamin/Cummings Pub. Co., 1991), 559 pp.
16. Harwood, J. "Fatty Acid Metabolism," *Annu. Rev Plant Physiol. Plant Mol. Biol.* 39:101-138 (1988).
17. Focke, M., A. Feld, and H. K. Lichtenthaler. "Inhibition of Early Steps of *de novo* Fatty-Acid Biosynthesis by Different Xenobiotica," *Physiol. Plant.* 81:251-255 (1991).
18. Gronwald, J. W. "Lipid Biosynthesis Inhibitors," *Weed Sci.* 39:435-449 (1991).
19. Burton, J. D., J. W. Gronwald, D. A. Somers, J. A. Connelly, B. G. Gengenbach, and D. L. Wyse. "Inhibition of Plant Acetyl-Coenzyme A Carboxylase by the Herbicides Sethoxydim and Haloxyfop," *Biochem. Biophys. Res. Commun.* 148:1039-1044 (1987).
20. Kobek, K., M. Focke, and H. K. Lichtenthaler. "Fatty-Acid Biosynthesis and Acetyl-CoA Carboxylase as a Target of Diclofop, Fenoxaprop and other Aryloxyphenoxypropionic Acid Herbicides," *Z. Naturforsch.* 43c:47-54 (1988).
21. Kobek, K., M. Focke, H. D. Lichtenthaler, G. Retzlaff, and B. Wurzer. "Inhibition of Fatty Acid Biosynthesis in Isolated Chloroplasts by Cycloxydim and other Cyclohexane-1,3-Diones," *Physiol. Plant.* 72:492-498 (1988).
22. Rendina, A. R., J. M. Felts, J. D. Beaudoin, A. C. Craig-Kennard, L. L. Look, S. L. Paraskos, and J. A. Hagenah. "Kinetic Characterization, Stereoselectivity, and Species Selectivity of the Inhibition of Plant Acetyl-CoA Carboxylase by the Aryloxyphenoxypropionic Acid Grass Herbicides," *Arch. Biochem. Biophys.* 265:219-225 (1988).
23. Rendina, A. R., and J. M. Felts. "Cyclohexanedione Herbicides are Selective and Potent Inhibitors of Acetyl-CoA Carboxylase from Grasses," *Plant Physiol.* 86:983-986 (1988).

24. Walker, K. A., S. M. Ridley, T. Lewis, and J. L. Harwood. "Fluazifop, a Grass-Selective Herbicide which Inhibits Acetyl-CoA Carboxylase in Sensitive Plant Species," *Biochem. J.* 254:307-310 (1988).
25. Burton, J. D., J. W. Gronwald, R. A. Keith, D. A. Somers, B. G. Gengenbach, and D. L. Wyse. "Kinetics of Inhibition of Acetyl-Coenzyme A Carboxylase by Sethoxydim and Haloxyfop." *Pestic. Biochem. Physiol.* 39:100-109 (1991).
26. Rendina, A. R., A. C. Craig-Kennard, J. D. Beaudoin, and M. K. Breen. "Inhibition of Acetyl-Coenzyme A Carboxylase by Two Classes of Grass-Selective Herbicides." *J. Agric. Food Chem.* 38:1282-1287 (1990).
27. Banaś, A., I. Johansson, G. Stenlid, and S. Stymne. "The Effect of Haloxyfop-Ethoxyethyl on Lipid Metabolism in Oat and Wheat Shoots," *Swedish J. Agric. Res.* 20:97-104 (1990).
28. Serrano, R. *Plasma Membrane ATPase of Plants and Fungi,* (Boca Raton, FL: CRC Press, Inc., 1985), 174 pp.
29. Sze, H. "H^+-Translocating ATPases of the Plasma Membrane and Tonoplast of Plant Cells," *Physiol. Plant.* 61:683-691 (1984).
30. Uribe, E. G., and U. Lüttge. "Solute Transport and the Life Functions of Plants," *Am. Sci.* 72:567-573 (1984).
31. Nicholls, D. G. *Bioenergetics: An Introduction to the Chemiosmotic Theory,* (New York: Academic Press, 1982), 190 pp.
32. Senn, A. P., and M. H. M. Goldsmith. "Regulation of Electrogenic Proton Pumping by Auxin and Fusicoccin as Related to the Growth of *Avena* Coleoptiles," *Plant Physiol.* 88:131-138 (1988).
33. Santoni, V., G. Vansuyt, and M. Rossignol. "The Changing Sensitivity to Auxin of the Plasma-membrane H^+-ATPase: Relationship Between Plant Development and ATPase Content of Membranes," *Planta* 185:227-232 (1991).
34. Reinhold, L., and A. Kaplan. "Membrane Transport of Sugars and Amino Acids," *Annu. Rev. Plant Physiol.* 35:45-83 (1984).
35. Tester, M. "Plant Ion Channels: Whole-cell and Single-channel Studies," *New Phytol.* 114:305-340 (1990).
36. Hedrich, R., and J. I. Schroeder. "The Physiology of Ion Channels and Electrogenic Pumps in Higher Plants," *Annu. Rev. Plant Physiol. Plant Mol. Biol.* 40:539-569 (1989).
37. Kurkdjian, A., and J. Guern. "Intracellular pH: Measurement and Importance in Cell Activity," *Annu. Rev. Plant Physiol. Plant Mol. Biol.* 40:271-303 (1989).
38. Grignon, C., and H. Sentenac. "pH and Ionic Conditions in the Apoplast," *Annu. Rev. Plant Physiol. Plant Mol. Biol.* 42:103-128 (1991).
39. Lucas, W. J., C. Wilson, and J. P. Wright. "Perturbation of *Chara* Plasmalemma Transport Function by 2[4(2',4'-dichlorophenoxy)phenoxy]propionic Acid," *Plant Physiol.* 74:61-66 (1984).
40. Wright, J. P., and R. H. Shimabukuro. "Effects of Diclofop and Diclofop-Methyl on the Membrane Potentials of Wheat and Oat Coleoptiles," *Plant Physiol.* 85:188-193 (1987).
41. Holtum, J. A. M., J. M. Matthews, R. E. Häusler, D. R. Liljegren, and S. B. Powles. "Cross-resistance to Herbicides in Annual Ryegrass (*Lolium rigidum*). III. On the Mechanism of Resistance to Diclofop-Methyl," *Plant Physiol.* 97:1026-1034 (1991).

42. Häusler, R. E., J. A. M. Holtum, and S. B. Powles. "Cross-resistance to Herbicides in Annual Ryegrass (*Lolium rigidum*). IV. Correlation Between Membrane Effects and Resistance to Graminicides," *Plant Physiol.* 97:1035-1043 (1991).
43. Shimabukuro, R. H., and B. L. Hoffer. "Effect of Diclofop on the Membrane Potentials of Herbicide-Resistant and -Susceptible Annual Ryegrass Root Tips," *Plant Physiol.* 98:1415-1422 (1992).
44. Ratterman, D. M., and N. E. Balke. "Herbicidal Disruption of Proton Gradient Development and Maintenance by Plasmalemma and Tonoplast Vesicles from Oat Root," *Pestic. Biochem. Physiol.* 31:221-236 (1988).
45. Ratterman, D. M., and N. E. Balke. "Diclofop-Methyl Increases the Proton Permeability of Isolated Oat Root Tonoplast," *Plant Physiol.* 91:756-765 (1989).
46. Cleland, R. E., G. Buckley, S. Nowbar, N. M. Lew, C. Stinemetz, M. L. Evans, and D. L. Rayle. "The pH Profile for Acid-Induced Elongation of Coleoptile and Epicotyl Sections is Consistent with the Acid-Growth Theory," *Planta* 186:70-74 (1991).
47. Renault, S., and M. D. Devine. Unpublished results (1992).
48. Shimabukuro, M. A., R. H. Shimabukuro, and W. C. Walsh. "The Antagonism of IAA-Induced Hydrogen Ion Extrusion and Coleoptile Growth by Diclofop-Methyl," *Physiol. Plant.* 56:444-452 (1982).
49. Shimabukuro, R. H., W. C. Walsh, and J. P. Wright. "Effect of Diclofop-Methyl and 2,4-D on Transmembrane Proton Gradient: A Mechanism for their Antagonistic Interaction," *Physiol. Plant.* 77:107-114 (1989).
50. Kafiz, B., J. P. Caussanel, R. Scalla, and P. Gaillardon. "Interaction Between Diclofop-methyl and 2,4-D in Wild Oat (*Avena fatua* L.) and Cultivated Oat (*Avena sativa* L.) and Fate of Diclofop-Methyl in Cultivated Oat," *Weed Res.* 29:299-305 (1989).
51. Peregoy, R. S., and S. Glenn. "Physiological Responses to Fluazifop-Butyl in Tissue of Corn (*Zea mays*) and Soybean (*Glycine max*)," *Weed Sci.* 33:443-446 (1985).
52. Hoppe, H. H. "Effect of Diclofop-Methyl on Protein, Nucleic Acid and Lipid Biosynthesis in Tips of Radicles from *Zea mays* L.," *Z. Pflanzenphysiol.* 102:189-197 (1981).
53. Cho, H.-Y., J. M. Widholm, and F. W. Slife. "Effects of Haloxyfop on Corn (*Zea mays*) and Soybean (*Glycine max*) Cell Suspension Cultures," *Weed Sci.* 34:496-501 (1986).
54. Hoppe, H. H. "Untersuchungen zum Wirkungsmechanismus von 'Diclofop-methyl'," *Z. Pflanzenkr. (Pflanzenpathol.) Pflanzenschutz Sonderh.* IX:187-195 (1981).
55. Ikai, T., K. Suzuki, K. Hattori, and H. Igarashi. "The Site of Action of Quizalofop-ethyl, NCI-96683," *British Crop Protection Conference — Weeds* (Farnham, U.K.: The British Crop Protection Council, 1985) pp. 163-169.
56. Sentenac, H., and C. Grignon. "Effect of H^+ Excretion on the Surface pH of Corn Root Cells Evaluated by Using Weak Acid Influx as a pH Probe," *Plant Physiol.* 84:1367-1372 (1987).
57. Shimabukuro, R. H., and B. L. Hoffer. "Perturbation of Transmembrane Proton Gradient and Inhibition of Fatty Acid Metabolism: Their Roles in the Mechanism of Action of Diclofop-Methyl," *Weed Science Society of America, Abstr.* No. 176 (1990).

58. Walker, K. A., S. M. Ridley, and J. L. Harwood. "Effects of the Selective Herbicide Fluazifop on Fatty Acid Synthesis in Pea (*Pisum sativum*) and Barley (*Hordeum vulgare*)," *Biochem. J.* 254:811-817 (1988).
59. Hoppe, H. H. "Differential Effect of Diclofop-Methyl on Fatty Acid Biosynthesis in Leaves of Sensitive and Tolerant Plant Species," *Pestic. Biochem. Physiol.* 23:297-308 (1985).
60. Boldt, L. D., and M. Barrett. "Effects of Diclofop and Haloxyfop on Lipid Synthesis in Corn (*Zea mays*) and Bean (*Phaseolus vulgaris*)," *Weed Sci.* 39:143-148 (1991).
61. Shimabukuro, R. H. Unpublished results (1988).
62. DiTomaso, J. M., P. H. Brown, A. E. Stowe, D. L. Linscott, and L. V. Kochian. "Effects of Diclofop and Diclofop-methyl on Membrane Potentials in Roots of Intact Oat, Maize, and Pea Seedlings," *Plant Physiol.* 95:1063-1069 (1991).
63. Devine, M. D., and R. Sasata. Unpublished results (1992).
64. Weber, A., E. Fischer, H. S. von Branitz, and U. Lüttge. "The Effects of the Herbicide Sethoxydim on Transport Processes in Sensitive and Tolerant Grass Species. I. Effects on the Electrical Membrane Potential and Alanine Uptake," *Z. Naturforsch.* 43c:249-256 (1988).
65. Weber, A., and U. Lüttge. "The Effects of the Herbicide Sethoxydim on Transport Processes in Sensitive and Tolerant Grass Species. II. Effects on Membrane-Bound Redox Systems in Plant Cells," *Z. Naturforsch.* 43c:257-263 (1988).
66. Crane, F. L. "Plasma Membrane Redox Reactions Involved in Signal Transduction" in *Second Messengers in Plant Growth and Development, Plant Biology,* Vol. 6, W. F. Boss, and D. J. Morre, Eds. (New York: Alan R. Liss, 1989), pp.115-144.
67. Banaś, A., I. Johansson, G. Stenlid, and S. Stymne. "Flavonoids, Pyridazinones and Salicylic Acid Counteract the Effects of Haloxyfop and Alloxydim," in *Proc. Int. Symp. Plant Lipids (9th), Plant Lipid Biochem., Struct. Util.*, Quinn, P. J., and Harwood, J. L., Eds. (London: Portland Press, 1990) pp. 407-409.
68. Nakahira, K., M. Uchiyama, T. Ikai, H. Igarashi, and K. Suzuki. "Effect of (R)-(+)- and (S)-(−)-Quizalofop-Ethyl on Lipid Metabolism in Excised Corn Stem-Base Meristems," *J. Pestic. Sci.* 13:269-276 (1988).
69. Aguero-Alvarado, R., A. P. Appleby, and D. J. Armstrong. "Antagonism of Haloxyfop Activity in Tall Fescue (*Festuca arundinacea*) by Dicamba and Bentazon," *Weed Sci.* 39:1-5 (1991).
70. Hoppe, H. H. "Einfluss von Diclofop-Methyl auf Wachstum und Entwicklung der Keimlinge von *Zea mays* L.," *Weed Res.* 20:371-376 (1980).
71. Hoppe, H. H., and H. Zacher. "Inhibition of Fatty Acid Biosynthesis in Tips of Radicles from *Zea mays* by Diclofop-Methyl," *Z. Pflanzenphysiol.* 106S:287-298 (1982).
72. Cooke, D. T., R. S. Burden, D. T. Clarkson, and C. S. James. "Xenobiotic Induced Changes in Membrane Lipid Composition: Effects on Plasma-Membrane AT-Pases," in *British Plant Growth Regulator Group, Monograph 18, Mechanism and Regulation of Transport Processes*, pp. 41-53 (1989).
73. Norberg, P., and C. Liljenberg. "Lipids of Plasma Membranes Prepared from Oat Root Cells: Effects of Induced Water-Deficit Tolerance," *Plant Physiol.* 96:1136-1141 (1991).

74. Hoppe, H. H., and H. Zacher. "Inhibition of Fatty Acid Biosynthesis in Isolated Bean and Maize Chloroplasts by Herbicidal Phenoxy-Phenoxypropionic Acid Derivatives and Structurally Related Compounds," *Pestic. Biochem. Physiol.* 24:298-305 (1985).
75. Donald, W. W., R. V. Parke, and R. H. Shimabukuro. "The Effects of Diclofop-Methyl on Root Growth of Wild Oat," *Physiol. Plant.* 54:467-474 (1982).
76. Shimabukuro, R. H., W. C. Walsh, and A. Jacobson. "Aryl-*O*-Glucoside of Diclofop: A Detoxification Product in Wheat Shoots and Wild Oat Cell Suspension Culture," *J. Agric. Food Chem.* 35:393-397 (1987).
77. Jacobson, A., and R. H. Shimabukuro. "Metabolism of Diclofop-Methyl in Root-Treated Wheat and Oat Seedlings," *J. Agric. Food Chem.* 32:742-746 (1984).
78. Zimmerlin, A., and F. Durst. "Xenobiotic Metabolism in Plants: Arylhydroxylation of Diclofop by a Cytochrome-P-450 Enzyme from Wheat," *Phytochemistry* 29:1729-1732 (1990).
79. Devine, M. D., S. A. MacIsaac, M. L. Romano, and J. C. Hall. "Investigation of the Mechanism of Diclofop Resistance in Two Biotypes of *Avena fatua*," *Pestic. Biochem. Physiol.* 42:88-96 (1992).
80. Buhler, D. D., B. A. Swisher, and O. C. Burnside. "Behavior of ^{14}C-Haloxyfop-Methyl in Intact Plants and Cell Cultures," *Weed Sci.* 33:291-299 (1985).
81. Coupland, D. "The Influence of Environmental Factors on the Metabolic Fate of ^{14}C-Fluazifop-Butyl in *Elymus repens*," *Proceedings British Crop Protection Conference — Weeds*, (Farnham, U.K.: The British Crop Protection Council, 1985) pp. 317-324.
82. Yaacoby, T., J. C. Hall, and G. R. Stephenson. "Influence of Fenchlorazole-Ethyl on the Metabolism of Fenoxaprop-Ethyl in Wheat, Barley, and Crabgrass," *Pestic. Biochem. Physiol.* 41:296-304 (1991).
83. Campbell, J. R., and D. Penner. "Sethoxydim Metabolism in Monocotyledonous and Dicotyledonous Plants," *Weed Sci.* 33:771-773 (1985).
84. Stanger, C. E., and A. P. Appleby. "Italian Ryegrass (*Lolium multiflorum*) Accessions Tolerant to Diclofop," *Weed Sci.* 37:350-352 (1989).
85. Heap, J., and R. Knight. "A Population of Ryegrass Tolerant to the Herbicide Diclofop-Methyl," *J. Aust. Inst. Agric. Sci.* 48:156-157 (1982).
86. Heap, I., and R. Knight. "The Occurrence of Herbicide Cross-Resistance in a Population of Annual Ryegrass, *Lolium rigidum*, Resistant to Diclofop-methyl," *Aust. J. Agric. Res.* 37:149-156 (1986).
87. Tardif, F. J., J. A. M. Holtum, and S. B. Powles. "Occurrence of a Herbicide-Resistant Acetyl-Coenzyme A Carboxylase Mutant in Annual Ryegrass (*Lolium rigidum*) Selected by Sethoxydim," *Planta* 190:176-181 (1993).
88. Joseph, O. O., S. L. A. Hobbs, and S. Jana. "Diclofop Resistance in Wild Oat (*Avena fatua*)," *Weed Sci.* 38:475-479 (1990).
89. Heap, I. M., B. G. Murray, H. A. Loeppky, and I. N. Morrison. "Resistance to Aryloxyphenoxypropionate and Cyclohexanedione Herbicides in Wild Oat (*Avena fatua*)," *Weed Sci.* 41:232-238 (1993).
90. Seefeldt, S. S., and D. R. Gealy. "Investigations of Diclofop-Methyl Resistant Wild Oat (*Avena fatua* L.) Biotypes from Oregon," *Weed Science Society of America, Abstr.* No. 266 (1992).
91. Mansooji, A. M., J. A. M. Holtum, P. Boutsalis, J. M. Matthews, and S. B. Powles. "Resistance to Aryloxyphenoxypropionate Herbicides in Two Wild Oat Species (*Avena fatua* and *Avena sterilis* ssp. *ludoviciana*)," *Weed Sci.* 40:599-605 (1992).

92. Heap, I. M., and I. N. Morrison, "Resistance to Aryloxyphenoxypropionate and Cyclohexanedione Herbicides in Green Foxtail (*Setaria viridis* (L.) Beauv.)." *Weed Science Society of America, Abstr.* No. 185 (1993).
93. Stoltenberg, D. E., and R. J. Wiederholt, "Giant Foxtail (*Setaria faberi* Herrm.) Resistance to Acetyl-CoA Carboxylase Inhibitors," *Weed Science Society of America, Abstr.* No. 183 (1993).
94. Stoltenberg, D. E. Personal communication (1992).
95. Smeda, R. J., W. L. Barrentine, and C. E. Snipes. "Johnsongrass (*Sorghum halepense* (L.) Pers.) Resistance to Postemergence Grass Herbicides," *Weed Science Society of America, Abstr.* No. 53 (1993).
96. Marshall, G., R. C. Kirkwood, and G. E. Leach. "Comparative Studies on Graminicide-Resistant and -Susceptible Biotypes of *Eleusine indica*," *Weed Res.* (in press, 1994).
97. Heap, I. M., and R. Knight. "Variation in Cross-Resistance among Populations of Annual Ryegrass (*Lolium rigidum*) Resistant to Diclofop-Methyl," *Aust. J. Agric. Res.* 41:121-128 (1990).
98. Matthews, J. "Management of Resistant Weed Populations" in *Herbicide Resistance in Plants: Biology and Biochemistry,* S. B. Powles and J. A. M. Holtum, Eds. (Lewis Publishers, Chelsea, MI, 1994), pp. 317-335.
99. Thai, K. M., S. Jana, and J. M. Naylor. "Variability for Response to Herbicides in Wild Oat (*Avena fatua*) Populations," *Weed Sci.* 33:829-835 (1985).
100. Devine, M. D., S. O. Duke, and C. Fedtke. *Physiology of Herbicide Action*, (Englewood Cliffs, New Jersey: PTR Prentice-Hall Inc., 1993), pp. 441.
101. Parker, W. B., L. C. Marshall, J. D. Burton, D. A. Somers, D. L. Wyse, J. W. Gronwald, and B. G. Gengenbach. "Dominant Mutations Causing Alterations in Acetyl-Coenzyme A Carboxylase Confer Tolerance to Cyclohexanedione and Aryloxyphenoxypropionate Herbicides in Maize," *Proc. Natl. Acad. Sci. U.S.A.* 87:7175-7179 (1990).
102. Parker, W. B., D. A. Somers, D. L. Wyse, R. A. Keith, J. D. Burton, J. W. Gronwald, and B. G. Gengenbach. "Selection and Characterization of Sethoxydim-Tolerant Maize Tissue Cultures," *Plant Physiol.* 92:1220-1225 (1990).
103. Tanaka, F. S., B. L. Hoffer, R. H. Shimabukuro, R. G. Wien, and W. C. Walsh. "Identification of the Isomeric Hydroxylated Metabolites of Methyl 2-[4-(2,4-Dichlorophenoxy)Phenoxy]-Propanoate (Diclofop-Methyl) in Wheat," *J. Agric. Food Chem.* 38:559-565 (1990).
104. Wink, O., E. Dorn, and K. Beyermann. "Metabolism of the Herbicide Hoe 33171 in Soybeans," *J. Agric. Food Chem.* 32:187-192 (1984).
105. Lefsrud, C., and J. C. Hall. "Basis for Sensitivity Differences Among Crabgrass, Oat, and Wheat to Fenoxaprop-Ethyl," *Pestic. Biochem. Physiol.* 34:218-227 (1989).
106. Koeppe, M. K., J. J. Anderson, and L. M. Shalaby. "Metabolism of [^{14}C]Quizalofop-Ethyl in Soybean and Cotton Plants," *J. Agric. Food Chem.* 38:1085-1091 (1990).
107. Devine, M. D., J. C. Hall, M. L. Romano, M. A. S. Marles, L. W. Thomson, and R. H. Shimabukuro. "Diclofop and Fenoxaprop Resistance in Wild Oat is Associated with an Altered Effect on the Plasma Membrane Electrogenic Potential," *Pestic. Biochem. Physiol.* 45:167-177 (1993).
108. Marles, M. A. S., M. D. Devine, and J. C. Hall. "Herbicide Resistance in *Setaria viridis* Conferred by a Less Sensitive Form of Acetyl Coenzyme A Carboxylase," *Pestic. Biochem. Physiol.* 46:7-14 (1993).

109. Marshall, L. C., D. A. Somers, P. D. Dotray, B. G. Gengenbach, D. L. Wyse, and J. W. Gronwald. "Allelic Mutations in Acetyl-Coenzyme A Carboxylase Confer Herbicide Tolerance in Maize," *Theor. Appl. Genet.* 83:435-442 (1992).
110. Stoltenberg, D. E., J. W. Gronwald, D. L. Wyse, J. D. Burton, D. A. Somers, and B. G. Gengenbach. "Effect of Sethoxydim and Haloxyfop on Acetyl-Coenzyme A Carboxylase Activity in *Festuca* Species," *Weed Sci.* 37:512-516 (1989).
111. Catanzaro, C. J., J. D. Burton, and W. A. Skroch. "Graminicide Resistance of Acetyl-CoA Carboxylase from Ornamental Grasses," *Pestic. Biochem. Physiol.* 45:147-153 (1993).
112. Hubbard, J., and T. Whitwell. "Ornamental Grass Tolerance to Postemergence Grass Herbicides," *HortScience* 26:1507-1509 (1991).
113. Gronwald, J. W., C. V. Eberlein, K. J. Betts, K. M. Roscow, N. J. Ehlke, and D. L. Wyse. "Diclofop Resistance in a Biotype of Italian Ryegrass," (Abstract No. 685), *Plant Physiol.* 89S:115 (1989).
114. Maneechote, C., J. A. M. Holtum, and S. B. Powles. "Resistant Acetyl CoA Carboxylase is the Mechanism of Herbicide Resistance in a Biotype of *Avena sterilis* spp. *ludoviciana*," *Plant Cell Physiol.* (Submitted for publication).
115. Marles, M. A. S., and M. D. Devine. Unpublished results (1992).
116. Matthews, J. M., J. A. M. Holtum, D. R. Liljegren, B. Furness, and S. B. Powles. "Cross-Resistance to Herbicides in Annual Ryegrass (*Lolium rigidum*). I. Properties of the Herbicide Target Enzymes Acetyl-Coenzyme A Carboxylase and Acetolactate Synthase," *Plant Physiol.* 94:1180-1186 (1990).
117. Shimabukuro, R. H., and B. L. Hoffer. "Metabolism of Diclofop-Methyl in Susceptible and Resistant Biotypes of *Lolium rigidum*," *Pestic. Biochem. Physiol.* 39:251-260 (1991).
118. Holtum, J. A. M., R. E. Häusler, M. D. Devine, and S. B. Powles. "The Recovery of Transmembrane Potentials in Plants Resistant to Aryloxyphenoxypropionate Herbicides: A Phenomenon Awaiting Explanation," *Weed Sci.* in press.
118a. Devine, M. D., S. Renault, and X. Wang. "Alternative Mechanisms of Resistance to Acetyl-CoA Carboxylase Inhibitors in Grass Weeds," *Proceedings British Crop Protection Conference — Weeds,* (Farnham, U.K.: The British Crop Protection Council, 1993) pp. 541-548.
119. Downey, J. A., and M. D. Devine. Unpublished results (1992).
120. Betts, K. J., N. J. Ehlke, D. L. Wyse, J. W. Gronwald, and D. A. Somers. "Mechanism of Inheritance of Diclofop Resistance in Italian Ryegrass (*Lolium multiflorum*)," *Weed Sci.* 40:184-189 (1992).
121. Tardif, F. J., and S. B. Powles. Unpublished results (1992).
122. Barr, A. R., A. M. Mansooji, J. A. M. Holtum, and S. B. Powles. "The Inheritance of Herbicide Resistance in *Avena sterilis* ssp. *ludoviciana*, Biotype SAS 1," *Proceedings of the First International Weed Control Congress,* (Melbourne: Weed Science Society of Victoria, 1992) pp. 70-75.
123. Morrison, I. N., I. M. Heap, and B. Murray. "Herbicide Resistance in Wild Oat — The Canadian Perspective," in *Proceedings of the Fourth International Oat Conference,* Adelaide, Australia (1992), pp. 36-40.
124. Warkentin, T. D., G. Marshall, R. I. H. McKenzie, and I. N. Morrison. "Diclofop-Methyl Tolerance in Cultivated Oats (*Avena sativa* L.)," *Weed Res.* 28:27-35 (1988).
125. Heap, I. M., and I. N. Morrison. Personal communication (1992).
126. Matthews, J. M., and S. B. Powles. Unpublished results (1992).

CHAPTER 6

Resistance to the Auxin Analog Herbicides

David Coupland

REVIEW OF THE AUXIN ANALOG HERBICIDES

Chemical Structure

The auxin analog herbicides were the first selective organic herbicides to be developed. MCPA and 2,4-D were discovered independently by English and American workers during the 1940s.[1-3] Auxin analog herbicides mimic the action of the natural plant growth substance IAA (indol-3-yl acetic acid) often more simply referred to as auxin. However, herbicides in this group are of diverse structure, belonging to several quite distinct chemical classes (Figure 1). By no means all auxin analog herbicides are shown in Figure 1 and numerous reviews provide further information on their chemistry.[4,5]

Modes of Action

Structure-Activity Relationships

In the 50 years since their discovery, there have been a succession of proposals to explain the structure-activity relationships of auxin analog herbicides. All of these theories are based on the premise that the herbicide molecule has to bind to a receptor (see below) in order to facilitate a biochemical response. However, the nature of this binding differs according to the different theories (Table 1). There have been several reviews of these structure-activity proposals.[6-9]

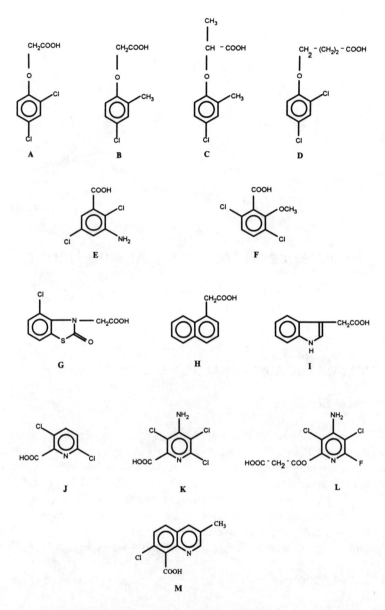

Figure 1. Chemical structures of IAA and some commonly used auxin analog herbicides. A. 2,4-D; B. MCPA; C. Mecoprop; D. 2,4-DB; E. Chloramben; F. Dicamba; G. Benazolin; H. 1-naphthylacetic acid; I. IAA; J. 3,6-dichloropicolinic acid; K. Picloram; L. Fluroxypyr; M. Quinclorac.

Physiological activity — Following uptake,[16] the physiological activity of these compounds depends largely on how closely their structures resemble those of endogenous auxin, and their persistence in the plant. Auxin mimicry

Table 1. Proposed auxin-receptor models

Theory	Description	Comments	Ref.
Charge separation Porter & Thimann (1965)	0.55 nm distance between a negative charge on the carboxyl group and a slight positive charge on a nitrogen atom or aryl group	Accommodated most of the auxin herbicides known at the time but unable to account for the activity of chiral pairs.[9] Lack of evidence for correct charge density on aryl ring[10]	11
Three-point attachment Wain & Fawcett (1969)	Three-point contact between the auxin molecule (carboxyl group, hydrogen atom and rest of the molecule) and a receptor	Explains enantiomeric activity but otherwise weak	12
Conformational change of receptor Kaethner (1977)	Binding only takes place if the ligand has the correct "recognition conformation"; after binding, there is a conformational change which then "modulates" further biochemical effects	Does not accommodate all compounds with auxin activity and poor "fit" for IAA	10, 13
Complementary binding site Katekar (1979)	Binding depends upon the correct charge localization and steric characteristics for the entire molecules	Accommodates most auxin herbicides	14, 15

and herbicide metabolism are considered specifically in more detail below. Physiological effects can result from two quite separate reactions of the plant to the applied xenobiotic. First, there are the growth-promoting effects of the molecule, dependent upon tissue type and cellular concentration. Auxin, and compounds having auxin-like activity, are known to stimulate cell growth; indeed this is used as an important criterion for characterizing such compounds.[8] This effect is usually considered to be due to the stimulation of plasma membrane-located ATPases (see below). A second action is the stimulation of ethylene production and auxins have long been known to enhance ethylene biosynthesis,[17,18] which is also a natural plant growth substance.[19] Separating the effects of these two individual processes is not always possible, but there have been cases where this has been done.[20,21] Table 2 summarizes the main physiological effects produced by auxin analog herbicides.

Biochemical activity and phytotoxicity — There is an enormous catalog of different effects on cell biochemistry produced by auxin analog herbicides, the list including virtually every kind of cellular metabolite.[4,29] The problem is that this literature describes "correlative"-type experiments in which herbicide is applied and some effect is measured. Hence many "auxin effects" have been documented. Brummell and Hall[30] have reviewed this extensive literature and a summary of their findings is given in Table 3.

Table 2. Summary of the physiological effects produced by compounds with auxin-like activity

Effect	Comments	Ref.
Ethylene stimulation	Common to all auxin analog herbicides	22
Chloroplast damage	Precedes chlorotic damage to leaves	23
Stomatal aperture closure	Precedes effects on transpiration and photosynthesis rate	24
Production of epinasty	Common to all auxin analogs; may be mediated by ethylene	25,26
Stem tissue proliferation	May result in tissue damage and blockage of vascular tissues	22, 23, 27
Root initiation	Both adventitious and "normal" roots can be produced	25
Changes in cell wall plasticity	Often regarded as an early event in the phytotoxic mode of action; primarily responsible for cell elongation	5, 28

Table 3. Biochemical responses to auxin

Response	Time (min)	Ref.
Respiration rate increased	5	31
Decrease in cytosolic pH	5	32
Increase in mRNA sequences	5	33
Increase in H^+ extrusion	7–8	34
Increase in plasma membrane potential	12	35
Increase in β-glucan synthase activity	10–15	36
Increase in K^+ uptake	30	37
Increase in protein synthesis	30	38

Note: these responses are from whole tissues, not *in vitro* assays. After Reference 16.

It is often difficult to separate the effects of auxin analog herbicides from those of IAA itself, in particular during the early stages when the concentration of herbicide within the cell is likely to be low, perhaps equivalent to that of IAA and its metabolites. However, as the concentration of xenobiotic increases in the plant, herbicidal effects will ensue which ultimately lead to plant death. This herbicidal effect has been described as an auxin-overdose,[5] that is, a continual over-stimulation of the plant's metabolism. Various workers have considered this phytotoxicity to occur in distinct stages: first, a stimulation of metabolic processes including photosynthesis, respiration, and nucleic acid biosynthesis; second, stem elongation accompanied by abnormal effects such as epinasty and tissue swelling; and finally, cell membrane damage and tissue collapse.[22] However, not all of these processes are necessarily affected, nor in this sequence, and it should be emphasized that plants vary in response according to species and age, and different types of auxin analogs have different effects.[23]

Auxin mimicry — Even though none of the receptor models can fully explain the growth-regulating activities of all of the auxin analog herbicides, it is clear that they mimic the activity of auxin because these compounds invoke responses so characteristic of IAA. Perhaps the most important features of the "mode of action" of IAA are its initial binding to a receptor and the cascade of biochemical events that follow.

Auxin binding: "signal reception" — Using a variety of techniques, investigators have identified several auxin-binding proteins (ABPs) that may have receptor functions (Table 4). Auxin-binding proteins have been located in the plasma membrane and within the cell, primarily associated with the endoplasmic reticulum (ER),[46] but also may be localized in the nucleus.[47] The bulk of the ABP is usually ER-located[48] and, for maize, several ABP isoforms have been found.[49] This bilocation of ABPs is somewhat paradoxical. There is considerable evidence supporting the intracellular location of ABPs. Auxin is a weak acid so will preferentially be located within the cytoplasm due to "acid-trapping". In addition, most plant cells have a proton/auxin symport uptake mechanism, therefore auxin is readily accessible to plant cells and will be located within the cell. Auxin is usually translocated symplastically.[50] Of the ABPs that have been sequenced to date, all show characteristics typical of ER polypeptides such as: a C-terminus signal peptide of 38 residues that contains sufficient hydrophobic amino acid residues to span a lipid bilayer; a C-terminus KDEL sequence which "targets" the protein for the ER lumen; and a single N-glycosylation site of the high mannose type, again characteristic of ER proteins.[51] Hence, it is not clear why ABPs occur on the outside of the plasma membrane. Conversely, all the immunological data[52,53] indicate ABPs are on the outside of the plasmalemma. In addition, impermeant auxin analogs have auxin activity,[54] and all of the physiological data are consistent with the existence of plasma membrane-located ABPs.

The crucial aspect of this issue is whether the ABPs have a true receptor function. While conclusive evidence is lacking, there is considerable indirect

Table 4. Auxin-binding proteins in plants

Plant tissue	Molecular mass (kDa)	Method of identification	Ref.
Maize shoots	22,24,43	Photoaffinity labeling	39
Maize shoots	22	Ion exchange chromatography	40
Maize shoots	21	Affinity chromatography	41
Maize shoots	20	Immunoaffinity chromatography	42
Zucchini hypocotyl	40*	Photoaffinity labeling	43
Tomato roots	40*	Photoaffinity labeling	44
Mung beans (seedling tissues)	15,47	Affinity chromatography	45
Several species including monocots and dicots	65	Anti-idiotypic antibodies	47

*Doublet in SDS-PAGE

evidence that these proteins do have receptor function. First, antibodies raised against ABPs specifically block auxin-induced responses, such as plasma membrane hyperpolarization.[46] Furthermore, antibodies raised against a synthetic peptide, mimicking the auxin-binding site, not only recognize all isoforms of the maize ABP but these antibodies hyperpolarize protoplast plasma membranes in an auxin-like manner.[55] Secondly, plants exhibiting the "hairy root" phenotype (after transformation by *Agrobacterium rhizogenes*),[56] exhibit increased auxin sensitivity correlated directly with an increased titer of anti-ABP IgG.[51] Lastly, adding purified maize ABP to "wild-type" protoplasts resulted in an enhanced sensitivity to auxin.[57] Thus, there is good evidence for ABPs at the plasmalemma and associated with internal membranes. Could there be a common explanation for this bilocation? One possibility is that ABPs are synthesized in the ER but then are secreted from the ER lumen, perhaps as a result of perturbed Ca^{2+} levels (see below), and subsequently migrate to the outer face of the plasma membrane.[46] Another possibility is that two types of ABP are needed, one type at the plasma membrane, ultimately regulating H^+ pumping, the other type sensing changes in auxin level within the cell and affecting changes in gene transcription.

Biochemical changes induced by binding: the signal transduction pathway — Irrespective of where the receptor is located, after auxin has become bound, the reception signal has to be transmitted and amplified to effect the multitude of biochemical events which precede growth stimulation (see Tables 2 and 3). Signal amplification is thought to be necessary due to the relatively small number of available receptors.[30] Calcium (as Ca^{2+}), along with other "second messengers", in particular phospholipids, are generally considered to play a key role in this signal transduction pathway.[57-60] However, how the functional integration of auxin action at the cell surface and at the ER is achieved has still to be resolved. Key issues are undoubtedly the activation and regulation of the enzymes involved in phospholipid metabolism.[59,60] Although the research on transmembrane signal transduction in plants is considered to be still in "its infancy",[59] several models have been published. The integration of the various components of plasma membrane, ER and tonoplast, involved in signal reception and transduction, is a complex topic and beyond the scope of this review. However, a summary of some of the main biochemical reactions involved is presented in Figure 2.

The control of gene expression by auxin — Studies into auxin regulation of gene expression have utilized mutants and transformed plants in which auxin metabolism is specifically affected resulting in changes to protein, RNA, and DNA metabolism. It has been recognized for nearly 40 years that auxins affect nucleic acid synthesis[62] and this fact alone formed the basis of a "gene activation" hypothesis of auxin action in the 1960s.[63] Changes in the protein content of plants treated with auxins have been demonstrated both *in vivo* and *in vitro*.[64]

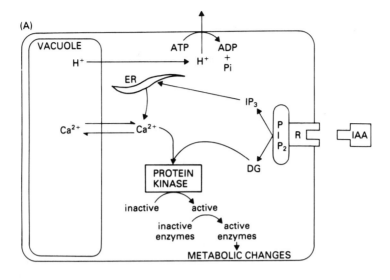

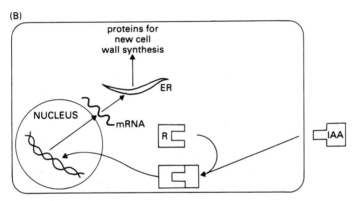

Figure 2. Biochemical interactions of auxin with a plasma membrane (A) or cytoplasmic (B) receptor. Key: PIP_2 (phosphatidyl inositol 4,5-bisphosphate); ER (endoplasmic reticulum); R (receptor); IP_3 (inositol 1,4,5-triphosphate); DG (diacylglycerol). After [61] (with permission).

In vitro studies have demonstrated rapid effects following auxin treatment, within 15 min.[65] How these effects are achieved is not known in detail, but both transcription and translation are affected. Specific auxin-regulated mRNAs have been identified using soybean tissues as the source of poly A^+-mRNA which has been used for the construction of the cDNA libraries. Table 5 shows a summary of the results from four such studies.

Increases in the specific mRNA sequences that are upregulated are only found after treatment with naturally occurring auxin (IAA) or synthetic auxins (naphthalene acetic acid [NAA], 2,4-D, and 2,4,5-T). No such induction is observed with non-auxin analogs, other endogenous plant growth substances,

Table 5. Effects of auxins on mRNA regulation

Source of poly A+-mRNA	Herbicide used	Size of mRNA showing largest change after auxin treatment[a]	Fold increase or decrease in regulation	Earliest detectable change	Ref.
Intact soybean hypocotyls	2,4-D	1000	66 (down)	4 h	66
Excised soybean hypocotyls	2,4-D	1000	8 (up)	15 min	67
Intact soybean hypocotyls	2,4-D	2400	30 (up)	30 min	68
Excised pea hypocotyls	IAA	850 and 950	50 (up)	10–15 min	69

[a] number of nucleotides.
After Reference 70.

or environmental stresses.[64] The fact that cycloheximide, emetine, and anisomycin (three protein synthesis inhibitors) are effective in inducing many of the mRNAs normally associated with auxin treatment[69] has led to the suggestion that auxin induction of RNA is regulated in vivo by a protein with a short half-life.[64] However, Theologis[70] suggests that the accumulation of mRNAs that are insensitive to protein inhibitors constitutes a primary hormonal response, whereas the induction of RNAs, which is dependent on protein synthesis, is a secondary response. Whatever the precise mechanism, clearly auxins can rapidly and specifically regulate mRNA biosynthesis. Increases in specific mRNA sequences could result from increased transcription rates, or posttranscriptional events, such as increased transport of mRNA from the nucleus to cytoplasm. Transcription rates for all auxin-regulated clones increased from 10- to 100-fold (depending on the clone analyzed) after auxin treatment. Responses were rapid, one clone showing increases in transcription rate after 5 min.[70] Interestingly, increases in transcription rates appeared approximately linear over a wide dose range for 2,4-D, from 10^{-8} to 10^{-3} M.[64,68] Furthermore, similar transcriptional responses occurred using nuclei prepared from different species, and from different tissues. All of the above effects were specific only for auxins.[64,72] These are convincing data that auxin and auxin analog herbicides have a direct (i.e., not dependent on protein synthesis), selective and rapid effect on nucleic acid biosynthesis and that the primary site of action is at the level of transcription.

Usage, Selectivity, and Metabolism

Agricultural Use

The auxin analog herbicides 2,4-D and MCPA were the first truly selective herbicides and their discovery was revolutionary for agriculture. Their unique ability to control dicot weeds in cereal crops came at a crucial time between the two World Wars when maximum food production was desperately needed. Since then, hundreds of analogs have been synthesized and many active molecules have been developed into commercially important herbicides.[73,74]

Selectivity and Metabolism

Selectivity should not be regarded in simplistic terms. For example, dosage will obviously influence phytotoxicity and it is easy to demonstrate herbicidal effects on so-called tolerant species just by increasing the amounts applied. Plant age is another factor; cucumbers are highly sensitive to 2,4-D as seedlings but are less so when mature.[75] In addition, there are the often overriding influences of weather factors, plant competition, and varietal differences within species.[5] Mechanisms of selectivity fall into two basic categories: first, differences in the amounts of active ingredient reaching the sites of action; and second, differences in sensitivity at the sites of action. The first of these two mechanisms involves herbicide uptake, translocation, metabolism, and sequestration en route to the sites of action.

Uptake — Even though different plant species vary considerably with respect to epidermis morphology, cuticle thickness, waxiness, etc., there is little evidence that differential uptake is primarily responsible for selectivity.[5,71,75]

Translocation — Pillmoor and Gaunt have reviewed the literature concerning the translocation of the phenoxyacetic acid herbicides.[5] Although they reported several instances where differences in translocation undoubtedly contributed to selectivity,[76] they concluded that since there were many exceptions, other factors must have had an overriding influence. Another factor affecting selectivity is the excretion of active ingredient from roots. There is a considerable literature on the excretion of herbicides from plants, and auxin analog herbicides feature prominently in this list.[77,78] As considerable amounts of herbicide may be lost in this way (e.g., up to 65% of absorbed dicamba was exuded from soybean[79]) losses via the roots could account for a significant part of the selectivity mechanism. The presence of herbicides in root exudates may occur naturally, due to diffusion from intact tissues, or the result of leakage caused by herbicide damage to root tissue.

Metabolism — There is an extensive literature on the metabolism and selectivity of the auxin analog herbicides and only the main reactions involved in the metabolism of these herbicides are described, as they are important in understanding the mechanisms involved in herbicide resistance (see Figures 3 and 4).

Deesterification — Like other herbicides containing carboxylic acid functions, the auxin analog herbicides are often applied as organic ester derivatives in a suitable formulation in order to maximize absorption through plant cuticles. These esters are generally considered not to be phytotoxic but are rapidly converted to the free acids by esterase activity within the plant. The rapidity of this conversion, and the lack of substrate specificity, indicate that hydrolytic cleavage is unlikely to confer any degree of species selectivity.[80]

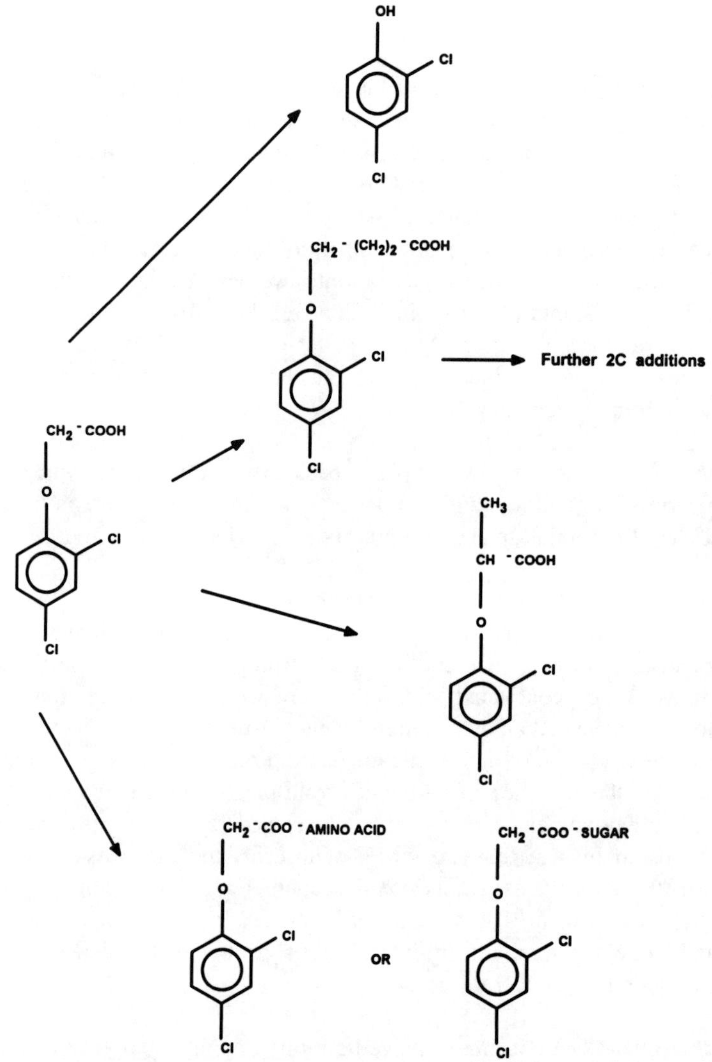

Figure 3. Side-chain metabolism of the auxin analogs. Metabolic reactions (from top to bottom):- Decarboxylation and further degradation of side-chain [80]; elongation [87]; alkylation [86]; conjugation.[80,93] See text for further details.

Formation of amides — Amide bonds are similar to ester linkages and are probably hydrolyzed by a similar group of esterases in plants, although there are amidases which specifically hydrolyze these types of bonds. Amino acid amides of the auxin analog herbicides are well documented (see below) and their formation provides several opportunities for potential selectivity mechanisms. For example, the rate of their formation could be different in different species.[81] Second, their formation could be a prerequisite for cellular (vacuolar?) compartmentation, although there is no direct evidence for this as a

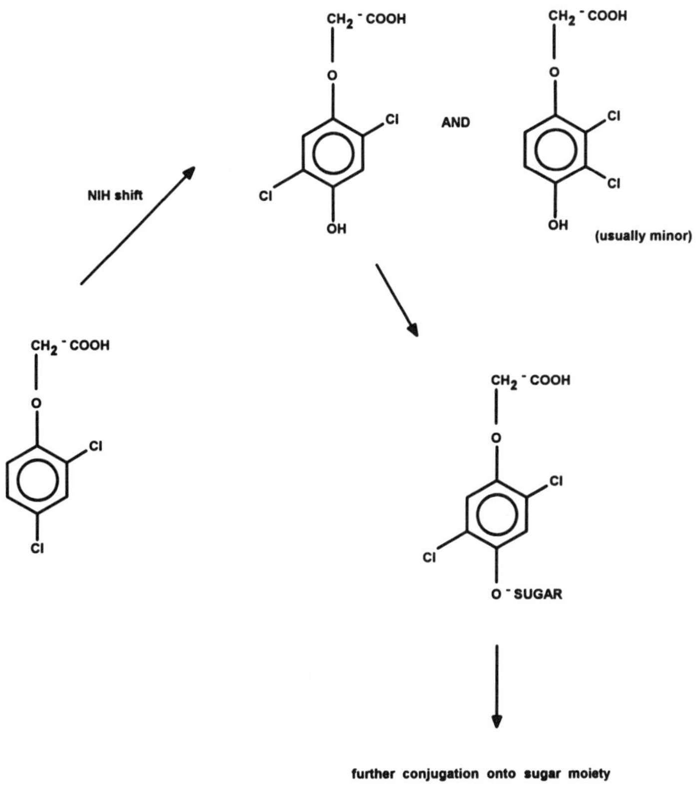

Figure 4. Ring hydroxylation and associated metabolism.

selectivity mechanism. Third, as these amino acid conjugates are sometimes the chemical form in which the herbicide is translocated in the phloem, their formation and subsequent movement could help to regulate the localized concentration of "active" herbicide.[80] Fourth, the rate of their hydrolysis may provide another means of regulating the amount of herbicide in plant tissues. The susceptibility of *Chenopodium album* to triclopyr is thought to be due to greater hydrolysis of aspartate conjugates than in *Stellaria media*.[82]

Decarboxylation and side chain degradation — *O*-dealkylation, resulting in the loss of the side chain, is important to selectivity in a number of species.[80] In particular, *O*-demethylation of 2,4-D, following decarboxylation, results in the production of the nonphytotoxic dichloroanisole in red currant (*Ribes sativum*) whereas in black currant (*Ribes nigrum*) and other plant species, these reactions occur only to a minor extent.[83] Selectivity between *Datura stramonium* and *Galium aparine* was also correlated with different rates of decarboxylation. In the more resistant *G. aparine*, the rate of decarboxylation was ten times faster in the first 6 h of treatment than in *D. stramonium*.[84] Pillmoor and Gaunt,[5] however, have doubted the direct involvement of decarboxylation of the phenoxyacetic acids as a major factor in determining selectivity. They consider an oxidative removal of the side chain more likely, resulting in the formation of glycolate which then gives rise to CO_2. Loss of the side chain can also occur by β-oxidation, that is, the removal of two carbon atoms at a time. This is a very effective detoxification mechanism for those plants that are capable of this type of oxidative catabolism, for example in legumes and in tomato.[85]

Side chain elongation — In a few species, elongation of the side chain by 1- or 2-carbon additions (see Figure 3) has been demonstrated for phenoxyalkanoic acid herbicides and this has conferred selectivity to some species. This selectivity of 2,4-D in the grasses *Bromus inermis*, *Phleum pratense,* and *Dactylis glomerata* appeared to be the result of rapid conversion of the parent molecule to 2,4-dichlorophenoxypropionic acid.[86] There was no mechanism offered to explain this unusual (in plants) conversion of acetate to propionate. The addition of 2-carbon units, however, is common in plants, for example, 2,4-DB is converted to homologues with longer aliphatic side chains[87] in resistant legumes, such as alfalfa.

Aryl hydroxylation — These are oxidative reactions resulting from the activities of mixed-function oxidases, probably cytochrome P450 (Cyt P450) dependent. Ring hydroxylation of 2,4-D occurs preferentially in the 4 position at the same time displacing the *para*-chlorine atom. However, this atom is usually not lost but moves to one of the adjacent positions on the aryl ring forming 4-OH-2,5-D and lesser amounts of 4-OH-2,3-D (see Figure 4). This molecular rearrangement has been named the "NIH" shift.[88] In some species, chlorine atoms can be eliminated forming 2-OH-4-chloro- or 4-OH-2-chlorophenoxyacetic acid.[89] Crop species that are resistant due to their capacity for ring hydroxylation include several monocots (wheat, barley, and oats) and some dicots, including cucumber and soybean.[80] This "NIH" shift has not been reported to occur in chemically similar herbicides which are alkylated on the aryl ring (e.g., MCPA and mecoprop). With these compounds, it is the alkyl substituent that becomes hydroxylated in preference to the aryl ring.[90,91] Figure 5 compares the proportions of metabolites formed in barley (resistant) compared with beet (susceptible) after treatment with MCPA for 48 h. As can be seen, barley effectively metabolizes the herbicide to nonphytotoxic metabolites, mainly OH-MCPA, while in beet, over 50% of the parent herbicide remained intact over the same period. With the benzoic acid herbicides, such as dicamba, hydroxylation of the aryl ring followed by glycosylation (see below) are the important reactions conferring selectivity in wheat and *Poa annua*.[92] With the structurally related chloramben, ring hydroxylation does not take place and detoxification in resistant species is more often associated with *N*-glycoside formation (see below).

Glycoside conjugate formation — The generation of hydroxylated metabolites invariably is followed by further reactions in which sugars (principally) are conjugated to the "herbicide" via ether linkages using the hydroxyl group(s). In addition, there is the possibility with acidic compounds for sugar esters to be formed directly.

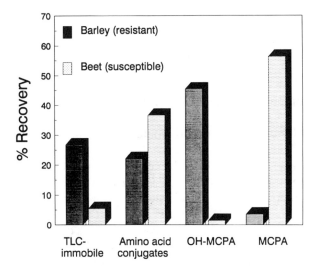

Figure 5. Metabolic fate of MCPA in barley and in beet. (Adapted from[90])

However, unless these are compartmentalized, away from the main metabolic activity within the cell, this form of metabolic change is not an effective detoxification mechanism since these esters are inherently labile, being hydrolyzed (chemically or enzymatically) to release the parent herbicide.[80] Less common is the formation of *N*-glycosides, and direct *N*-glucosylation of chloramben is a significant detoxification step in resistant but not in susceptible plants.[93]

Linked with metabolism is the process of compartmentation. This is defined as "a sequestration of herbicide, or its metabolites, in a specific cellular compartment". Herbicides are thought to be compartmentalized predominantly in the cell vacuole, although the evidence for this is mainly circumstantial.[80,91] Nevertheless, 2,4-D (as the β-D-glucoside) has been shown to accumulate in cell vacuoles[94] and the sequence: herbicide ⇒ conjugate metabolite ⇒ vacuolar sequestration, is considered responsible for resistance to 2,4-D in a tobacco cell culture variant.[95] This mechanism is discussed in more detail below. The second way in which selectivity can be achieved involves differential sensitivity of the target sites to the herbicide molecule. As discussed above, the plant cell probably responds to auxin analog herbicides in many ways. However, the early stages in the mode of action of these herbicides may be common for all species and for all types of cell; namely, receptor binding and subsequent effects on nucleic acid biosynthesis. If so, it should be possible to perform structure/activity studies for receptor binding, or for effects on nucleic acid biosynthesis. Unfortunately, these two aspects of the mode of action of auxin analog herbicides have not been studied in sufficient detail, using plants with contrasting susceptibilities to these herbicides. ABPs have been isolated from different plants, but the best evidence for a receptor function comes mainly from experiments with monocots, in particular maize. Conversely, most of the work on nucleic acid biosynthesis has been performed with dicots, especially soybean. A goal for future studies, would be to compare receptor binding and gene expression with molecular structure in both monocots and dicots.

THE DEVELOPMENT OF RESISTANCE

Weeds Resistant to the Auxin Analog Herbicides

There are many examples in the literature of inherent varietal and interspecific differential effects from the auxin analog herbicides.[96] It is often very difficult to determine from the literature whether a plant is resistant to these herbicides. This is invariably because all possible causes for an apparently "poor" herbicidal effect have not been considered, this detail being especially relevant when different genera are being compared. In such cases, there could be a multitude of reasons why one plant responds differently from another. Accordingly, only variation *within* a given species is considered in detail here. Examples for which only limited information was readily available are summarized in Table 6; other case histories are discussed more fully.

***Daucus carota* L.** — Biotypes of *D. carota* resistant to 2,4-D were first reported in 1957 in Ontario, Canada.[97,98] Resistance correlated with the continual use of 2,4-D over a number of seasons for roadside vegetation management. Resistant plants were morphologically identical to susceptible plants. Cross resistance to chemically similar herbicides, such as 2,4-DB and MCPB, was also found but both R and S biotypes were equally susceptible to 2,4,5-T. It was estimated that the R biotype represented approximately 1% of the natural population.[97]

***Carduus nutans* L.** — In the 1970s, 2,4-D was reported as ineffective against *C. nutans* in New Zealand,[99,100] with cross resistance to MCPB but not to the related auxin analogs picloram, clopyralid, and dicamba.[101] The development of resistance to MCPA was found to correlate directly with a previous history of 2,4-D application. From limited studies there was no evidence that R plants were any less fit ecologically than the "normal" S plants. Both biotypes produced similar amounts of shoot material when grown at high or low densities.[102]

***Stellaria media* (L.) Vill.** — *S. media* plants at a site near Bath (U.K.) were found to be strongly resistant to mecoprop with some resistance to MCPA and dichlorprop.[103] It was not evident how this resistance had developed as there was little evidence of a long history of mecoprop usage.[104] It appeared that resistance had been present for perhaps a long time, but at a low level at several locations. In studies with plants from 18 separate sites, biotypes could be grouped into three categories according to sensitivity, from susceptible through to resistant. Individual biotypes can sometimes be distinguished morphologically, the resistant ones generally having a more trailing habit and smaller leaves. However, these differences are small and vary according to environmental conditions. Superficial observations in a series of five experiments[104] indicated that the R plants

Table 6. Differential response of weed biotypes to auxin analog herbicides

Plant species	Comments	Ref.
Commelina diffusa Erechtites hieracifolia Burm.	Both these plants were found as weeds in sugar cane in Hawaii. One biotype of *C. diffusa* could withstand 5 times the dose of 2,4-D which effectively controlled susceptible plants.	97
Sphenoclea zeylanica Gaertn.	Reported as being resistant to 2,4-D (no other details available).	185
Polygonum lapathifolium L.	Five biotypes were treated with dichlorprop and LD_{50} values calculated. Three biotypes were approximately twice as tolerant as the other two, over a wide range of growth stages.	128
Cardia chalepensis (L.) Hand.-Maz	Biotypes varied in their responses to several auxin analog herbicides including 2,4-D, 2,3,6-TBA, and a 1:2 mixture of 2,3,6-TBA and MCPA. Susceptibility was not correlated with plant vigor, growth rate or size of leaf.	129
Kochia scoparia (L.) Schrad.	Thirteen selections were self-pollinated for four generations and treated with 2,4-D or dicamba. There was an approximate 2-fold difference in response to both herbicides from the most tolerant to the most susceptible. Differences in response could not be explained by plant size or maturity.	130
Sonchus arvensis L.	Poor control was thought to be due to a proportion of individuals in the population which had cross-pollinated with annual sowthistle (*S. oleraceus* L.), *S. oleraceus* is inherently more tolerant of 2,4-D and dicamba.	131
Chenopodium album L.	Reported as 2,4-D resistant, no further information available.	132
Taraxacum officinale, Ranunculus spp., Trifolium repens L.	In the 1950s, increased tolerance to 2,4-D and MCPA was reported in these three species in Belgium after nine annual herbicide treatments.	133

grew as vigorously as the S plants. Inheritance studies with this self-pollinated species[106,107] have yielded equivocal results and further work is needed to clarify the genetics relating to resistance in this species.[191]

Ranunculus acris L. — In the early 1980s, poor field control of *R. acris* with MCPA suggested that some populations in the North Island of New Zealand had developed resistance to this herbicide.[108,109] Plants from two populations with contrasting responses to MCPA showed a fivefold difference in susceptibility.[110] Leaf morphology was similar between biotypes and a chromosome analysis did not reveal any translocations, deletions, or additions in the nuclear chromosomes of R plants, nor were there any changes in ploidy.[109] Further investigations, using plants obtained from the North and South Islands, showed that the variation in resistance correlated with a history of MCPA use, with those fields receiving the most MCPA containing the most resistant plants.[111,112] As with *S. media*, there was no evidence that resistant plants were less "fit", as plant competition studies, using a replacement series technique, clearly showed that the two most contrasting biotypes achieved similar monoculture yields and competed equally with each other.[112,113]

***Convolvulus arvensis* L.** — Substantial variation in response to 2,4-D was noted between different strains of *C. arvensis*.[114,115] Resistant plants had the ability to resprout, especially from the roots after herbicide treatment, whereas susceptible plants showed complete necrosis of shoot tissues and no resprouting. There was no correlation between the resistance trait and plant vigor, root/shoot ratios, or initial knock-down effect. Interestingly, in some biotypes, there was only moderate injury to shoots but extensive root damage, implying that in these plants, translocation of the herbicide was more extensive than in others. No differences in leaf properties or foliar uptake and translocation were evident.[116] With stem callus tissue, no differences were found in the amounts of herbicide absorbed by the cells or in metabolism of 2,4-D.[117] The most obvious effects of the herbicide were increases in the levels of nitrate reductase, soluble protein, and RNA in the S biotype, while these components remained at a constant level (or decreased slightly) in the R biotype. Differential binding within the cells was offered as the only explanation for the differences in response to 2,4-D.

***Sinapis arvensis* L.** — In 1990, resistance in *S. arvensis* to several auxin analog herbicides such as 2,4-D, MCPA, dichlorprop, mecoprop, and dicamba was reported in western Canada in fields that had been treated regularly with these herbicides for over 20 years.[118] In contrast to the species described above, resistant *S. arvensis* plants did appear to be less competitive than the susceptible plants.

***Cirsium arvense* (L.) Scop.** — Several independent reports of resistance to 2,4-D and/or MCPA in *C. arvense* have been made. Resistance to 2,4-D and MCPA was reported in Hungary in regions where these herbicides had been used for about 15 to 20 years.[119,120] Although poor control from MCPA in Sweden was first reported in the 1950s,[121] it was not realized that these effects were due to herbicide resistance until approximately 20 years later. Sixty clones of *C. arvense* from areas of intensive farming, and greater herbicide use, were found to be resistant at some level.[122]

***Matricaria perforata* (*Tripleurospermum maritimum* spp. inodorum) (L.) Schultz Bip.** — Several independent studies, from 1956 onwards in three countries, reported that the spread of *M. perforata* had increased. This was apparently due to a reduction in herbicide efficacy, notably in regions where there was increased usage of auxin analog herbicides, such as 2,4-D, MCPA and mecoprop.[123-125] In later studies,[126] the most resistant population was found to be 2.5 times less affected by MCPA than the most susceptible, based on ED_{50} values. These differences in susceptibility were independent of plant growth stage and there was no indication that the resistant populations were any less vigorous or different in morphology. A survey of 48 populations, gathered in Britain and in France, showed that resistance to the herbicide was significantly greater in populations with a history of intensive herbicide usage than in other populations.[127] An interesting observation was that MCPA resistance correlated with resistance to ioxynil, indicating that, whatever mechanisms were involved, resistance was not specific to the chemistry of the herbicides involved, nor to their modes of action.

Other species — Other species with reduced sensitivity to the auxin analogs, but for which only limited information is available from the literature, are described in Table 6.

Crops Resistant to the Auxin Analog Herbicides

Tobacco (*Nicotiana tabacum* L.) — Several attempts have been made to develop resistance to auxin analog herbicides in tobacco.[134,135] Usually this has involved cell culture techniques and *in vitro* screening for resistant cell lines. Stable genetic changes were found in one cell line which maintained its resistance to 2,4-D under nonselective growth conditions.[135] Cross resistance to IAA, NAA, and picloram was also evident. In a separate study, cell lines were selected for resistance to picloram and regenerated plants were characterized genetically.[138] Of seven resistant cell lines isolated initially, four gave rise to plants in which the resistance trait proved stable and heritable. Other cell lines exhibited resistance but were either unable to produce viable plants (perhaps IAA synthesis mutants, see below) or gave rise to plants from which only sensitive callus was produced, i.e., genetically unstable. In the four cases in which picloram resistance was transmitted across sexual generations, three appeared as dominant alleles and one as a semidominant allele of single nuclear genes, estimated to occur at a frequency of 10^{-5} in the cell population.[138]

Clover (*Trifolium* spp. L.) — White clover (*T. repens*) cell suspension cultures have yielded cell lines with resistance to various auxin analogs.[139] All cell lines transmitted the trait to succeeding generations by asexual propagation. Recurrent half-sib family selection techniques were used to select 2,4-D resistance in red clover (*T. pratense*).[140] Starting with a broad-based germplasm bank based on three varieties, 20,000 plants were treated with a range of five doses of 2,4-D (up to 4.5 kg a.i. ha^{-1}). Selection criteria were based on plant survival and the extent of regrowth. Plants surviving the first "selection cycle" were intercrossed with untreated half-sib plants, thereby creating the seed for the next cycle The results after four cycles indicated a 35% change in sensitivity to 2,4-D (Figure 6). The data also indicated that further progress in increasing the level of resistance should be possible by increasing the number of selection cycles, although the improvement per cycle appeared to be diminishing.[140] After four cycles, plants were tolerant enough to 2,4-D that the use of untreated, half-sib plants for intercrossing could be abandoned.

Birdsfoot Trefoil (*Lotus corniculatus* L.) — As early as 1975, phenotypic recurrent selection techniques were being used to develop plant lines that could withstand 2,4-D.[141] Progeny testing selected 34 clones from an original population of 75; these 34 were then intercrossed and the progeny treated with 2,4-D over five cycles, yielding an approximate twofold reduced sensitivity to the herbicide. Cross resistance to other auxin analogs, 2,4-DB and 2,4,5-T, was also found. Cell tissue culture techniques have yielded cell lines tolerant to 2,4-D.[142,143]

Cucumber (*Cucumis sativus* L.) — Intraspecific variability in response to chloramben has been observed in several cucumber varieties.[144] There is no evidence that this arose due to herbicide selection pressure, but probably resulted from inherent physiological and metabolic differences between cultivars.[145] Cucumber cultivars with contrasting sensitivity, and their progenies, were evaluated for sensitivity to chloramben.[146] Gene action was primarily additive, although partial dominance of genes controlling tolerance was found.

Physiological Studies into Resistance to IAA and Exogenous Auxins

As well as the investigations above, describing resistance in both weeds and crops to auxin analog herbicides, studies have also been carried out into resistance to IAA. There are many examples of IAA synthesis mutants; most exhibit abnormal phenotypes, as would be expected, and many are not viable. Resistance to IAA is unlikely to be due to reduced uptake, and there are no mutants known with this single trait.[147] Plants with over- and under-production of IAA are known, the latter type probably being the most useful in enabling plants to withstand exogenously applied auxins. The *iba1* mutant of *Nicotiana plumbaginifolia* contains very low amounts of IAA and can withstand growth media containing high concentrations of auxins.[148] Uptake or metabolism of

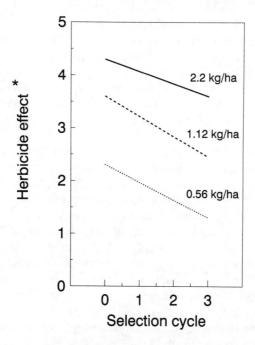

Figure 6. Decrease in 2,4-D injury over four cycles of recurrent selection in red clover. * = Visual rating: 0 = as controls, 5 = dead. Adapted from Reference 140.

NAA or IBA was not different between the mutant cell lines and wild types. Auxin mutants were also resistant to paclobutrazol (a gibberellic acid [GA] biosynthesis inhibitor) and to abscisic acid, which may indicate that the *IBA1* gene is involved more with the metabolism of ABA and GA than with IAA. The fact that the mutants appeared less severely affected in their morphology than other auxin-resistant mutants may support this idea.[149] However, most IAA-resistant mutants are affected by the way in which the cell responds to IAA. Perhaps the first of these responses is a hyperpolarization of the plasma membrane. Protoplasts prepared from the *Rac*− mutant of tobacco required a 10-fold higher concentration of auxin to achieve the same hyperpolarization as wild-type protoplasts.[150] The auxin-resistant mutant of tomato, *dgt*, was originally identified because of its diageotropic growth habit.[151] Auxin-binding proteins of M_r 42 and 44 can be detected in membrane preparations using the photoaffinity probe ^{3}H-azido-IAA.[152] When this was used to probe membrane preparations from *dgt* tomato plants, ABPs were not detected in the shoots but were present in roots.[153] The level of ABP in two tobacco cell lines was also correlated with resistance to auxin.[154] In these studies, the resistant line retained the ability to differentiate shoots but could not form roots. Similarly, tobacco mutants resistant to NAA were unable to form roots, due to a single, dominant nuclear mutation.[155] Recessive mutants resistant to 2,4-D in *Arabidopsis thaliana* also displayed altered root geotropism[156] or unusual root morphology.[157]

Collectively, these data imply an interrelationship between root differentiation/growth and auxin resistance, although the precise mechanism remains to be elucidated. The *aux1*, *axr1*, and *axr2* mutants of *A. thaliana* were isolated by screening for resistance to auxin.[157] The recessive *aux1* mutation confers resistance to ethylene and its precursor ACC.[158] It was concluded that the *AUX1* gene functions primarily in the root but, in contrast to above,[154] it is not required for root differentiation or root development but rather functions specifically in the regulation of hormonal responses, perhaps in the signal transduction pathway.[158] Multiple hormone resistance is not unique to the *aux1* mutant of *A. thaliana*. The *axr2* mutation in *A. thaliana* confers resistance to auxin, ethylene, and ABA,[159] and in tobacco, the *iba1* mutant is resistant to auxin and ABA.[148] All of these mutants provide valuable genetic material for the elucidation of hormone interactions within the plant cell as well as providing important information regarding the biochemical mechanisms for resistance to the auxin analog herbicides.

MECHANISMS OF RESISTANCE — CASE STUDIES

2,4-D

Herbicide-Resistant Weeds

Despite widespread use of 2,4-D for over 50 years, there are relatively few detailed studies on the mechanisms of resistance to this herbicide in weeds.

Perhaps the first report, with *Cardaria chalepensis* (L.) Hand.-Maz. biotypes having considerable variability in leaf morphology, revealed no correlation between leaf size and herbicide effect for 2,4-D, indicating that spray retention was not an important factor in determining herbicide resistance.[129] Resistant and susceptible biotypes of *C. arvensis* were equally susceptible to 2,4-D when stem segments were treated directly with the herbicide, and cultured cells of the R and S biotypes absorbed 2,4-D at identical rates.[117] Both biotypes metabolized the herbicide to the same extent (although the analysis was not thorough). The authors concluded, without much evidence, that differential binding to structural components within the cells, rendering the herbicide inactive, was the most likely explanation for resistance in this species.[117] Similarily, resistance to 2,4-D in *S. arvensis* L. was not due to differences in herbicide absorption and there were no significant differences between R and S biotypes in the amounts of herbicide translocated within the plants, nor in the amounts that were exuded from the roots following foliar application.[160] Herbicide metabolism studies showed that resistance was not due to differences in the extent or pattern of metabolism within the two biotypes, including the amounts of bound herbicide. However, the metabolism studies were based on whole plant extracts which would not have revealed the intracellular location of any metabolites and extent of any sequestered herbicide.

Studies to determine the resistance mechanism in *Carduus nutans* revealed that 2,4-D uptake or translocation to the root tissues in R and S biotypes was not responsible for the difference in susceptibility to the herbicide.[161] However, the amounts of herbicide that were exuded from the roots were significantly greater in the S biotype, which could reflect a greater amount of herbicidal damage to the S plants or more herbicide being immobilized (i.e., bound) within the shoots of R plants. The metabolism of 2,4-D within these plants showed clear differences between S and R biotypes as the S plants had approximately twice as much radioactivity in the ether fraction containing 2,4-D acid and the potentially phytotoxic amino acid conjugates of 2,4-D (Table 7).[161] When this ether fraction was analyzed further, S plants were found to contain six times the amount of free 2,4-D than R plants. The final, ethanol-insoluble fraction, showed that R plants had less "bound" ^{14}C than the S plants. These data provide good evidence that herbicide resistance in this case is due to differential metabolism of 2,4-D. However, if resistance depends upon the conversion of 2,4-D to amino acid conjugates, these metabolites must be nonphytotoxic. Several reports have indicated that these compounds are either inherently herbicidal themselves, or are readily hydrolyzed, releasing parent 2,4-D.[5,89] If this is the case in *C. nutans*, then three further possibilities exist. First, that these conjugate metabolites are sequestered within the cells, so effectively detoxifying the herbicide. Second, that conversion to these conjugates results in temporary storage which is sufficient to prevent translocation to the sites of action at a critical time of herbicide application. Third, that this temporary storage allows the conversion to more water-soluble metabolites through further metabolism of the 2,4-D moiety, in particular by ring hydroxylation and that this results in more permanent detoxification.

Herbicide-Resistant Crops

The mechanisms of resistance in crops are generally better understood than in weeds because these characteristics have been deliberately introduced as a result of genetic engineering. However, where resistance traits have been produced by random induction of mutations, reasons for the difference in susceptibility are not always clear. Resistance has been successfully introduced into four species: birdsfoot trefoil, cotton, soybean, and tobacco.

Resistance to 2,4-D was developed in birdsfoot trefoil (*Lotus corniculatus* L.) by *in vitro* selection. Studies of translocation and metabolism of ^{14}C-2,4-D showed that R plants appeared to have significantly more "bound" radioactivity and translocated less herbicide out of the treated leaf than the S plants.[162] It is important to realize that these two results may be related. Radioactivity was found in the solutions bathing the roots, and although no statistical analysis was made, the S biotype exuded 6% more of applied ^{14}C than did the R. Unlike most other studies, only free 2,4-D was found in the alcohol extracts, no amino acid conjugates or other labeled compounds were detected. However, a study was made of the release of ^{14}CO$_2$ from treated seedlings of the two biotypes and this showed five times as much metabolism in the R, than in the S plants. Taken together, these data indicate that differential metabolism plays an important part in the resistance mechanism to 2,4-D in *L. corniculatus*.

Another species in which resistance to 2,4-D has been introduced by growing callus on selective media is soybean (*Glycine max* L. Merrill). In early studies,[163] three 2,4-D resistant root callus cell lines were examined for differences in herbicide uptake or metabolism. Not only did R cells absorb less herbicide during the 24-h experimental period, but a greater proportion of 2,4-D was converted to hydroxylated glycosides in the R than in the S biotype. It may also be significant that the resistant cells contained smaller amounts of ether-soluble metabolites, including the potentially phytotoxic amino acid conjugates. The authors concluded that variations in uptake and metabolism of 2,4-D partially explained herbicide resistance, although differences in target site sensitivity or differential compartmentation of metabolites could also be involved.

Table 7. Metabolism of ^{14}C-2,4-D in resistant and susceptible *C. nutans* biotypes

Sample	Probable metabolites	Resistant biotypes	Susceptible biotypes
Ether fraction	Free 2,4-D 2,4-D amino acid conjugates 2,4-dichlorophenol	15.1[a]	37.4*
Water fraction	Glycosides of 2,4-D and its metabolites	84.9[a]	62.2
"Bound" fraction	Unknown	8.3[b]	13.9*

[a] % of recovered ^{14}C.
[b] % of applied ^{14}C.
* significant difference between resistant and susceptible biotypes.

Perennial *Glycine* species have been found to withstand 2,4-D.[186] Herbicide uptake and metabolism were compared in several perennial species and in soybean; data for the most tolerant perennial accession and soybean are shown in Table 8. Herbicide uptake was substantially higher in S soybean than in the R perennial *Glycine*. This, together with a significantly greater metabolism of 2,4-D (at least one order of magnitude), strongly supports a resistance mechanism based on the quantity of active ingredient within the cells. However, the type of metabolism is also important because of the differences in toxicity of the various metabolites, as discussed above. Thus, soybean contained no hydroxylated derivatives of 2,4-D, while this was the most predominant metabolite occurring in *Glycine* (approximately 67% of the total). More rapid and extensive metabolism of 2,4-D was also thought responsible for the resistance of other perennial *Glycine* accessions in a separate study.[164]

Recently, 2,4-D resistance has been engineered into cotton (*Gossypium hirsutum* L.).[165] In order to accomplish this, a 2,4-D monooxygenase gene *tfdA* from the *Alcaligenes eutrophus* plasmid pJP5 was used. The *tfdA* gene was incorporated into a chimeric construct pBIN19::pRO17 which enabled its introduction into cotton using *Agrobacterium*. Shoots of putative transformed plants were screened for herbicide resistance using either a rooting bioassay or by applying 2,4-D directly to shoot meristems. Plants rooting in the 2,4-D-containing medium, or those unaffected by the herbicide treatment of the meristems, were vegetatively propagated and screened further. Transformants were unaffected by up to three times the concentration of herbicide recommended for weed control in cotton. Transformants were further characterized by assaying monooxygenase activity, the highest level obtained being some 38 times greater than in nontransformed plants. Progeny derived from self-fertilization of three of the transformants showed a 3:1 segregation pattern, indicating a single resistance locus.

Use of *Alcaligenes eutrophus* as a source of genes for degrading 2,4-D has also been successful for engineering resistance into tobacco.[166-168] Strategies employed for producing transformed plants have been similar. For example, the *tfdA* gene was first isolated, fused with a suitable promoter plus the nopaline synthase (*nos*) 3' polyadenylation signal and then cloned into a

Table 8. Uptake and metabolism in herbicide-resistant and -susceptible cell cultures of *glycine* 48 h after treatment with 10 μ*M* 2,4-D

Species	Uptake (nmol)	Metabolism (% of applied)	Composition of metabolites (nmol)		
			Hydroxy metabolites	Amino acid metabolites	Free 2,4-D
Soybean (*Glycine max*) (susceptible)	517	6	0	18	485
Glycine tomentella (resistant)	369	85	232	64	52

After Reference 186.

suitable vector system for plant transformation, e.g., using *Agrobacterium tumefaciens* [167] (Figure 7). Resistance to 2,4-D was monitored in several ways: (1) by measuring monooxygenase activity in plant homogenates by assay of 2,4-dichlorophenol production from the herbicide, (2) using a leaf disc assay in which the ability to regenerate shoots in the presence of the herbicide was measured, and (3) by using a germination assay in which seed produced from selfed transformants was grown in the presence of various concentrations of 2,4-D. The seed germination test was also used to investigate cross resistance to other auxin analogs, 2,4,5-T and MCPA.

Two transformants showing the highest resistance to the herbicide were selected from the above tests. The transgenic F1 progeny survived up to eight times normal field application rates, were moderately resistant to MCPA but were susceptible to 2,4,5-T. The authors suggested that this may imply a degree of substrate specificity by the monooxygenase for 2,4-D. In another study, transgenic tobacco was produced using the same *tfdA* gene as above,[166] although the procedure used for the construction of the plasmid for transformation was different. Analysis of monooxygenase-specific RNA showed a substantial variation in the level of expression of up to 30-fold between different transformants; the highest level of gene expression in any of the transformed plants corresponded to only three copies of the T-DNA. Nevertheless, plants with this level of expression could withstand up to four times the commercially applied dose. Figure 8 shows the response of control and transformed plants to other auxin analog herbicides. Resistance to 2,4-D and 4-chlorophenoxyacetic acid, and to a lesser extent to MCPA was found, but not to any of the phenoxypropionic acids or the tri-substituted phenoxyacetic acids. As discussed above, this could be due to substrate specificity of the monooxygenase.

Mecoprop

Investigations into the mechanisms of resistance to mecoprop have followed two distinct paths: those in which various physiological processes have been studied (absorption, translocation, and metabolic fate of the herbicide); and those in which the biochemical action of the herbicide has been determined. There is no reason why either of these approaches should be mutually exclusive. Herbicide uptake and translocation studies with S and R biotypes of *S. media* revealed no quantitative differences[169] (Figure 9). Furthermore, there were no qualitative differences in herbicide translocation between the two biotypes that could explain the difference (Figure 9C, D). Further evidence that differences in herbicide uptake and movement are not responsible for resistance, comes from a second study where mecoprop was applied to *S. media* biotypes via the roots and the leaves.[170] Plants were resistant to both foliar and root treatment, although the differences in response between the two biotypes were smaller when treated via the roots.

In all studies with S and R biotypes of *S. media*, there is a common pattern of symptom development. All plants, irrespective of their sensitivity to mecoprop,

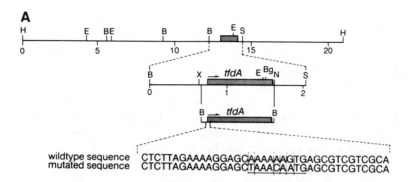
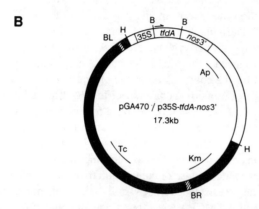

Figure 7. Cloning of the gene for 2,4-D monooxygenase. A. The 960 bp *Bam* HI-*Bam* HI fragment was produced from a 21 kb *Hind* II B fragment of plasmid pJP4 (see [167] for further details). This clone was subjected to oligomutagenesis to produce the ATG start codon (underlined). Scales beneath each map are in kb pairs; the arrow indicates the direction of translation of *tfdA*. B. Schematic representation of the tfdA gene cloned between the 35S promoter and the (nos) polyadenylation signal of pJ35SN and inserted into the binary vector plasmid pGA470. Km designates the gene for kanamycin resistance; Ap and Tc designate bacterial genes for ampicillin and tetracycline resistance respectively. BL and BR are the left and right borders respectively of the T-DNA. Reproduced from Reference 167 (with permission).

initially show symptoms characteristic of the auxin analog herbicides, but these symptoms gradually disappear from the R plants while remaining, or becoming more severe, in S biotypes. These symptoms include epinasty,[20,104,169,170] enhanced ethylene production,[20,171,172] and inducement of stomatal closure, with concomitant effects on photosynthesis and transpiration.[24] This symptomology is not unique to mecoprop; a similar response has been observed with biotypes of *Ranunculus acris* treated with MCPA.[110] This is *prima facie* evidence that similar amounts of herbicide are absorbed and translocated at similar rates in both S and R plants. One explanation for the disappearance of symptoms in R biotypes is that the herbicide may be detoxified faster and to a greater extent than in the S plants. This has been investigated in two *S. media* biotypes with

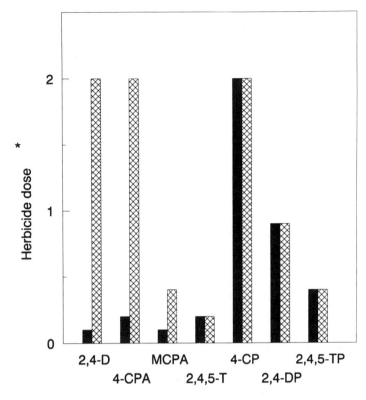

Figure 8. Cross resistance to auxin analogs in 2,4-D-resistant tobacco. * = the highest herbicide dose at which explants could differentiate shoots
Key: 4-CP (4-chlorophenoxypropionic acid); 2,4-DP (2,4-dichlorophenoxypropionic acid); 2,4,5-TP (2,4,5-trichlorophenoxypropionic acid); other abbreviations as in the text. Adapted from Reference 166.

contrasting sensitivities to mecoprop. Shoots were treated with ^{14}C-mecoprop and analysis by radio-TLC revealed that the extracts contained two main types of components, one being immobile on silica gel developed with an ether/formic acid solvent system, the other corresponding to mecoprop acid.[169] The TLC-immobile material was considered to be polar metabolites, probably conjugates. In general, more free mecoprop was recovered from the tissues of the S plants than from the R biotype (Figure 10). However, these differences were small in relation to the doses of mecoprop needed to control these two biotypes under normal conditions. This raises further questions involving the chemical identity of the metabolites, their location within the cells, and whether there is a significant and inherent difference in sensitivity between R and S plants. Since the earlier report,[169] further metabolic studies have been carried out. The metabolic profile, as described above, was confirmed for both biotypes and an identical distribution of metabolites was obtained from both biotypes when the ethanolic extracts were further analyzed by partitioning into

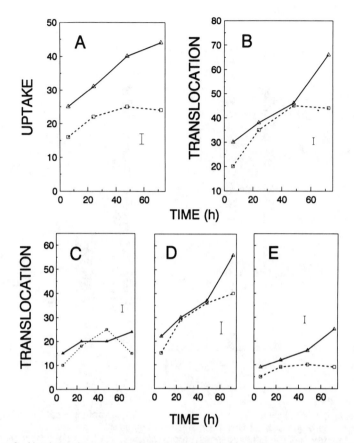

Figure 9. Uptake and translocation of ^{14}C-mecoprop in herbicide-resistant and -susceptible biotypes of *Stellaria media*. A: Foliar absorption (as % of applied ^{14}C) B: Translocation of ^{14}C from ^{14}C-mecoprop-treated leaves (sq. root transformed data). C-E: Distribution of ^{14}C in ^{14}C-mecoprop-treated plants. C = apical shoots; D = basal shoots; E = roots. All vertical bars are SED values. See [169] for further experimental details.

ether (Figure 11, right-hand side). However, when the extracts were treated with β-glucosidase, or a mild acid hydrolysis, there was a difference between the R and S biotypes. With the S biotype, only one ^{14}C-aglycone, mecoprop, was released after hydrolysis. With the R biotype, a second, minor component was released which co-chromatographed with an authentic sample of hydroxymethyl mecoprop (Figure 12). As ether-linked conjugates are generally recognized as being less phytotoxic and more stable (resistant to hydrolysis) *in vivo*, the occurrence of compounds conjugated to hydroxy derivatives of mecoprop may indicate a possible resistance mechanism. If the resistance mechanism is due to a greater ability to metabolize mecoprop via the formation of hydroxylated derivatives, then it may be possible to detect this by using inhibitors of the Cyt P450 mixed-function oxidases which mediate the oxidative metabolism of the auxin analogs. One such inhibitor is ABT (1-aminobenzotriazole), which has

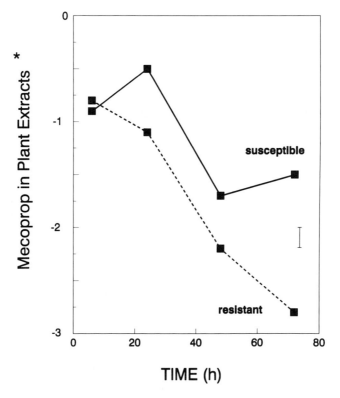

Figure 10. Amounts of mecoprop in leaves of herbicide-resistant and -susceptible biotypes of *Stellaria media*. * Values are DPM as % of recovered ^{14}C, logit transformed; vertical bar = SED.

been shown to inhibit phenoxyacetic acid degradation in several species.[173] ABT applied to *S. media* biotypes prior to mecoprop application had no additional effect on the S biotype but consistently increased the sensitivity of the R biotype by approximately 30%. However, the data were variable and not statistically significant. The possibility remains that cellular compartmentation of the conjugated metabolites of mecoprop may be a crucial part of the resistance mechanism (Figure 13), as other studies have concluded.[91] However, there is very little supporting evidence in the literature for this concept. Despite many years of research, there is only one report that demonstrates the specific vacuolar localization of conjugates of an auxin analog herbicide.[94] In this, soybean cell suspension cultures were treated with ^{14}C-2,4-D, protoplasts were then formed and their vacuoles released. The purified vacuoles contained β-D-glucosides of 2,4-D, no amino acid conjugates, and only traces of 2,4-D.

The other line of investigation mentioned earlier concerns the biochemical action of mecoprop. Two of the most obvious and early symptoms of treatment with auxin analog herbicides are the stimulation of ethylene and the production of epinasty, especially in leaves and petioles,[18,25] and mecoprop is

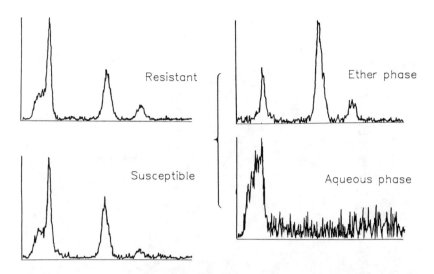

Figure 11. Radio-TLC analysis of the metabolic profile of mecoprop in *Stellaria media*. Left-hand side: total leaf extracts from resistant and susceptible biotypes. Right-hand side: metabolic profile after partitioning into ether (same profile irrespective of biotype). NB. The most mobile component on each TLC plate (right-hand side of each chromatogram) is the ethyl ester of mecoprop, an artifact of the ethanolic extraction solution.

no exception.[20,104,169,170] The epinastic response is generally assumed to result directly from herbicide-induced ethylene production.[187] Recently, some doubt has been cast on this assumption and there is convincing evidence that with *S. media,* the epinastic response is due to a combination of the growth-promoting (auxin) properties of the molecule and its ethylene-stimulating effect.[20] In fact, irrespective of herbicide dose, there were no differences in leaf epinasty observed between two biotypes with contrasting susceptibilities to mecoprop. The epinastic response, therefore, is not a direct indicator of mecoprop action in this species. The ethylene-stimulating response, on the other hand, may give some clues on the resistance mechanism because there appears to be a direct relationship between herbicide resistance and the stimulation of ethylene.[20,172] A possible explanation involves increases in malonyltransferase activity as a result of elevated levels of ethylene.[174] For many compounds, malonylation of herbicide conjugate metabolites is a prerequisite for their vacuolar sequestration, an important stage in the detoxification mechanism.[91,175] If this is the case in *S. media*, it could explain the resistance mechanism, enhanced ethylene synthesis, especially during the first few hours after herbicide treatment, promoting greater detoxification of mecoprop in resistant biotypes.

The few studies that have investigated mecoprop resistance at the biochemical level have examined H^+ transport. One experimental technique measures H^+ efflux, presumably due to the action of plasma membrane

Metabolic profile after treatment with β-glucosidase

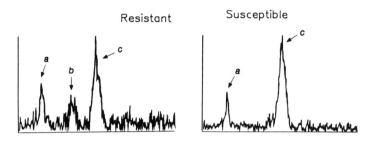

Metabolic profile after treatment with mild acid

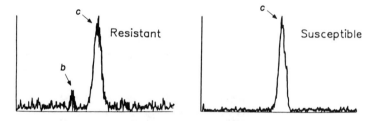

Figure 12. Metabolic profiles of ether-soluble ^{14}C-compounds from herbicide-resistant and -susceptible biotypes of *Stellaria media* treated with ^{14}C-mecoprop. a = TLC-immobile compounds (unknown chemistry); b = hydroxymethyl mecoprop; c = mecoprop.

ATPases, and this has been used to determine sensitivity to auxins in plants. The rate of acidification of the solution bathing abraded shoot explants was used to measure H^+ efflux in *S. media* and the doses producing half-maximal responses (H_{50}) for the two biotypes were calculated.[171] The R biotype was approximately five orders of magnitude less responsive to the herbicide than the S biotype, indicating a reduced binding affinity for mecoprop in R plants. However, the technique does not measure binding *per se* and suffers from the major disadvantage of having to use immature, etiolated and abraded tissues. The large difference between the two biotypes in H_{50} values is also at variance with the level of resistance in this

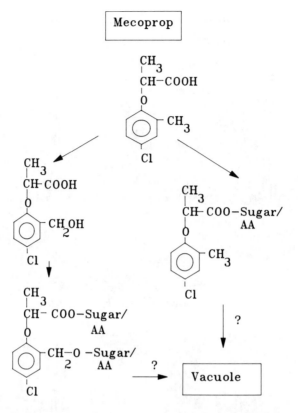

Figure 13. Proposed metabolism of mecoprop in resistant *Stellaria media*.

Table 9. H⁺ Pumping Activity in mecoprop-resistant and -susceptible biotypes of *Stellaria media*

Biotype	Rate of H⁺-pumping*	
	Vegetative plants	Flowering plants
Resistant	45	160
Susceptible	83	101

* = measured by a fluorescence quenching technique,[177] fluorescence units mg^{-1} protein min^{-1}.

species towards mecoprop. A more direct approach measured the H⁺ pumping activity in plasma membrane vesicles prepared from shoot tissues of R and S biotypes treated with mecoprop. With flowering plants, H⁺ pumping activity was significantly greater for the R biotype,[177] whereas the reverse was found for vegetative tissues (Table 9).[178] The only conclusion that can be made from the above work is that resistance is unlikely to result from an effect on H⁺ pumping. The changes in H⁺ efflux observed in plants are more likely to reflect the amounts of active ingredient in the tissues rather than an inherent difference in response.

Other Herbicides

MCPA

Uptake of MCPA was similar in R and S biotypes of *Ranunculus acris*, averaging 18.1% and 20.8% (of applied), respectively.[109] However, the distribution of ^{14}C-MCPA was different between biotypes, R plants translocating 5% ^{14}C to the stolons compared with 15% for S plants. Certain aspects of metabolism were studied, and decarboxylation of MCPA was significantly greater in R compared with S biotypes.[109]

Chloramben

Differential sensitivity in several lines of cucumber was related to differential uptake and translocation. One line appeared resistant due to a unique distribution of ^{14}C from ^{14}C-chloramben after root application, in that the label was confined in and around the vascular bundles, thus preventing any further transport into the leaves. Herbicide metabolism also affected resistance, the more S lines containing more parent herbicide and less of the non-phytotoxic *N*-glucosyl metabolite.[146]

Dicamba

R and S biotypes of *Sinapis arvensis* were studied to determine the mechanism of resistance to dicamba.[160] The two biotypes showed similar patterns of uptake, translocation, and exudation from the roots following foliar application of herbicide. High-performance liquid chromatography (HPLC) analysis showed no differences in herbicide metabolism or in the amounts of bound radioactivity after treatment with ^{14}C-dicamba. The authors concluded that differential sensitivity at the target sites of action is the mechanism of resistance, but no other details were presented.

Recent measurements of the binding characteristics of ^{3}H-IAA to ABPs prepared from resistant and susceptible *S. arvensis* biotypes have added support to this conclusion.[188] Scatchard analysis indicated a fundamental physiological difference between the two biotypes, in that two populations of IAA binding sites were associated with ABPs from susceptible plants, whereas only one was revealed in protein from the resistant biotype. IAA binding in susceptible plants was more sensitive to dicamba than in herbicide resistant plants, highlighting a further difference.

Picloram

Four picloram-tolerant cell lines of tobacco were isolated using *in vitro* techniques.[189] Tolerance was not due to impaired uptake, in fact, more picloram was absorbed by the tolerant mutants than wild-type cells. Similarly, there was no indication that the tolerant cells metabolized more of the herbicide than the susceptible lines. The authors speculated on the possibility that picloram was sequestered within mutant cells preventing the active ingredient from reaching the sites of action.

Picloram was evaluated on the same two biotypes of *Sinapis arvensis* as described above.[188] Essentially the same results were obtained as with dicamba, herbicide resistance not being a result of differential uptake, translocation, or metabolism.[190] Similarly, IAA binding to ABPs derived from the susceptible biotype was more sensitive to picloram inhibition than in resistant plants.[188] The significance of these results remains to be determined.

FINAL DISCUSSION

Despite the occurrence of weeds resistant to the auxin analog herbicides for nearly 40 years, very little research has been done on the potential impact that this might have on agriculture. The reason for this is probably related to the relative ease, currently, with which these weeds can be controlled using other types of herbicides. Plants that are resistant to the auxin-type compounds still remain sensitive to compounds with other modes of action. In fact, some weeds resistant to one herbicide can be controlled with other compounds belonging to the same chemical class, probably with similar modes of action. Furthermore, it appears that weeds resistant to the auxin analogs are not widely dispersed in comparison with those resistant to other types of herbicide. The auxin analog herbicides are considered as "low-risk" compounds as they do not affect a single target site and they do not have long-term residual activity. The fact that they are often applied several times throughout the year and are frequently applied to the same areas, over several seasons, has provided the "base" from which resistance has developed.

It is difficult not to sense a certain degree of irony when considering the information in this chapter. It is often quoted that herbicide resistance develops directly in response to prolonged herbicide usage,[179-182] yet, even though the auxin analog herbicides have been used for over a half century and frequently at high rates of application, there are relatively few reports of resistance to these compounds. It is also ironic that the modes of action of these herbicides are not entirely understood, considering how long they have been in commercial use. Also, the auxin analogs are generally regarded as herbicides yet some, in particular 2,4-D and NAA, are essential ingredients in tissue culture media, as replacements for the naturally occurring plant growth "hormone", IAA. Furthermore, 2,4-D and NAA have been instrumental in the development of affinity-chromatographic techniques for the isolation of auxin-binding proteins,[41,45] which have subsequently been used to investigate the mode of action of IAA and the mechanisms of resistance to compounds with auxin-like activity.[154-156]

The reason why it has been so difficult to resolve the modes of action of these auxin analogs is undoubtedly due to the multiplicity of biochemical effects that they have within the cell. This multifaceted mode of action may be the very reason why there has not been any extensive evolution of resistance to these herbicides. It is generally assumed that because the auxin analogs do not inhibit just one biosynthetic pathway, but affect several, often unrelated metabolic processes, it is far less likely for biotypes to evolve possessing all the genetic changes required for resistance.[180]

Biotypes that are able to withstand high concentrations of auxin analogs (or, indeed, IAA) as a result of a fundamental change in their physiology, are likely to have abnormal phenotypes and most are unable to survive under normal conditions. However, those biotypes that are viable and herbicide resistant have one characteristic in common: herbicide metabolism is significantly different in R compared with S biotypes. The ways in which this detoxification occurs differ according to species. There are examples where resistance is due to more extensive degradative metabolism,[80,83] greater production of nonphytotoxic metabolites,[92,93] greater binding of herbicide or its metabolites to structural components (e.g., lignin),[5] or greater sequestration of the active ingredient, or any potentially phytotoxic metabolites, into cell vacuoles.[94] However, in any given situation, resistance to one auxin analog never confers resistance to all others. This may not be too surprising, considering the chemical diversity of this class of herbicides (Figure 1). Nevertheless, it could also reflect an element of substrate specificity, perhaps with enzymes involved in herbicide degradation.[166,167] Furthermore, resistance to the auxin analog herbicides is never absolute and, "at best", can be measured in tenfold differences in sensitivity between contrasting biotypes. Compare this with resistance to the triazines, for example, where thousandfold differences in target site binding have been reported.[96] Again, this might imply that the mechanisms of resistance to the auxin analogs are more concerned with reducing the concentration of active ingredient within the cell rather than involving any target site modification. An additional piece of evidence in support of the "differential metabolism" theory concerns the pattern of herbicide symptom development. Most plants, irrespective of their sensitivity, initially show identical herbicide symptoms characteristic of the auxin analogs but these symptoms gradually disappear from R plants while remaining, or becoming more severe, in S plants. If target site alteration was the primary mechanism of resistance, it is most unlikely that herbicide-treated plants would recover in this way. It should be emphasized that there are many examples of such differential metabolism, both in weed species as well as in crops.

What can be done to delay the spread of resistance to the auxin analogs? The management of herbicide resistance has received a considerable amount of attention in the last few years[181,182,185] and is considered fully in other chapters. The salient message is "diversity": crops and herbicides should be rotated whenever possible; reliance should not be placed on herbicides that have a single site of action; and methods of weed control should be varied, using mechanical methods as well as chemical ones.[185] It is not difficult to see that these recommendations are at variance with current trends in herbicide usage. One possibility to counter resistance is the use of synergists that reduce herbicide detoxification.[183] This may be particularly relevant to the auxin analogs where enhanced metabolism is a key feature of the resistance mechanism. However, these compounds are not selective in their action and increased activity will probably occur in both the crop and the weed. Finally, there may be opportunities for developing biological control systems for herbicide-resistant weeds. A pathogenic fungus, *Sclerotinia sclerotiorum*, has shown potential as a selective mycoherbicide for the control of MCPA-resistant *R. acris* in pasture.[184]

To the author's knowledge, this is the first review published specifically on resistance to auxin analog herbicides. While this review is by no means complete, it is hoped that it is comprehensive enough to satisfy both expert and newcomer alike in this exciting and ever-changing area of weed science.

ACKNOWLEDGMENTS

I wish to thank Drs. A. H. Cobb, and D. Llewellyn for allowing me to use their previously published diagrams. Special thanks go to Drs. P. J. W. Lutman (Rothamsted Experimental Station, U.K.) and K. C. Harrington (Massey University, NZ) for giving me permission to use some of their research data prior to publication.

REFERENCES

1. Zimmerman, P. W., and A. E. Hitchcock. "Substituted Phenoxy and Benzoic Acid Growth Substances and the Relation of Structure to Physiological Activity," *Contrib. Boyce Thompson Inst.* 12:321-344 (1942).
2. Nutman, P. S., H. G. Thornton, and J. H. Quastel. "Inhibition of Plant Growth by 2,4-dichlorophenoxyacetic Acid and other Plant Growth Substances," *Nature* 155:488-500 (1945).
3. Kirby, C. "*The Hormone Weedkillers*," (Farnham: British Crop Protection Council, 1980), p. 54.
4. Loose, M. A. "Phenoxyalkanoic Acids," in *Herbicide Chemistry, Degradation and Mode of Action*, P. C. Kearney, and D. D. Kaufman, Eds. (New York: Dekker, 1975), pp. 1-128.
5. Pillmoor, J. B., and J. K. Gaunt. "The Behaviour and Mode of Action of the Phenoxyacetic Acids in Plants," in *Progress in Pesticide Biochemistry Vol. 1*, D. H. Hutson, and T. R. Roberts, Eds. (Chichester: Wiley & Sons, 1981), pp. 147-218.
6. Venis, M. A. "*Hormone Binding Sites in Plants*," (Harlow: Longman, 1985), p. 191.
7. Rubery, P. H. "Auxin Receptors," *Annu. Rev. Plant Physiol.* 32:569-596 (1981).
8. Corbett, J. R., K. Wright, and A. C. Bailie. "*The Biochemical Mode of Action of Pesticides*," (London: Academic Press, 1984), p. 382.
9. Lehmann, P. A. "Stereoselectivity and Affinity in Molecular Pharmacology. III Structural Aspects in the Mode of Action of Natural and Synthetic Auxins," *Chem. Biol. Interact.* 20:239-249 (1978).
10. Farrimond, J. A., M. C. Elliott, and D. W. Clack. "Charge Separation as a Component of the Structural Requirement for Hormone Activity," *Nature* 267:401-402 (1978).

11. Porter, W. L., and K. V. Thimann. "Molecular Requirements for Auxin Action," *Phytochemistry* 4:229-243 (1965).
12. Wain, R. L., and C. H. Fawcett. "Chemical Plant Growth Regulation," in *Plant Physiology*, F. C. Steward, Ed. (London: Academic Press, 1969), pp. 231-296.
13. Kaethner, T. M. "Conformational Change Theory for Auxin Structure-Activity Relationships," *Nature* 267:19-23 (1977).
14. Katekar, G. F., and A. E. Geissler. "Auxins II. The Effect of Chlorinated Indolylacetic Acids on Pea Stems," *Phytochemistry* 21:257-260 (1982).
15. Katekar, G. F. "Auxins: on the Nature of the Receptor Site and Molecular Requirements for Auxin Activity," *Phytochemistry* 18:223-233 (1979).
16. Fletcher, W. W., and R. C. Kirkwood. "*Herbicides and Plant Growth Regulators*," (London: Granada, 1982), p. 408.
17. Zimmerman, P. W., and F. Wilcoxon. "Several Chemical Growth Substances which Cause Initiation of Roots and Other Responses in Plants," *Contrib. Boyce Thompson Inst.* 7:209-229 (1935).
18. Morgan, P. W., and W. C. Hall. "Effects of 2,4-dichlorophenoxyacetic Acid on the Production of Ethylene by Cotton and Grain Sorghum," *Physiol. Plant.* 15:420-427 (1962).
19. Mattoo, A. K., and J. C. Suttle, Eds. "*The Plant Hormone Ethylene*," (Boca Raton: CRC Press, 1991).
20. Coupland, D., and M. B. Jackson. "Effects of Mecoprop (an Auxin Analogue) on Ethylene Evolution and Epinasty in Two Biotypes of *Stellaria media*," *Ann. Bot.* 68:167-172 (1991).
21. Ursin, V. W., and K. J. Bradford. "Auxin and Ethylene Regulation of Petiole Epinasty in Two Developmental Mutants of Tomato, *Diageotropica* and *Epinastic*," *Plant Physiol.* 90:1341-1346 (1989).
22. Fedtke, C. "*Biochemistry and Physiology of Herbicide Action*," (Berlin: Springer-Verlag, 1982), p. 202.
23. Sanders, G. E., and K. E. Pallett. "Physiological and Ultrastructural Changes in *Stellaria media* Following Treatment with Fluroxypyr," *Ann. Appl. Biol.* 111:385-398 (1987).
24. Breeze, V. G., D. Coupland, P. J. W. Lutman, and A. Hutchings. "Initial Effects of the Herbicide Mecoprop on Photosynthesis and Transpiration by a Mecoprop-resistant Biotype of *Stellaria media*," in *Herbicide Resistance in Weeds and Crops*, J. C. Caseley, G. W. Cussans, and R.K. Atkin, Eds. (Oxford: Butterworths-Heinemann, 1991), p. 513.
25. Fisher, A. D., D. E. Bayer, and T. E. Weier. "Morphological and Anatomical Effects of Picloram on *Phaseolus vulgaris*," *Bot. Gaz.* 129:67-70 (1968).
26. Hall, J. C., P. K. Bassi, M. S. Spencer, and W. H. Vanden Born. "An Evaluation of the Role of Ethylene in Herbicide Injury by Picloram and Clopyralid in Rapeseed and Sunflower Plants," *Plant Physiol.* 79:18-23 (1985).
27. Eames, A. J. "Destruction of Phloem in Young Bean Plants After Treatment with 2,4-D," *Am. J. Bot.* 37:840-847 (1950).
28. Hoson, T., and Y, Masuda. "Relationship Between Polysaccharide Synthesis and Cell Wall Loosening in Auxin-induced Elongation of Rice Coleoptile Segments," *Plant Sci.* 83:149-154 (1992).
29. Penner, D., and F. M. Ashton. "Biochemical and Metabolic Changes in Plants Induced by Chlorophenoxy Herbicides," *Res. Rev.* 14:39-113 (1966).
30. Brummell, D. A., and J. L. Hall. "Rapid Cellular Responses to Auxin and the Regulation of Growth," *Plant Cell Environ.* 10:523-543 (1987).

31. Anderson, P. C., R. E. Lovrien, and M. L. Brenner. "Energetics of the Response of Maize Coleoptile Tissue to Indoleacetic Acid. Characterisation by Flow Calorimetry as a Function of Time, IAA Concentration, and pH," *Planta* 151:499-505 (1981).
32. Brummer, B., A. Bertl, I. Potrykus, H. Felle, and R. W. Parish. "Evidence that Fusicoccin and Indole-3-acetic Acid Induce Cytosolic Acidification of *Zea mays* Cells," *FEBS Lett.* 189:109-114 (1985).
33. Hagen G., and T. J . Guilfoyle. "Rapid Induction of Selective Transcription by Auxins," *Mol. Cell. Biol.* 5:1197-1203 (1985).
34. Evans, M. L., and M. J. Vesper. "An Improved Method for Detecting Auxin-Induced Hydrogen Ion Efflux from Corn Coleoptile Segments," *Plant Physiol.* 66:561-565 (1980).
35. Bates, G. W., and M. H. M. Goldsmith. "Rapid Response of the Plasma-membrane Potential in Oat Coleoptiles to Auxin and other Weak Acids," *Planta* 159:213-237 (1983).
36. Ray, P. M. "Regulation of β-glucan Synthetase Activity by Auxin in Pea Stem Tissue. I. Kinetic Aspects," *Plant Physiol.* 51:601-608 (1973).
37. Cleland, R. E., and T. Lomax. "Hormonal Control of H^+-excretion from Oat Cells," in *Regulation of Cell Membrane Activities in Plants*, E. Marre and O. Ciferri, Eds. (Amsterdam: North-Holland Publishing Co., 1977), pp. 161-171.
38. Meyer, T., L. Aspart, and Y. Chartier. "Auxin-induced Regulation of Protein Synthesis on Tobacco Mesophyll Protoplasts Cultivated *in vitro*. II. Time Course and Level of Auxin Control," *Plant Physiol.* 75:1034-1039 (1984).
39. Jones, A. M., and M. A. Venis. "Photoaffinity Labelling of Indole-3-acetic Acid-binding Proteins in Maize," *Proc. Natl. Acad. Sci. U.S.A.*, 86:6153-6156 (1989).
40. Napier, R. M., M. A. Venis, M. A. Bolton, L. I. Richardson, and G.W. Butcher. "Preparation and Characterisation of Monoclonal and Polyclonal Antibodies to Maize Membrane Auxin-binding Proteins," *Planta* 176:519-526 (1988).
41. Shimomura, S., S. Sotobayashi, M. Futai, and T. Fukui. "Purification and Properties of an Auxin-binding Protein from Maize Shoot Membranes," *J. Biochem.* 99:1513-1524 (1986).
42. Löbler, M., and D. Klämbt. "Auxin-binding Proteins of Corn (*Zea mays*). I. Purification by Immunological Methods and Characterization," *J. Biol. Chem.* 260:9848-9853 (1985).
43. Hicks, G. R., D. L. Rayle, A. M. Jones, and T. L. Lomax. "Specific Photoaffinity Labeling of Two Plasma Membrane Polypeptides with an Azido Auxin," *Proc. Natl. Acad. Sci. U.S.A.* 86:4948-4952 (1989).
44. Hicks, G. R., D. L. Rayle, and T. L. Lomax. "The *Diageotropica* Mutant of Tomato Lacks High Specific Activity Auxin Binding Sites," *Science* 245:52-54 (1989).
45. Sakai, S., and T. Hanagata. "Purification of Auxin-binding Protein from Etiolated Mung Bean Seedlings by Affinity Chromatography," *Plant Cell Physiol.* 24:685-693 (1983).
46. Jones, A. M. "Do We Have the Auxin Receptor Yet?" *Physiol. Plant.* 80:154-158 (1990).
47. Barbier-Brygoo, H., G. Ephritikhine, D. Klambt, M. Ghislain, and J. Guern. "Functional Evidence for an Auxin Receptor at the Plasmalemma of Tobacco Mesophyll Protoplast." *Proc. Natl. Acad. Sci. U.S.A.* 86:891-895 (1989).
48. Ray, P. M. "Auxin-binding Sites of Maize Coleoptiles are Localised in Membranes of the Endoplasmic Reticulum," *Plant Physiol.* 59:594-599 (1977).

49. Hesse, T., J. Feldwisch, D. Balshüemann, G. Bauw, M. Pupe, J. Vandekerckchove, M. Löbler, D. Klämbt, J. Schell, and K. Palme. "Molecular Cloning and Structural Analysis of a Gene from *Zea mays* (L.) Coding for a Putative Receptor for the Plant Hormone Auxin," *EMBO J.* 8:2453-2461 (1989).
50. Goldsmith, M. H. M. "The Polar Transport of Auxins," *Annu. Rev. Plant Physiol.* 28:439-478 (1977).
51. Napier, R. M., and M. A. Venis. "From Auxin-binding Protein to Plant Hormone Receptor?" *TIBS* 16:72-75 (1991).
52. Barbier-Brygoo, H., G. Ephritikhine, D. Klämbt, C. Maurel, K. Palme, J. Schell, and J. Guern. "Perception of the Auxin Signal at the Plasma Membrane of Tobacco Mesophyll Protoplasts," *Plant J.* 1:83-93 (1991).
53. Napier, R. M., and M. A. Venis. "Monoclonal Antibodies Detect an Auxin-induced Conformational Change in the Maize Auxin-binding Protein," *Planta* 182:313-318 (1990).
54. Venis, M. A., E. W. Thomas, H. Barbier-Brygoo, G. Ephritikhine, and J. Guern. "Impermeant Auxin Analogues have Auxin Activity," *Planta* 182:232-235 (1990).
55. Venis, M. A., R. M. Napier, H. Barbier-Brygoo, C. Maurel, C. Perrot-Rechenmann, and J. Guern. "Antibodies to a Peptide from the Maize Auxin-Binding Protein Have Auxin Agonist Activity," *Proc. Natl. Acad. Sci. USA* 89:7208-7212 (1992).
56. Shen, W. H., A. Petit, J. Guern, and J. Tempé. "Hairy Roots are More Sensitive to Auxins than Normal Roots," *Proc. Natl. Acad. Sci. U.S.A.* 85:3417-3421 (1988).
57. Hepler, P. K., and R. O. Wayne. "Calcium and Plant Development," *Annu. Rev. Plant Physiol.* 36:397-439 (1985).
58. André, B., and G. F. E. Scherer. "Stimulation by Auxin Phospholipase A in Membrane Vesicles from an Auxin-sensitive Tissue is Mediated by an Auxin Receptor," *Planta* 185:209-214 (1991).
59. Morré, D. J. "Transmembrane Signal Transduction and the Control of Plant Growth," *Curr. Top. Plant Biochem. Physiol.* 9:47-65 (1990).
60. Hepler, P. K. "Does Calcium Regulate Events Through Amplitude Modulation?" *Curr. Top. Plant Biochem. Physiol.* 9:1-9 (1990).
61. Cobb, A. *Herbicides and Plant Physiology* (London: Chapman & Hall, 1992), p. 176.
62. Silberger, J., and F. Skoog. "Changes Induced by Indoleacetic Acid in Nucleic Acid Content and Growth of Tobacco Pith Tissue," *Science* 118:443-444 (1953).
63. Key, J. L. "Hormones and Nucleic Acid Metabolism," *Annu. Rev. Plant Physiol.* 20:449-474 (1969).
64. Hagen, G. "The Control of Gene Expression by Auxin," in *Plant Hormones and Their Role in Plant Growth and Development*, P. Davies, Ed. (Dordrecht: Martinus Nijhoff, 1987), p. 681.
65. Zurfluh, L. L., and T. J. Guilfoyle. "Auxin-induced Changes in the Population of Translatable Messenger RNA in Elongating Sections of Soybean Hypocotyl," *Plant Physiol.* 69:332-337 (1982).
66. Baulcombe, D. C., P. A. Kroner, and J. L. Key. "Auxin and Gene Regulation," in *Levels of Genetic Control in Development*, S. Subtelny and U. K. Abbott, Eds. (New York: Liss, 1986), pp. 83-97.
67. Walker, J. C., and J. L. Key. "Isolation of Cloned cDNAs to Auxin-Responsive Poly(A)$^+$ RNAs of Elongating Soybean Hypocotyl," *Proc. Natl. Acad. Sci. U.S.A.* 79:185-189 (1982).

68. Hagen, G., A. Kleinschmidt, and T. J . Guilfoyle. "Auxin-regulated Gene Expression in Intact Soybean Hypocotyl and Excised Hypocotyl Sections," *Planta* 162:147-153 (1984).
69. Theologis, A., T. V. Huynh, and R. W. Davis. "Rapid Induction of Specific mRNAs by Auxin in Pea Epicotyl Tissue," *J. Mol. Biol.* 183:53-68 (1985).
70. Theologis, A. "Rapid Gene Regulation by Auxin," *Annu. Rev. Plant Physiol.* 37:407-438 (1986).
71. Dexter, A. G., F. W. Slife, and H. S. Butler. "Detoxification of 2,4-D by Several Plant Species," *Weed Sci.* 19:721-726 (1971).
72. Guilfoyle, T. J ., and J. L. Key. "Purification and Characterization of Soybean DNA-dependent RNA Polymerases and the Modulation of their Activities during Development," in *Nucleic Acids and Protein Synthesis in Plants*, L. Bogorad and J. Weil, Eds. (New York: Plenum, 1977), pp. 37-63.
73. Matthews, L. J. *Weed Control by Chemical Methods* (Wellington, NZ: Government Printers, 1975), p. 710.
74. Fletcher, W. W., and R. C. Kirkwood, Eds. *Herbicides and Plant Growth Regulators*, (London: Granada, 1982), p. 408.
75. Chkanikov, D. I., N. N. Pavlova, A. M. Makeev, T. A. Nazarova, and A. Y. Makoveichuk. "Paths of Detoxification and Immobilisation of 2,4-D in Cucumber Plants," *Sov. Plant Physiol.* 24:457-463 (1977).
76. Hallem, U. "Translocation and Complex Formation of Root-applied 2,4-D and Picloram in Susceptible and Tolerant Species," *Physiol. Plant.* 34:266-272 (1975).
77. Hale, M. G., C. L. Foy, and F. J. Shay. "Factors Affecting Root Exudation," *Adv. Agron.* 23:89-109 (1971).
78. Coupland, D. "Factors Affecting the Phloem Translocation of Foliage-applied Herbicides," in *British Plant Growth Regulator Group, Monograph 18 — 1989. Mechanisms and Regulation of Transport Processes*, R. K. Atkin and D. R. Clifford, Eds. (Bristol: British Plant Growth Regulator Group, 1989), pp. 85-112.
79. Peterson, P. J., L. C. Haderlie, R. H. Hoefer, and R. S. McAllister. "Dicamba Absorption and Translocation as Influenced by Formulation and Surfactant," *Weed Sci.* 33:717-720 (1985).
80. Owen, W. J. "Herbicide Metabolism as a Basis for Selectivity," in *Target Sites for Herbicide Action*, R.C. Kirkwood, Ed. (New York: Plenum Press, 1991), pp. 285-314.
81. Scheel, D., and H. Sandermann. "Metabolism of 2,4-dichlorophenoxyacetic Acid in Cell Suspension Cultures of Soybean (*Glycine max* L.) and Wheat (*Triticum aestivum* L.). I. General Results," *Planta* 152:248-252 (1981).
82. Lewer, P., and W. J. Owen "Selective Action of the Herbicide Triclopyr," *Pestic. Biochem. Physiol.* 36:187-200 (1990).
83. Luckwell, L. C., and C. P. Lloyd-Jones. "Metabolism of Plant Growth Regulators. I. 2,4-dichlorophenoxyacetic Acid in Leaves of Red and Black Currant," *Ann. Appl. Biol.* 48:613-625 (1960).
84. Sanad, A. J., and F. Mäller. "Untersuchungen über den Metabolismus von ^{14}C-markiertem 2,4-D bei Verschiedenen Resistenten Unkräutern," *Z. Pflanzenkr. Pflanzenschutz* 79:651-658 (1972).
85. Wain, R. L., and M. S. Smith. "Selectivity in Relation to Metabolism," in *Herbicides. Physiology, Biochemistry, Ecology*. Vol. 2, L. J. Audus, Ed. (London: Academic Press, 1976) pp. 279-302.
86. Hagin, R. D., D. L. Linscott, and J. E. Dawson. "2,4-D Metabolism in Resistant Grasses," *J. Agric. Food Chem.* 18:848-850 (1970).

87. Linscott, D. L., R. D. Hagin, and J. E. Dawson. "Conversion of 4-(2,4-dichlorophenoxy)butyric Acid to Homologs by Alfalfa," *J. Agric. Food Chem.* 16:844-848 (1968).
88. Guroff, G., J. Daly, D. Jerina, J. Renson, and B. Witkop. "Hydroxylation-induced Migration: the NIH shift," *Science* 157:1524-1527 (1967).
89. Feung, C., R. H. Hamilton, and R. O. Mumma. "Metabolism of 2,4-dichorophenoxyacetic acid. VII. Comparison of Metabolites from Five Species of Plant Callus Tissue Cultures," *J. Agric. Food Chem.* 23:373-376 (1975).
90. Cole, D. J., and B. C. Loughman. "The Metabolic Fate of (4-chloro-1-methylphenoxy) acetic Acid in Higher Plants," *J. Exp. Bot.* 34:1299-1310 (1983).
91. Coupland, D. "The Role of Compartmentation of Herbicides and their Metabolites in Resistance Mechanisms," in *Herbicide Resistance in Weeds and Crops*, J. C. Caseley, G. W. Cussans, and R. K. Atkin, Eds. (Oxford: Butterworths-Heinemann, 1991), p. 513.
92. Broadhurst, N. A., M. L. Montgomery, and V. H. Freed. "Metabolism of 2-methoxy-3,6-dichlorobenzoic acid (Dicamba) by Wheat and Blue Grass Plants," *J. Agric. Food Chem.* 14:585-588 (1966).
93. Frear, D. S., H. R. Swanson, E. R. Mansager, and R. G. Wien. "Chloramben Metabolism in Plants: Isolation and Identification of Glucose Ester," *J. Agric. Food Chem.* 26:1347-1351 (1978).
94. Schmitt, R., and H. Sandermann. "Specific Localisation of ß-D-glucoside Conjugates of 2,4-dichlorophenoxyacetic Acid in Soybean Vacuoles," *Z. Naturforsch.* 37:772-777 (1982).
95. Nakakmura. C., M. Nakata, M. Shioji, and H. Ono. "2,4-D Resistance in a Tobacco Cell Culture Variant: Cross-resistance to Auxin and Uptake, Efflux and Metabolism of 2,4-D," *Plant Cell Physiol.* 26:271-280 (1985).
96. LeBaron, H. M., and J. Gressel, Eds. *Herbicide Resistance in Plants*, (New York: John Wiley & Sons, 1982), p. 401.
97. Bandeen, J., G. R. Stephenson, and E. R. Cowett. "Discovery and Distribution of Herbicide-Resistant Weeds in North America," in *Herbicide Resistance in Plants*, H. M. LeBaron, and J. Gressel, Eds. (New York: John Wiley & Sons, 1982), pp. 9-30.
98. Whitehead, C. W., and C. M. Switzer. "The Differential Response of Strains of Wild Carrot to 2,4-D and Related Herbicides," *Can. J. Plant Sci.* 43:255-262 (1967).
99. Bourdôt, G. W., K. C. Harrington, and A. I. Popay. "The Appearance of Phenoxy-Herbicide Resistance in New Zealand Pasture Weeds," *Proceedings of the Brighton Crop Protection Conference — Weeds*, pp. 309-316 (1989).
100. Harrington, K. C., and A. I. Popay. "Differences in Susceptibility of Nodding Thistle Populations to Phenoxy Herbicides," *Proceedings of the 8th Australian Weeds Conference*, pp. 126-129 (1989).
101. Harrington, K. C. "Distribution and Cross-tolerance of MCPA-tolerant Nodding Thistle," *Proceedings of the 42nd New Zealand Weed and Pest Control Conference*, pp. 39-42 (1989).
102. Harrington, K. C. "Spraying History and Fitness of Nodding Thistle, *Carduus nutans*, populations resistant to MCPA and 2,4-D," *Proceedings of the 9th Australian Weeds Conference*, pp. 201-204 (1990).
103. Lutman, P. J. W., and A. W. Lovegrove. "Variations in the Tolerance of *Galium aparine* (cleavers) and *Stellaria media* (chickweed) to Mecoprop," *Proceedings of the 1985 British Crop Protection Conference — Weeds*, pp. 411-418 (1985).

104. Lutman, P. J. W. "Further Investigations into the Resistance of Chickweed (*Stellaria media*) to Mecoprop," *Proceedings of the British Crop Protection Conference — Weeds*, pp. 901-908 (1987).
105. Radosevich, S. R. and J. S. Holt. "Physiological Responses and Fitness of Susceptible and Resistant Weed Biotypes to Triazine Herbicides," in *Herbicide Resistance in Plants*, H. M. LeBaron, and J. Gressel, Eds. (New York: John Wiley & Sons, 1982), pp. 163-182.
106. Sobney, D. G. "Biological Flora of the British Isles, No. 150: *Stellaria media* (L.) Vill.," *J. Ecol.* 69:311-335 (1981).
107. Briggs, D., H. Hodgkinson, and M. Block. "Precociously Developing Individuals in Populations of Chickweed *Stellaria media* (L.) Vill. from Different Habitat Types, with Special Reference to the Effects of Weed Control Measures," *New Phytol.* 117:153-164 (1991).
108. Popay, A. I., D. K. Edmonds, L. A. Lyttle, and H. T. Phung. "Timing of MCPA Applications for Control of Giant Buttercup," *Proceedings of the 37th New Zealand Weed and Pest Control Conference*, pp. 17-19 (1984).
109. McNaughton, A. S. "Physiological and Genetic Aspects of Herbicide Resistance in Giant Buttercup (*Ranunculus acris* L. subsp. *acris*)," M.Sc. Thesis, Lincoln College, Canterbury, New Zealand (1991).
110. Bourdôt, G. W., and G. A. Hurrell. "Differential Tolerance of MCPA among Giant Buttercup (*Ranunculus acris*) biotypes in Takaka, Golden Bay," *Proceedings of the 41st New Zealand Weed and Pest Control Conference*, pp. 231-234 (1988).
111. Bourdôt, G. W., G. A. Hurrell, and D. J. Saville. "MCPA-resistance in Giant Buttercup (*Ranunculus acris* L.) in North Island Dairy Pastures," *Proceedings of the 43rd New Zealand Weed and Pest Control Conference*, pp. 229-232 (1990).
112. Bourdôt, G. W., G. A. Hurrell, and D. J. Saville. "Variation in MCPA-resistance in *Ranunculus acris* L. subsp. *acris* and its Correlation with Historical Exposure to MCPA," *Weed Res.* 30:449-457 (1990).
113. Bourdôt, G. W., and G. A. Hurrell. "Evidence for the Heritability of MCPA-resistance in Giant Buttercup (*Ranunculus acris*)," *Proceedings of the 44th New Zealand Weed and Pest Control Conference*, pp. 270-274 (1991).
114. Whitworth, J. W., and J. A. Long. "The Association of Plant Characters with the Differential Tolerance of Field Bindweed to Growth Regulator Herbicides," *Annu. Rep. Co-op. Regional Project W-11, Mimeo. New Mexico Exp. Station* (1956).
115. Whitworth, J. W. "The Reaction of Strains of Field Bindweed to 2,4-D," *Weeds* 12:57-58 (1964).
116. Whitworth, J. W., and T. J. Muzik. "Differential Response of Selected Clones of Bindweed to 2,4-D," *Weeds* 15:275-280 (1967).
117. Harvey, R. G., and T. J. Muzik. "Effects of 2,4-D and Amino Acids on Field Bindweed *in vitro*," *Weed Sci.* 21:135-138 (1973).
118. Heap, I. M., and I. N. Morrison. "Resistance to Auxin-type Herbicides in Wild Mustard (*Sinapis arvensis* L.) Populations in Western Canada," *Abstracts of the Weed Science Society of America*, 32:55 (1992).
119. Solymosi, P., Z. Kostyál, and A. Gimesi. "Resistance of *Cirsium arvense* (L.) to Phenoxyacetic Acid Derivatives," *Növényvedelem* 23:301-305 (1987).
120. Hodgson, J. M. "The Response of Canada Thistle Ecotypes to 2,4-D, Amitrole and Intensive Cultivation," *Weed Sci.* 18:253-255 (1970).

121. Abel, A. L. "The Rotation of Weed Killers," *Proceedings of the British Weed Control Conference* 1:249-255 (1954).
122. Fogelfors, H. "Changes in the Flora of Farmland" *Swed. Univ. Agric. Sci. Dept. Ecol. Environ. Res. Rep. No. 5, Uppsala* (1979), p. 66.
123. Aberg, E. "Weed Control Research and Development in Sweden," *Proceedings of the 3rd British Weed Control Conference*, pp. 141-146 (1956).
124. Mukula, J., M. Raatikainen, R. Lallukka, and T. Raatikainen. "Composition of Weed Flora in Spring Cereals in Finland," *Ann. Agric. Fenn.* 8:61-110 (1969).
125. Thurston, J. M. "Weed Studies in Broadbalk," *Report of the Rothamsted Experimental Station for 1968* (1969), pp. 186-208.
126. Ellis, M., and Q. O. N. Kay. "Genetic Variation in Herbicide Resistance in Scentless Mayweed (*Tripleurospermum inodorum* (L.) Schultz Bip.). I. Differences Between Populations in Response to MCPA," *Weed Res.* 15:307-315 (1975).
127. Ellis, M., and Q. O. N. Kay. "Genetic Variation in Herbicide Resistance in Scentless Mayweed (*Tripleurospermum inodorum* (L.) Schultz Bip.). II. Intraspecific Variation in Response to Ioxynil and MCPA and the Role of Spray Retention Characteristics," *Weed Res.* 15:317-326 (1975).
128. Hammerton, J. L. "Studies on Weed Species of the Genus *Polygonum* L. III. Variation in Susceptibility to 2-(2,4-dichlorophenoxy)propionic Acid within *P. lapathifolium*," *Weed Res.* 6:132-141 (1966).
129. Sexsmith, J. J. "Morphological and Herbicide Susceptibility Differences among Strains of Hoary Cress," *Weeds* 12:19-22 (1964).
130. Bell, A. R., J. D. Nalewaja, and A. B. Schooler. "Response of *Kochia* Selection to 2,4-D, Dicamba and Picloram," *Weed Sci.* 20:458-462 (1972).
131. Bell, A. R., J. D. Nalewaja, A. B. Schooler, and T. S. Hsieh. "Herbicide Response and Morphology of Interspecific Sowthistle Crosses," *Weed Sci.* 21:189-193 (1973).
132. Beauge, A. "*Chenopodium album* et Spèces Affinés," *SEDES*, p. 409 (1974).
133. Stryckers, J. "OnderzoeKingen Naar de toePassings-mogelijkheden van Synthetische Groeistoffen als Selektieve Herbicidem in Ggrasland en Akkerbougeswassen," *Gent Rijkslandbouwhogsch. Rep.* (1958), p. 100.
134. Chaleff, R. S. "Further Characterisation of Picloram-tolerant Mutants of *Nicotiana tabacum*," *Theor. Appl. Genet.* 58:91-95 (1980).
135. Ono, H. "Genetical and Physiological Investigations of a 2,4-D Resistant Cell Line Isolated from the Tissue Cultures in Tobacco. I. Growth Responses to 2,4-D and IAA," *Sci. Rep. Fac. Agric. Kobe Univ.* 13:272-277 (1979).
136. Ono, H., and M. Nakano. "The Regulation of Expression of Cellular Phenotypes in Cultural Tissues of Tobacco," *Jpn. J. Genet.* 53:241-250 (1978).
137. Nakamura, C., M. Nakata, M. Shioji, and H. Ono. "2,4-D Resistance in a Tobacco Cell Culture Variant: Cross-resistance to Auxins and Uptake, Efflux and Metabolism of 2,4-D," *Plant Cell Physiol.* 26:271-280 (1985).
138. Chaleff, R. S., and M. F. Parsons. "Direct Selection *in vitro* for Herbicide-resistant Mutants of *Nicotiana tabacum*," *Proc. Natl. Acad. Sci. U.S.A.* 75:5104-5107 (1978).
139. Oswald, T. H., A. E. Smith, and D. V. Phillips. "Herbicide Tolerance Developed in Cell Suspension Cultures of Perennial White Clover," *Can. J. Bot.* 55:1351-1358 (1977).
140. Taylor, S. G., D. D. Baltensperger, and K. H. Quesenberry. "Recurrent Half-sib Family Selection for 2,4-D Tolerance in Red Clover," *Crop Sci.* 29:1109-1114 (1989).

141. Devine, T. E., R. R. Seaney, D. L. Linscott, R. D. Hagin, and B. Brace. "Results of Breeding for Tolerance to 2,4-D in Birdsfoot Trefoil," *Crop Sci.* 15:721-724 (1975).
142. Swanson, E. B., and D. T. Tomes. "*In vitro* responses of Tolerant and Susceptible Lines of *Lotus corniculatus* L. to 2,4-D," *Crop Sci.* 20:792-795 (1980).
143. Swanson, E. B., and D. T. Tomes. "Plant Regeneration from Cell Cultures of *Lotus corniculatus* and the Selection and Characterisation of 2,4-D Tolerant Cell Lines," *Can. J. Bot.* 58:1205-1209 (1980).
144. Miller, J. C. Jr., and L. R. Baker. "Differential Phytotoxicity of Amiben Methyl Ester to *Cucumis sativus* Lines," *Hortic. Sci.* 6:276 (1971).
145. Miller, J. C. Jr., D. Penner, and L. R. Baker. "Basis for Variability in the Cucumber for Tolerance to Chloramben Methyl Ester," *Weed Sci.* 21:207-211 (1973).
146. Miller, J. C. Jr., L. R. Baker, and D. Penner. "Inheritance of Tolerance to Chloramben Methyl Ester in Cucumber," *J. Am. Soc. Hortic. Sci.* 98:386-389 (1973).
147. Klee, H. J., and M. Estelle. "Molecular Genetic Approaches to Plant Hormone Biology," *Annu. Rev. Plant Physiol. Plant Mol. Biol.* 42:529-551 (1991).
148. Bitoun, R., P. Rousselin, and M. Caboche. "A Pleiotropic Mutation Results in Cross-Resistance to Auxin, Abscisic Acid and Paclobutrazol," *Mol. Gen. Genet.* 220:234-239 (1990).
149. Muller, J.-F., C. Missionier, and M. Caboche. "Low Density Growth of Cells Derived from *Nicotiana* and *Petunia* Protoplasts: Influence of the Source of Protoplasts and Comparison of the Growth Promoting Activity of Various Auxins," *Plant Physiol.* 57:35-41 (1983).
150. Ephritikhine, G., H. Barbier-Brygoo, J.-F. Muller, and J. Guern. "Auxin Effect on the Transmembrane Potential Difference of Wild-type and Mutant Tobacco Protoplasts Exhibiting a Differential Sensitivity to Auxin," *Plant Physiol.* 83:801-804 (1987).
151. Kelly, M. A., and K. J. Bradford. "Insensitivity of the Diageotropica Tomato Mutant to Auxin," *Plant Physiol.* 82:713-717 (1986).
152. Hicks, G. R., D. L. Rayle, A. Jones, and T. L. Lomax. "Specific Photoaffinity Labeling of Two Plasma Membrane Polypeptides with an Azido Auxin," *Proc. Natl. Acad. Sci. U.S.A.* 86:4948-4952 (1989).
153. Hicks, G. R., D. L. Rayle, and T. L. Lomax. "The Diageotropica Mutant of Tomato Lacks High Specific Activity Auxin Binding Sites," *Science* 245:51-54 (1989).
154. Nakamura, C., H. J. van Telgen, M. Mennes, H. Ono, and K. R. Libbenga. "Correlation Between Auxin Resistance and the Lack of a Membrane-bound Auxin Binding Protein and a Root-specific Peroxidase in *Nicotiana tabacum*," *Plant Physiol.* 88:845-849 (1988).
155. Muller, J.-F., J. Goujaud, and M. Caboche. "Isolation *in vitro* of Naphthaleneacetic Acid-tolerant Mutants of *Nicotiana tabacum* which are Impaired in Root Morphogenesis," *Mol. Gen. Genet.* 199:194-200 (1985).
156. Mirza, J. I., G. M. Olsen, T.-H. Iversen, and E. P. Maher. "The Growth and Gravitropic Responses of Wild-type and Auxin-resistant Mutants of *Arabidopsis thaliana*," *Plant Physiol.* 60:516-522 (1984).
157. Estelle, M. A., and C. R. Somerville. "Auxin Resistant Mutants of *Arabidopsis thaliana* with an Altered Morphology," *Mol. Gen. Genet.* 206:200-206 (1986).
158. Pickett, F. B., A. K. Wilson, and M. Estelle. "The *aux1* Mutation of *Arabidopsis* Confers both Auxin and Ethylene Resistance," *Plant Physiol.* 94:1462-1466 (1990).

159. Wilson, A. K., F. B. Pickett, J. C. Turner, and M. Estelle. "A Dominant Mutation in *Arabidopsis* Confers Resistance to Auxin, Ethylene and Abscisic acid," *Mol. Gen. Genet.* 222:377-383 (1990).
160. Penuik, M. G., M. L. Romano, and J. C. Hall. "Absorption, Translocation and Metabolism are not the Basis for Differential Selectivity of Wild Mustard (*Sinapis arvensis* L.) to Auxinic Herbicides," *Abstracts Weed Science Society of America* p. 55 (1992).
161. Harrington, K. C. "Aspects of Resistance to Phenoxy Herbicides in Nodding Thistle (*Carduus nutans* L.)," Ph.D. Thesis, Massey University, New Zealand, (1992).
162. Davis, C., and D. L. Linscott. "Tolerance of Birdsfoot Trefoil (*Lotus corniculatus*) to 2,4-D," *Weed Sci.* 34:373-376 (1986).
163. Davidonis, G. H., R. H. Hamilton, and R. O. Mumma. "Metabolism of 2,4-dichlorophenoxyacetic Acid in 2,4-dichlorophenoxyacetic Acid-resistant Soybean Callus Tissue," *Plant Physiol.* 70:104-107 (1982).
164. Hart, S. E., S. Glenn, and W. W. Kenworthy. "Tolerance and the Basis for Selectivity to 2,4-D in Perennial *Glycine* species," *Weed Sci.* 39:535-539 (1991).
165. Bayley, C., N. Trolinder, C. Ray, M. Morgan, J. E. Quisenberry, and D.W. Ow. "Engineering 2,4-D Resistance into Cotton," *Theor. Appl. Genet.* 83:645-649 (1992).
166. Streber, W. R., and L. Willmitzer. "Transgenic Tobacco Plants Expressing a Bacterial De-toxifying Enzyme are Resistant to 2,4-D," *Biotechnology* 7:811-816 (1989).
167. Lyon, B. R., D. J. Llewellyn, J. L. Huppatz, E. S. Dennis, and W. J. Peacock. "Expression of a Bacterial Gene in Transgenic Tobacco Plants Confers Resistance to the Herbicide 2,4-dichlorophenoxyacetic Acid," *Plant Mol. Biol.* 13:533-540 (1989).
168. Perkins, E. J., C. M. Stiff, and P. F. Lurquin. "Use of *Alcaligenes eutrophus* as a Source of Genes for 2,4-D Resistance in Plants," *Weed Sci.* 35:12-18 (1987).
169. Coupland, D., P. J. W. Lutman, and C. Heath. "Uptake, Translocation and Metabolism of Mecoprop in a Sensitive and a Resistant Biotype of Stellaria media." *Pestic. Biochem. Physiol.* 36:61-67.
170. Lutman, P. J. W., and C. R. Heath. "Variations in the Resistance of *Stellaria media* to Mecoprop due to Biotype, Application Method and 1-aminobenzotriazole," *Weed Res.* 30:129-137 (1990).
171. Barnwell, P., and A. H. Cobb. "Physiological Studies of Mecoprop-resistance in Chickweed (*Stellaria media* L .)," *Weed Res.* 29:135-140 (1989).
172. Coupland, D., G. M. Arnold, T. Haynes, and M. Hindle. "Ethylene Production in Relation to the Resistance of *Stellaria media* (chickweed) to Mecoprop," *Pestic. Sci.* 34:365-367 (1992).
173. Gonneau, M., B. Paquette, F. Cabanne, and R. Scalla. "Transformation of Phenoxyacetic Acid and Chlorotoluron in Wheat, Barren Brome, Cleavers and Speedwell. Effects of an Inactivator of Monooxygenases," *Proceedings of the 1987 British Crop Protection Conference — Weeds*, pp. 329-336 (1987).
174. Liu, Y., L.-Y. Su, and S. F. Yang. "Ethylene Promotes the Capability to Malonylate 1-aminocyclopropane-1-carboxylic Acid and D-amino Acids in Pre-climacteric Tomato Fruits," *Plant Physiol.* 77:891-895 (1985).
175. Lamoureux, G. L., and D. G. Russness. "Xenobiotic Conjugation in Higher Plants," in *Xenobiotic Conjugation Chemistry*, G. D. Paulson, J. Caldwell, D. H. Hutson, and J. J. Mann, Eds. *American Chemical Society Symposium Series No. 299*, (Washington, D.C.: American Chemical Society, 1986), pp. 62-105.

176. Weyers, J. B. D., N. W. Paterson, and R. A. Brook. "Towards a Quantitative Definition of Plant Hormone Sensitivity," *Plant Cell Environ.* 10:1-10 (1987).
177. Coupland, D., D. T. Cooke, and C. S. James. "Effects of 4-chloro-2-methylphenoxypropionate (an Auxin Analogue) on Plasma Membrane ATPase Activity in Herbicide-resistant and Herbicide-susceptible Biotypes of *Stellaria media* L.," *J. Exp. Bot.* 42:1065-1071 (1991).
178. Coupland, D., D. T. Cooke, C. S. James, and B. James. "Effects of Plant Age on H^+-ATPase Activity in Two Biotypes of *Stellaria media* L. with Different Sensitivities to Mecoprop, an Auxin Analogue," Abstracts of the 8th Congress of the Federation of European Societies of Plant Physiology, *Physiol. Plant.* 85:A60 (1992).
179. Shaner, D., A. Sinha, and R. Braddock. "Designing Strategies to Delay Development of Resistance to Herbicides," *Proceedings of the First Int. Weed Control Conference* 1:236-239 (1992).
180. Schulz, A., F. Wengenmayer, and H. M. Goodman. "Genetic Engineering of Herbicide Resistance in Higher Plants," *Crit. Rev. Plant Sci.*, 9:1-15 (1990).
181. Putwain, P. D., and A. M. Mortimer. "The Resistance of Weeds to Herbicides: Rational Approaches for Containment of a Growing Problem," *Proceedings of the Brighton Crop Protection Conference — Weeds*, pp. 285-294 (1989).
182. Roush, M. L., S. R. Radosevich, and B. D. Maxwell. "Future Outlook for Herbicide-Resistance Research," *Weed Technol.* 4:208-214 (1990).
183. Caseley, J. C. "Improving Herbicide Performance with Synergists that Modulate Metabolism," *Proceedings of the First Int. Weed Control Congress,* Melbourne, Australia, 2:113-115 (1992).
184. Bourdôt, G. W., and I. Harvey. "A Potential Mycoherbicide for the Control of Phenoxy-herbicide-resistant Giant Buttercup (*Ranunculus acris*) in Dairy Pastures," *Proceedings of the First Int. Weed Control Congress,* Melbourne, Australia, 2:94 (1992).
185. LeBaron, H. M., and J. McFarland. "Herbicide Resistance in Weeds and Crops. An Overview and Progress," in *Managing Resistance in Agrochemicals. From Fundamental Research to Practical Strategies.* M. B. Green, H. M. LeBaron, and W. K. Moberg, Eds. ACS Symposium Series No 421, (Washington, D.C.: American Chemical Society, 1990), pp. 336-352.
186. White, R. H., R. A. Liebl, and T. Hymowitz "Examination of 2,4-D Tolerance in Perennial Glycine Species," *Pest. Biochem. Physiol.* 38:153-161 (1990).
187. Bovey, R. W. "Physiological Effects of Phenoxy Herbicides in Higher Plants," *The Science of 2,4,5-T and Associated Phenoxy Herbicides,* R. W. Bovey and A. L. Young, Eds. (New York: John Wiley & Sons, 1980) pp. 217-238.
188. Webb, S. R., and J. C. Hall "Indole-3-acetic Acid Binding Characteistics in Susceptible and Resistant Biotypes of Wild Mustard (*Sinapis arvensis* L.)," *Abstract Weed Science Society of America,* p. 66 (1993).
189. Shaner, D., A. Sinha, and R. Braddock "Designing Strategies to Delay Development of Resistance to Herbicides," *Proceedings First Int. Weed Control Congress,* Melbourne, Australia, 1:236-239 (1992).
190. Peniuk, M. G., M. L. Romano, and C. J. Hall "Physiological Investigations into the Resistance of a Wild Mustard (*Sinapis arvensis* L.) Biotype to Auxinic Herbicides," *Weed Sci.,* 33:431-440 (1993).
191. Lutman, P., personal communication.

CHAPTER 7

Resistance to Dinitroaniline Herbicides

Reid J. Smeda and Kevin C. Vaughn

INTRODUCTION

Between a quarter and a third of all of the herbicides marketed affect mitosis as a primary mechanism of action,[1] although many of them are of limited use or have been replaced by more effective herbicides. Within a class of mitotic disrupters termed dinitroanilines, trifluralin, oryzalin, and pendimethalin are still widely used in cotton, soybeans, and wheat, as well as other agronomic, tree, vine, and ornamental crops.[2] In use for over two decades, dinitroaniline herbicides selectively inhibit the growth of many annual grasses and some dicot weeds. Because herbicide absorption occurs through roots or shoot tissue emerging from the soil, dinitroaniline herbicides are applied prior to weed emergence. They provide greater than 90% control of target weeds for up to 10 weeks in many environments and season-long control in some situations. Their efficacy and low cost has contributed to their continuous use in many cropping systems. Despite the extensive and sustained use of dinitroaniline herbicides, only a few examples of resistance have been reported. In this chapter, we will describe the laboratory and field studies on the resistant biotypes that have arisen, as well as discuss the mechanisms by which these weed biotypes and various crops resist these herbicides.

Selectivity of Dinitroaniline Herbicides

Dinitroaniline herbicides are more phytotoxic to grass species than dicot species. Selectivity may be based on the lipid content of seeds, because dinitroanilines are quite lipophilic, and lipid content in seeds of grasses is generally much lower than in dicots. Hilton and Christiansen[3] examined this hypothesis and found a close correlation between trifluralin sensitivity and percent lipid content of seeds from various weed and crop species. Further experiments demonstrated that increasing the seed lipid content of a sensitive species by impregnating seeds with α-tocopherol acetate prior to planting, moderately improved shoot and root growth in the presence of trifluralin.[3] It is believed that the dinitroanilines are partitioned preferentially into lipid bodies in the seed and thereby sequestered away from the site of action ascribed to these herbicides (microtubules, see below).

Although seed of resistant weed and crop species will germinate in soil incorporated with dinitroanilines, development of roots, especially lateral roots, is often adversely affected. Dinitroaniline herbicides are typically incorporated into the top 10 cm of soil, and do not readily leach downward. Therefore, large-seeded crops and weeds with greater levels of stored lipids and proteins are capable of growing through the zone where the herbicide is incorporated, whereas small-seeded weeds cannot. Hilton and Christiansen[3] have found that lateral root development of cotton seedlings was inhibited in the top 8.5 cm of soil which contained trifluralin. However, below this level, trifluralin was absent from the soil and lateral root development was normal. In crop species with established root systems, such as tree, vine, or ornamental crops, dinitroaniline herbicides are effective in suppressing weed species in the upper soil profile with no effect on crop roots. Carrots and other members of the Umbelliferae are notable exceptions to selectivity based on seed lipid content or seed size.[4] They are small-seeded and are relatively low in lipid content, yet have a high level of resistance to dinitroaniline herbicides. The weed *Abutilon theophrasti* Medik. was also shown to be exceptionally resistant, despite having relatively low quantities of seed lipids.[3] The resistance mechanisms in *A. theophrasti* and Umbelliferae are not known.

MODE OF ACTION

The most striking symptoms of dinitroaniline treatment are a reduction in root length and the swelling of seedling roots into a characteristic club shape.[5] This is an effect identical to that obtained by treating seedlings with the classic mitotic disrupter colchicine,[6] and was one of the first indicators as to the mechanism of action of these compounds. Within treated roots, the cells are able to proceed through prometaphase normally, but no later stages of mitosis are observed, resulting in a number of cells arrested in this C (colchicine-type)-mitosis at prometaphase. Eventually, a nuclear membrane reforms around the chromosomes, but because of the doubling in chromosome number and the spreading of the chromosomes throughout the cytoplasm, the nucleus takes on a highly lobed morphology.[1,5] Electron micrographs reveal that no or few spindle and kinetochore microtubules, the cellular structures responsible for chromosome movement during mitosis, occur in cells treated with dinitroaniline herbicides.[1,5] Colchicine also causes the loss of plant spindle microtubules, but at concentrations 100 to 1000 times higher compared to dinitroanilines.[7] Cortical microtubules, which determine the shape of root cells in the zone of elongation by directing the deposition of cell wall components, are also affected by the dinitroaniline herbicides. The end results are isodiametric (square-shaped) rather than elongate (rectangular) cells, contributing to the swollen appearance of the root tips.

Biochemical results with dinitroaniline herbicides parallel those with colchicine. Upon addition of dinitroaniline herbicides to plant extracts enriched for tubulin, the major protein constituent of microtubules, polymerization of tubulin into microtubules is inhibited.[8,9] Although the exact molecular mechanism for this inhibition is not known, it believed that the dinitroanilines react with free tubulin heterodimers (composed of α and β subunits) in the cytoplasm and, when attached to the growing (or plus) end of the microtubule, prevent further polymerization. Colchicine also prevents plant tubulin polymerization in microtubules by binding to tubulin.[10,11] Microtubules are dynamic structures, extending at one end through polymerization of tubulin and either gradually losing subunits from the minus ends or cataclysmically losing whole microtubules or segments thereof. By blocking polymerization, the depolymerization process continuously shortens the microtubules until they eventually are undetectable. Unlike colchicine, dinitroaniline herbicides do not inhibit *in vitro* polymerization of animal tubulin,[7,9,12] nor do these compounds disrupt mitosis in animal cells.[13] This indicates that the binding sites for dinitroaniline herbicides on plant tubulin are not present on animal tubulin.

DEVELOPMENT/DISCOVERY OF RESISTANCE

Resistant Plant Species

Since the early 1970s, the number of herbicide-resistant weed species has gone from 1 to over 100.[14] Resistance has evolved predominantly where a herbicide, with a single site of action, has imposed a high selection pressure and has been used as the sole or primary herbicide for many years.[15]

Dinitroaniline herbicides have been used annually for 10 to 20 years in fields dedicated to producing cotton in the Southeastern U.S. and for approximately the same length of time in oilseed and small grain crops on the Canadian prairies and elsewhere in the world. It was not unexpected, therefore, that weed resistance to dinitroaniline herbicides was discovered in these two areas. Resistance to dinitroaniline herbicides has been reported in *Eleusine indica*

(L.) Gaertn. (goosegrass),[16,17] *Setaria viridis* (green foxtail),[18] and more recently in *Sorghum halepense* (L.) Pers. (Johnsongrass)[19] and *Amaranthus palmeri* S. Wats.(Palmer amaranth).[20] To our knowledge, there are no reports of normally sensitive crop species that have been selected for resistance to dinitroanilines. However, Davis et al.[21] found a wide range of responses to trifluralin among 52 lines of corn, indicating genetic variability for sensitivity to trifluralin.

In two cases, resistance to dinitroaniline herbicides has been discovered in weed species with multiple herbicide resistance. As discussed in Chapter 9, in Australia, biotypes of *Lolium rigidum* (annual ryegrass) which exhibit multiple resistance to the ACCase and ALS inhibiting herbicides are also resistant to dinitroanaline herbicides.[22] Biotypes of *Alopecurus myosuroides* Huds. (slender foxtail) in the United Kingdom are resistant to chlortoluron and isoproturon, and to other chemically unrelated herbicides including pendimethalin.[23]

In addition to reports of higher plants resistant to dinitroaniline herbicides, various researchers have documented numerous *Chlamydomonas reinhardtii* mutants exhibiting varying levels of resistance.[24,25] Mutants were selected *in vitro* by exposing chemically mutagenized cells to sublethal herbicide concentrations. The level of resistance to oryzalin was 3- to 6-fold for selected mutants and the resistance characteristic remained stable through five generations of progeny backcrossed to a dinitroaniline-sensitive, nonmutagenized strain,[24] indicating it is not an epigenetic modification.

Discovery of dinitroaniline-resistant weed species has sparked research to determine whether resistance is widespread and whether all resistant types are the same. In 1984, biotypes of *E. indica* resistant to field use rates of trifluralin were found in seven adjacent counties in South Carolina.[16] In each case, dinitroaniline herbicides had been used for over 10 years annually in fields devoted to cotton production. One of the four most highly resistant biotypes was not controlled by any of seven different dinitroaniline herbicides at up to twice the recommended dose.[16] The level of trifluralin resistance in this highly resistant R biotype was 1,000- to 10,000-fold compared to the S biotype, based upon changes in the mitotic index (a measure of the number of cells in mitosis at a given time point).[26] Growth of three of the seven biotypes was suppressed by trifluralin at twice the recommended dose,[16] with a level of resistance to trifluralin about 50-fold, based upon elevation of mitotic indices in treated seedlings.[17] Biotypes exhibiting this lower level of resistance were described as intermediate-resistant (I). Recently, biotypes of *E. indica* from other cotton-producing areas in Alabama, Georgia, and Tennessee were identified as dinitroaniline resistant. Comparing root growth and development with a known susceptible and a highly resistant biotype from South Carolina, the Alabama and Georgia biotypes exhibited a high level of resistance to oryzalin, trifluralin, and pendimethalin. The Tennessee biotype exhibited an intermediate level of resistance to these herbicides.[27]

Biotypes of *S. viridis* from three sites (in Manitoba, Canada) were identified as resistant, following annual applications of trifluralin for over 15 years.[18] Under greenhouse conditions, the R populations were 5-fold more resistant to trifluralin than a representative S population. Additional research in our laboratory determined the I_{50} R:S value (concentration of herbicide required to inhibit root growth of the R biotype by 50% divided by the concentration inducing the same effect on the S biotype) was 3.2 for trifluralin (Figure 1).[28] Resistant *S. viridis* was also resistant to five other dinitroaniline herbicides with I_{50} R:S values ranging from 1.6 to 14.8.[28] The level of resistance in *S. viridis* to dinitroaniline herbicides is clearly lower than the resistance exhibited by biotypes of *E. indica*.

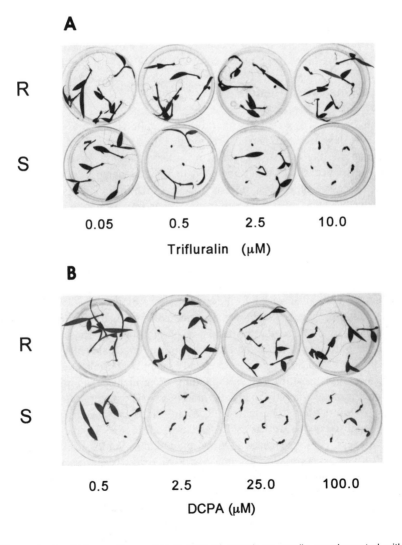

Figure 1. *S. viridis* seedlings after 7 days of growth on media supplemented with trifluralin (A) or DCPA (B). Note the inhibition of both root and shoot growth in the S biotype with a correspondingly lower effect in the R biotype.

Another grass species, *S. halapense*, is the first reported perennial with resistance to dinitroaniline herbicides.[19] The R biotype was discovered in cotton fields which had been treated for 12 consecutive years with trifluralin. Resistance was found to both trifluralin and pendimethalin, although the level of resistance appears to be low (less than 2-fold).

Biotypes of *A. palmeri* with resistance to trifluralin were reported recently in eight separate locations in South Carolina.[20] Trifluralin had been applied continuously for many years, similar to the scenario for development of other dinitroaniline-resistant weeds. Interestingly, the R *A. palmeri* was discovered in areas similar to where R *E. indica* biotypes were reported. Under field conditions, R *A. palmeri* exhibited variable levels of resistance to six dinitroaniline herbicides, with acceptable weed control obtained with trifluralin at rates six times those required for control of the susceptible biotype.[20]

Target Site Cross Resistance

The discovery of weeds resistant to dinitroaniline herbicides has prompted studies to determine cross resistance to compounds structurally dissimilar from dinitroaniline herbicides, but which also disrupt mitosis. In addition, some cross resistance studies have included herbicides which selectively disrupt plant processes other than mitosis. These studies provide insight into the mechanism(s) that might underlie resistance, as well as alternative herbicides for control of the resistant biotypes. In *E. indica*, the R biotype exhibits target site cross resistance to all of the dinitroaniline herbicides, as well as dithiopyr and amiprophosmethyl, herbicides which cause effects similar to the dinitroaniline herbicides. However, both the R and S biotypes are equally sensitive to the herbicides pronamide, terbutol, and DCPA.[29] Interestingly, the R biotype is more sensitive to the carbamate herbicides IPC (propham) and CIPC (chlorpropham).[29] Both propham and chlorpropham disrupt spindle microtubules by directing the movement of chromosomes during anaphase into multiple poles, resulting in the formation of multinucleate cells.[1] Under field conditions, R and S *E. indica* were equally sensitive to fluometuron, alachlor, atrazine, butylate, MSMA and sethoxydim, all herbicides with mechanisms of action not related directly to mitotic inhibition.[16] The pattern of resistance for R *S. viridis* varies from that of R *E. indica*. Resistance is exhibited to amiprophosmethyl as well as DCPA and terbutol (Figures 1 and 2).[28] These are the first reports of resistance to DCPA and terbutol. Similar to R *E. indica*, R *S. viridis* is cross resistant to dithiopyr. R *S. viridis* is not cross resistant to pronamide, and is more sensitive than the S biotype to propham and chlorpropham (Figure 2).[28] Resistance of R biotypes of *S. halapense* and *A. palmeri* to mitotic disrupters other than dinitroanilines has not been investigated. Field applications of numerous nonmitotic disrupting herbicides on R and S biotypes of *A. palmeri*, demonstrate no appreciable differences in control.[20] No comprehensive laboratory studies of cross resistance have been

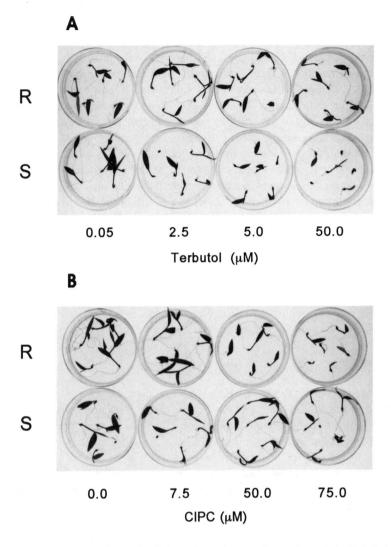

Figure 2. *S. viridis* seedlings after 7 days of growth on media supplemented with terbutol (A) or CIPC (B). The R biotype exhibits resistance to terbutol, but is actually more sensitive to CIPC.

attempted with R *A. palmeri*. Trifluralin-resistant biotypes of *L. rigidum* in Australia are also resistant across a range of dinitroaniline herbicides.[22]

Factors Limiting Resistance to Dinitroaniline Herbicides

The number of reports documenting resistance to dinitroanilines is relatively small given the long-term and widespread use of these herbicides. In contrast, more than 57 different weed species have evolved resistance to triazine herbicides,[14] which have been used widely for approximately the same duration as dinitroanilines (see Chapter 2). One underlying reason for the limited number of weeds resistant to dinitroaniline herbicides may involve overall use of these herbicides. In general, dinitroaniline herbicides provide short-term control of selected weeds prior to weed emergence. Growers typically apply any of a number of registered herbicides on emerged weeds throughout the season, as needed. Under these situations, weeds resistant to the dinitroanilines may be controlled by other herbicides with varying sites of action. Use of different herbicides for control of a weed resistant to some class of herbicides assumes that effective alternative herbicides are available. For R biotypes of *E. indica*, *S. viridis*, and *S. halapense*, this assumption is met, although growers continue to use dinitroaniline herbicides because of their low cost and effectiveness. However, broadcasted postemergence herbicides are not effective on populations of R *A. palmeri* in cotton, and directed postemergence herbicides are effective only on small plants.[20] Therefore, the loss in efficacy of dinitroaniline herbicides on *A. palmeri* has contributed to the proliferation of resistant biotypes. Effective control of R *A. palmeri* can be achieved by rotating infested fields to maize or soybeans, for which effective herbicides are available.

Fitness

One aspect of herbicide-resistant weeds often studied is the relative fitness of the R and S biotypes (see Chapter 11). Survival, plant competition, and fecundity influence relative fitness. Repeated use of the same herbicide imposes a continuous selection pressure which favors the survival of the R biotype. In the absence of herbicide use, however, survival of the R and S biotypes depends upon plant competition and fecundity, as well as other factors which contribute to fitness and, ultimately, survival.[30] Relative fitness studies abound for weed species exhibiting triazine resistance, but little information is available on dinitroaniline-resistant weeds. Under noncompetitive conditions, nonisogenic lines of R and S *E. indica* appear similar in growth and development, with total seed dry weight (representing fecundity) reduced in the R biotype.[31] In laboratory studies with R and S *S. viridis* under noncompetitive conditions, shoot and root biomass of 7-day-old seedlings was similar.[32] Shoot dry weights of R and S *A. palmeri* were not significantly different in field-grown plants.[20] Similarly, a multiple resistant biotype of *L. rigidum*[33] with resistance to dinitroaniline herbicides is equally fit to the S biotype. Although more conclusive fitness studies are needed, the lack of apparent differences in fitness among R and S biotypes indicates that abandoning the use of dinitroaniline herbicides in fields infested with resistant weeds will not in itself lead to an immediate reduction in populations of R biotypes.

MECHANISMS OF RESISTANCE

Large differences in the level of resistance and the spectrum of target site cross resistance to other herbicides among the various R weeds indicate that the mechanism of resistance may vary between species, and possibly within a species.

Uptake, Translocation, Metabolism, Calcium Effects

Chernicky[34] investigated the I and R biotypes of *E. indica* with respect to altered uptake and translocation of trifluralin, pendimethalin, and oryzalin. Based upon both autoradiography of whole plants and scintillation counting of the plant parts, there was no significant difference between the translocation patterns or the amount of radioactivity recovered in the I and R versus S biotypes. These data indicate there is no gross difference in translocation but, of course, do not preclude differences in subcellular distribution. Dinitroaniline herbicides are very lipid soluble, therefore measurements of total lipids and lipid composition might account for potential differences in cellular compartmentation between R and S biotypes. The lipid content of *E. indica* seed was one of the lowest of any plant measured and the R biotype actually had less lipid than the S biotype.[34] This is opposite to the result expected if resistance was based on an increase in lipid content, and actually might render the R biotype more susceptible to the effects of these herbicides.

No studies of herbicide metabolism have been done, to our knowledge, on the R *A. myosuroides*, *E. indica*, *S. viridis*, *S. halapense*, *L. rigidum*, or *A. palmeri* biotypes. It should be noted that dinitroaniline herbicides are not rapidly metabolized by any plant,[35] even highly resistant crops such as carrot. Thus, it is unlikely that herbicide metabolism has a critical role in resistance to dinitroanilines. An exception to this statement might be the resistance in multiple resistant weeds such as *L. rigidum* (see Chapter 9). Although the mechanism of multiple resistance is not fully elaborated, some biotypes have accelerated metabolism of some but not all herbicides.

One of the proposed mechanisms for dinitroaniline herbicide action involves stimulation of the movement of calcium from mitochondria to the cytoplasm.[36] This increase in cellular calcium would lead to the depolymerization of microtubules. The concentrations of oryzalin and trifluralin required to affect mitochondrial calcium stores are very high[5,6] compared to concentrations needed to disrupt mitosis.[29] To determine if calcium efflux from the mitochondria was differentially affected by dinitroaniline herbicides in R versus S *Eleusine*, pyroantimonate-precipitable calcium in the mitochondria was examined by electron microscopic histochemistry.[6] Treatment with 10 μM trifluralin was required to cause a reduction in pyroantimonate-precipitable calcium, and this effect was similar in both biotypes.

Differences in Cytoskeletal Proteins

Because dinitroanilines affect tubulin polymerization into microtubules, we examined the ability of oryzalin to inhibit the polymerization of tubulin-enriched (>85% tubulin based upon densitometric analysis of stained) protein fractions of the R and S *Eleusine* biotypes in the presence and absence of herbicide. Extracts from both biotypes polymerized into microtubules in the absence of oryzalin, although only the R biotype formed microtubules in the presence of oryzalin.[6] These data indicate that an intrinsic change in tubulin protein (or another co-purifying protein) was responsible for the difference.

An initial Western blot utilizing one R *E. indica* biotype and one S biotype probed with a polyclonal antiserum that recognized both α and β tubulin, resulted in an additional polypeptide detected in a position similar to β tubulin, indicating that this R biotype was a β tubulin mutant.[6] Biotypes collected from other regions sometimes demonstrated the same immunoreaction by this serum, although others did not.[37] Likewise, probing blots with monoclonal β tubulin antisera sometimes resulted in detectable differences while others did not, although there were some blots that showed changes with one sera and not another.[37] Most of the *E. indica* seed collections made by Gossett and colleagues were bulked seed from fields and different seed batches would be likely to show such variation. Thus, although there are some electrophoretic variants in the population, it is not clear whether any are related to herbicide resistance. No differences were found in the I biotype.

A similar screen was initiated on the R biotype of *S. viridis*. No differences between R and S biotypes were found in Western blots from single dimensional sodium dodecyl sulfate (SDS) gel separations of total cell homogenates probed with antisera specific for α and β tubulin, actin, or a mixture of microtubule-associated proteins (Figure 3).[38]

Both the R biotypes of *E. indica* and of *C. reinhardtii* are supersensitive to the microtubule stabilizer taxol.[24,39] That is, lower concentrations of taxol are required to produce mitotic irregularities or inhibit growth in the R compared to the S biotype. We interpreted this data to mean that the microtubules of the R biotype are hyperstabilized, thus rendering them less affected by compounds such as trifluralin which destabilize microtubules. Alternatively, the mechanism conferring resistance also confers hyperstability to the microtubules, although the two are unrelated. Both the I biotype of *E. indica* and R biotype of *S. viridis* are dinitroaniline resistant, but no increased sensitivity to taxol was detectable.[28]

GENETIC STUDIES

Initial crossing studies between the R and S biotypes of *E. indica* indicate that the dinitroaniline resistance trait is dominant.[37] The F_1 hybrids from S X R crosses are resistant, but less so than the R parent. For example, the R parent

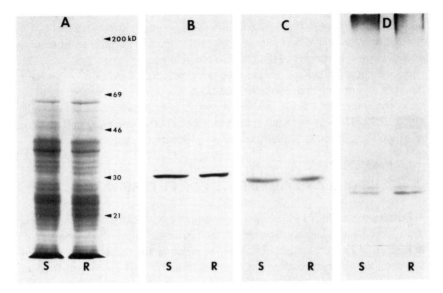

Figure 3. Coomassie blue-stained gel (A) of total protein homogenates of dinitroaniline-resistant (R) and -susceptible (S) biotypes of *S. viridis* separated on a 7.5% (w/v) acrylamide gel and corresponding Western blots of α (B) and β tubulin (C) from a similar protein separation. No qualitative or quantitative differences are noted in the tubulin subunit proteins or proteins separated on a 10% acrylamide gel and probed with anti-MAPs (D).

exhibited no root tip swelling even after treatment with 10 μM trifluralin, whereas the F_1 hybrids exhibit some swelling after 10 μM but not after 1 μM treatments. F_2 hybrids from the self-pollination of the F_1s resulted in progeny with a broad spectrum of sensitivity, some individuals being similar to both R and S parents and all ranges in between, with a mean sensitivity similar to the F_1s. The complex segregation indicates that several genes are responsible for resistance, although the numbers of offspring were too small to substantiate any genetic model.[37]

Baird et al.[40] examined various populations of R and S *E. indica* using isozyme protocols to determine if all of the R biotypes arose from a single progenitor (i.e., a founder effect). The R biotypes collected from a given site were more like the S biotypes at that site than the R biotypes at other sites. These data, coupled with the high percentage of self-pollination and the widespread occurrence of the R biotypes in a relatively short time period, indicate that there were several independent occurrences of resistance. This may also be true of R biotypes of *S. viridis*[18] and *A. palmeri*,[20] where biotypes were isolated from numerous, geographically separate areas, and where self-pollination is also likely.

The *Chlamydomonas* mutants that have been genetically mapped are found on the *uni* linkage group.[24] This extranuclear cluster of genes contains many of the loci involved in flagellar functions, probably microtubule-associated proteins, although none of the mutants are associated with tubulin loci. The level of resistance of the *Chlamydomonas* mutants is generally lower than that of the R weed biotypes, but this is partially due to the relatively high level of natural resistance of this species to dinitroaniline herbicides.

No genetic studies have been carried out on the *S. halapense* or *A. palmeri* biotypes, to our knowledge. Inheritance of resistance in *S. viridis* appears to be under the control of a single, nuclear recessive gene, however.[41] This strongly implicates a different mechanism of resistance than that operating in *E. indica*, in which the preliminary inheritance data indicates a multigenetic control.

CONCLUSIONS AND PROSPECTS FOR FUTURE WORK

Despite the widespread use and high efficacy of dinitroaniline herbicides, there have been relatively few reports of resistance to this herbicide group. Indeed, in our laboratory, selection of *Arabidopsis* for resistance to oryzalin failed to identify one resistant biotype in screening over 10^6 individuals from an M2 population mutagenized with ethylmethane sulfonate;[37] similar negative results were obtained by others.[42] From these results, one might predict that there would be few other instances of dinitroaniline resistance. Compared with some other examples of resistance to other herbicide groups (for example, see Chapters 2 and 5) resistance in weed species in the field is relatively rare. However, as documented in this chapter, there are now a number of dinitroaniline-resistant weed biotypes and given the continued widespread use of dinitroaniline herbicides there will be further cases of resistance. Those resistant biotypes that have been selected offer an intriguing view as to what perturbations to the cytoskeletal proteins can occur and their consequences. As of yet, we know so little of the identity or roles of proteins in the plant cytoskeleton other than tubulins and actin, that such mutants could become useful in elucidating these other critical proteins.

REFERENCES

1. Vaughn, K. C., and L. P. Lehnen, Jr. "Mitotic Disrupter Herbicides," *Weed Sci.* 39:450-457 (1991).
2. Ashton, F. M., and A. S. Crafts. "Dinitroanilines," in *Mode of Action of Herbicides* (New York: John Wiley & Sons, 1981), pp. 201-223.
3. Hilton, J. L., and M. N. Christiansen. "Lipid Contribution to Selective Action of Trifluralin," *Weed Sci.* 20:290-294 (1972).
4. Vaughan, M. A., and K. C. Vaughn. "Carrot Microtubules Are Dinitroaniline Resistant. I. Cytological and Cross-resistance Studies," *Weed Res.* 28:73-83 (1988).
5. Hess, F. D. "Herbicide Effects on the Cell Cycle of Meristematic Plant Cells," *Rev. Weed Sci.* 3:183-203 (1987).
6. Vaughn, K. C., and M. A. Vaughan. "Structural and Biochemical Characterization of Dinitroaniline-resistant *Eleusine*," in *Managing Resistance to Agrochemicals*, M. B. Green, H. M. LeBaron, and W. K. Moberg, Eds. (Washington, D.C.: The American Chemical Society, 1990), 421:364-375.

7. Pickett-Heaps, J. D. "The Effects of Colchicine On the Ultrastructure of Dividing Plant Cells, Xylem Wall Differentiation and Distribution of Cytoplasmic Microtubules," *Dev. Biol.* 15:206-236 (1967).
8. Morejohn, L. C., T. E. Bureau, J. Mole-Bajer, A. S. Bajer, and D. E. Fosket. "Oryzalin, a Dinitroaniline Herbicide, Binds to Plant Tubulin and Inhibits Microtubule Polymerization In Vitro," *Planta* 172:252-264 (1987).
9. Strachan, S. D., and F. D. Hess. "The Biochemical Mechanism of Action of the Dinitroaniline Herbicide Oryzalin," *Pestic. Biochem. Physiol.* 20:141-150 (1983).
10. Wilson, L. "Minireview: Action of Drugs On Microtubules," *Life Sci.* 17:303-310 (1975).
11. Morejohn, L. C., T. E. Bureau, L. P. Tocchi and D. E. Fosket. "Resistance of *Rosa* Microtubule Polymerization to Colchicine Results From a Low-Affinity Interaction of Colchicine and Tubulin," *Planta* 170:230-241 (1987).
12. Bartels, P. G., and J. L. Hilton. "Comparison of Trifluralin, Oryzalin, Pronamide, Propham, and Colchicine Treatments on Microtubules," *Pestic. Biochem. Physiol.* 3:462-472 (1973).
13. Hess, F. D., and D. E. Bayer. "Binding of the Herbicide Trifluralin to *Chlamydomonas* Flagellar Tubulin," *J. Cell Sci.* 24:351-360 (1977).
14. LeBaron, H. M. "Distribution and Seriousness of Herbicide-resistant Weed Infestations Worldwide," in *Herbicide Resistance in Weeds and Crops*, J. C. Caseley, G. W. Cussans, and R. K. Atkin, Eds. (Oxford, U.K.: Butterworth-Heinemann, Ltd., 1991), pp. 27-43.
15. Gressel, J., and L. A. Segel. "The Paucity of Plants Evolving Genetic Resistance to Herbicides: Possible Reasons and Implications," *J. Theor. Biol.* 75:349-371 (1978).
16. Mudge, L. C., B. J. Gossett, and T. R. Murphy. "Resistance of Goosegrass (*Eleusine indica*) to Dinitroaniline Herbicides," *Weed Sci.* 32:591-594 (1984).
17. Vaughn, K. C., M. A. Vaughan, and B. J. Gossett. "A Biotype of Goosegrass (*Eleusine indica*) With an Intermediate Level of Dinitroaniline Herbicide Resistance," *Weed Technol.* 4:157-162 (1990).
18. Morrison, I. N., B. G. Todd, and K. M. Nawolsky. "Confirmation of Trifluralin-resistant Green Foxtail (*Setaria viridis*) in Manitoba," *Weed Technol.* 3:544-551 (1989).
19. Wills, G. D., J. D. Byrd, Jr., and H. R. Hurst. "Herbicide Resistant and Tolerant Weeds," in *Proceedings of the Southern Weed Science Society*, 45:43 (1992).
20. Gossett, B. J., E. C. Murdock, and J. E. Toler. "Resistance of Palmer Amaranth (*Amaranthus palmeri*) to the Dinitroaniline Herbicides," *Weed Technol.* 6:587-591 (1992).
21. Davis, J. L., J. R. Abernathy, and A. F. Wiese. "Tolerance of 52 Corn Lines to Trifluralin," in *Proceedings of the Southern Weed Science Society*, 31:123 (1978).
22. McAlister, F. M., J. A. M. Holtum, and S. B. Powles. "Dinitroaniline Herbicide Resistance in Rigid Ryegrass (*Lolium rigidum*)," *Weed Sci.* (submitted).
23. Moss, S. R. "Herbicide Cross-Resistance in Slender Foxtail (*Alopecurus myosuroides*)," *Weed Sci.* 38:492-496 (1990).
24. James, S. W., L. P. W. Ranum, C. D. Silflow, and P. A. Lefebvre. "Mutants Resistant to Anti-microtubule Herbicides Map to a Locus On the *uni* Linkage Group in *Chlamydomonas reinhardtii*," *Genetics* 118:141-147 (1988).
25. Bolduc, C., V. D. Lee, and B. Huang. "β-Tubulin Mutants of the Unicellular Green Alga *Chlamydomonas reinhardtii*," *Proc. Natl. Acad. Sci. U.S.A.* 85:131-135 (1988).

26. Vaughn, K. C. "Cytological Studies of Dinitroaniline-resistant *Eleusine*," *Pestic. Biochem. Physiol.* 26:66-74 (1986).
27. Whitwell, T., V. Baird, J. Wells, and K. Tucker. "Comparison of Dinitroaniline Resistant Goosegrass Biotypes From Alabama, Tennessee, Georgia and South Carolina," in *Proceedings of the Southern Weed Science Society*, 45:297 (1992).
28. Smeda, R. J., K. C. Vaughn, and I. N. Morrison. "A Novel Pattern of Herbicide Cross-resistance in a Trifluralin-resistant Biotype of Green Foxtail (*Setaria viridis* (L.) Beauv.)," *Pestic. Biochem. Physiol.* 42:227-241 (1992).
29. Vaughn, K. C., M. D. Marks, and D. P. Weeks. "A Dinitroaniline-resistant Mutant of *Eleusine indica* Exhibits Cross-resistance and Supersensitivity to Antimicrotubule Herbicides and Drugs," *Plant Physiol.* 83:956-964 (1987).
30. Radosevich, S. R., B. D. Maxwell, and M. L. Roush. "Managing Herbicide Resistance Through Fitness and Gene Flow," in *Herbicide Resistance in Weeds and Crops*, J. C. Caseley, G. W. Cussans, and R. K. Atkin, Eds. (Oxford, U.K.: Butterworth-Heinemann Ltd., 1991), pp. 129-143.
31. Murphy, T. R., B. J. Gossett, and J. E. Toler. "Growth and Development of Dinitroaniline-Susceptible and -Resistant Goosegrass (*Eleusine indica*) Biotypes Under Noncompetitive Conditions," *Weed Sci.* 34:704-710 (1986).
32. Smeda, R. J. Unpublished results (1991).
33. Powles, S. B., and J. M. Matthews. "Multiple Herbicide Resistance in Annual Ryegrass (*Lolium rigidum*): A Driving Force for the Adoption of Integrated Weed Management," in *Resistance '91: Achievements and Developments in Combatting Pesticide Resistance*, I. Denholm, A. L. Devonshire, and D. W. Hollomon, Eds. (London: Elsevier Applied Science, 1992), pp. 75-87.
34. Chernicky, J. P. "An Investigation Into the Resistance of Goosegrass (*Eleusine indica*) to Dinitroaniline Herbicides," Ph. D. Thesis, University of Illinois at Urbana-Champaign, Urbana, IL (1985).
35. Probst, G. W., T. Golab, and W. L. Wright. "Dinitroanilines," in *Herbicides, Chemistry, Degradation, and Mode of Action*, P. C. Kearney, and D. D. Kaufman, Eds. (New York: Marcel Dekker, Inc., 1975), pp. 453-500.
36. Hertel, C., and D. Marme. "Herbicides and Fungicides Inhibit Ca^{2+} Uptake by Plant Mitochondria: A Possible Mechanism of Action," *Pestic. Biochem. Physiol.* 19:282-290 (1983).
37. Vaughn, K. C. Unpublished results.
38. Smeda, R. J., K. C. Vaughn, and I. N. Morrison. "Trifluralin-Resistant Green Foxtail (*Setaria viridis* (L.) Beauv.) Exhibits Cross-Resistance to Mitotic Disrupter Herbicides," *Plant Physiol. Suppl.* 96(1):114 (1992).
39. Vaughn, K. C., and M. A. Vaughan. "Dinitroaniline Resistance in *Eleusine indica* May Be Due to Hyperstabilized Microtubules," in *Herbicide Resistance in Weeds and Crops*, J. C. Caseley, G. W. Cussans, and R. K. Atkin, Eds. (Oxford, U.K.: Butterworth-Heinemann, Ltd., 1991), pp. 177-186.
40. Baird, V., W. M. Vance, C. A. Langner, J. Wells, K. Tucker, T. Whitwell, and C. R. Werth. "Dinitroaniline Resistant Goosegrass (*Eleusine indica* (L.) Gaertn.) From the Southeastern United States: Characterization of Biotypes and Population Genetic Analyses," *Am. J. Bot.* 79:89-90 (1992).
41. Jasieniuk, M., A. L. Brule-Babel, and I. N. Morrison. "The Genetics of Trifluralin Resistance in Green Foxtail (*Setaria viridis* (L.) Beauv.)," in *Weed Science Society of America Abstracts*, 33:61 (1993).
42. Siflow, C., and P. Snustad. Personal communication.

CHAPTER 8

Resistance to Glyphosate

William E. Dyer

CHEMISTRY

The herbicidal activity of glyphosate (*N*-[phosphonomethyl]glycine) and its salts was first described in 1971.[1] The compound was identified through a conventional screening program of tertiary aminomethyl phosphonic acids.[2] Interestingly, only one other compound from this screening program (glyphosine) has shown sufficient unit activity for development, but it is marketed as a plant growth regulator, not a herbicide. All subsequent efforts to identify herbicidal analogues, homologues, or derivatives have been unsuccessful, in contrast to the situation in several other herbicide families. The molecule is synthesized by the reaction of excess glycine with chloromethylphosphonic acid under basic conditions.[2] Due to the limited solubility of the acid form in water (1.2% at 25°C), glyphosate is formulated and marketed as the isopropylamine salt.

USAGE IN AGRICULTURE

Because glyphosate is a nonselective, nonresidual, and environmentally benign herbicide, it is used for total vegetation control in a number of situations. In horticultural and orchard industries, glyphosate is used for weed control around container nursery stock and beds, and for ground cover removal in plantations. Glyphosate is applied selectively using rope wick applicators, recirculating sprayers, and rollers in field crops wherever there is a size or location differential between crops and weeds (e.g., weeds projecting above the crop canopy). The herbicide has found wide use in fallow and industrial weed control programs, often in tankmix combinations with a residual herbicide. Home use of glyphosate for spot weed control (chemical hoeing), sucker removal from some tree species, and lawn renovation is widespread. Glyphosate is widely used to remove weeds prior to seeding crops.

The desirable characteristics of glyphosate mentioned above have given the herbicide a comfortable niche in total vegetation management programs. However, its nonselective nature precludes its use as a broadcast treatment in field crops, a potentially much more lucrative market. Expansion into this market has been one of several driving forces behind a considerable effort to introduce glyphosate resistance into crop plants, as discussed below.

MECHANISM OF ACTION

The intense and widespread interest in glyphosate has led to a substantial body of literature on its biochemical mechanism of action. This information has been reviewed previously,[3-5] and therefore this chapter will provide only a short overview of the subject.

Early studies by Jaworski[6] showed that the inhibitory effects of glyphosate in *Lemna* spp. could be reversed by the addition of aromatic amino acids. Similar results were shown for *Rhizobium japonicum* and several plant cell cultures.[7] Direct evidence implicating a specific enzyme target came from studies on *Aerobacter aerogenes* cells and buckwheat shoots.[8,9] After glyphosate treatment, both species accumulated high levels of shikimic acid, and conversion of shikimic acid to chorismate was inhibited 50% by 5 mM glyphosate in *A. aerogenes* cell-free extracts.[9] These and subsequent studies confirmed that the primary mechanism of action of glyphosate is the inhibition of 5-enolpyruvylshikimate-3-phosphate synthase (EPSPS) (EC 2.5.1.19), the penultimate enzyme of the aromatic amino acid biosynthetic (shikimate) pathway. The enzyme catalyzes the formation of EPSP and inorganic phosphate from phosphoenolpyruvate (PEP) and shikimate-3-phosphate in an unusual carboxyvinyl transfer reaction. Glyphosate is a competitive inhibitor of EPSPS with respect to PEP, and interacts with the enzyme:shikimate-3-phosphate complex.[10] Even though a number of other enzymes also utilize PEP, only EPSPS is inhibited by glyphosate, probably because only EPSPS interacts with PEP as an enzyme-substrate complex and not as the free enzyme.[11] Therefore, glyphosate is thought to act as a transition-state analogue of PEP. Plant death results from starvation for the aromatic amino acids, although there is evidence that glyphosate also alters carbon allocation patterns between starch and sucrose, indirectly affecting a number of other plant processes.[12]

Site-directed mutagenesis of a conserved region of the EPSPS enzyme has identified the active site, and shows that the active site is similar in bacteria and plants.[11] Amino acid residues 90 to 102 (numbering based on *E. coli* EPSPS sequence) flank a region that is highly conserved among six plants and five microorganisms. Substitution of an alanine for glycine in the petunia EPSPS at a position corresponding to *E. coli* residue 96 causes an increase in the $K_{m(app)}$ (PEP) of about 40-fold. Alternatively, substitution of a serine for the same glycine abolishes EPSPS activity but confers a novel hydrolase activity which converts EPSP to shikimate-3-phosphate and pyruvate.[11] These and other results confirm that the conserved region is centrally involved in the active site, specifically in EPSPS binding of the phosphate moiety of PEP. Mutations in this domain also confer glyphosate resistance, as discussed below, further confirming the idea that glyphosate binds to the EPSPS active site.

RESISTANCE TO GLYPHOSATE

Resistance in Bacteria

Early studies on the mechanism of action of glyphosate were closely followed by efforts to select resistant strains of enteric bacteria. A resistant strain of *Salmonella typhimurium* was identified after two cycles of chemical mutagenesis and the mutation conferring resistance was shown to be in the *aroA* gene, which encodes EPSPS.[13] The mutant gene was subsequently isolated and shown to confer resistance when transferred to *E. coli*. DNA sequencing revealed the presence of two independent mutations in the mutant gene: a promoter mutation conferred low levels of resistance due to elevated *aroA* expression, and a point mutation in the coding sequence caused a proline to serine substitution at residue 101 of the protein.[13,14] Selection in *E. coli*[15] and *Klebsiella pneumoniae*[16] has also resulted in mutants with highly glyphosate-resistant EPSPS enzymes. In both species, the mutation conferring resistance caused a substitution of alanine for glycine at position 96.[11,16] Overexpression of the wild-type *E. coli aroA* gene on a multicopy plasmid also conferred glyphosate resistance.[17] Glyphosate-resistant mutants selected in *Euglena gracilis* were shown to possess two mechanisms for resistance: an insensitive EPSPS in one line and elevated levels of the enzyme in another line.[18]

Resistance in Plants

In plants, glyphosate resistant lines have been selected from a number of species using various laboratory procedures. Suspension-cultured or callus lines of carrot,[19] tobacco,[20,21] tomato,[22] petunia,[23] chickory,[24] pea,[25] *Catharanthus roseus*,[26] and *Corydalis sempervirens*[27] were isolated either by single-step[20] or stepwise selection methods.[28] In all of these cases, resistance was shown to be due to elevated levels of EPSPS activity, although the enzymes were still sensitive to glyphosate inhibition. Overproduction of EPSPS effectively increases the number of enzyme molecules that must be inhibited in order to block carbon flow through the pathway. Further studies of the mutant tobacco,[29,30] carrot,[31] and petunia[32] cell lines showed that enzyme overproduction was due to amplification of the gene(s) encoding EPSPS. In some cell lines, EPSPS specific activity, gene copy number, and glyphosate resistance level were positively correlated,[30,33] and in many cases resistance was stable in the absence of herbicide. However, overproduction of EPSPS in *C. sempervirens* cells was not due to gene amplification, even though mRNA levels were elevated.[34] Elevated resistance levels[20-22,24] and gene amplification[29] have been demonstrated in plants regenerated from selected cell cultures.

In general, it is fairly straightforward to select glyphosate-resistant plant cells, although the levels of resistance obtained are not exceptionally high. With one exception (see below), resistance obtained in cultured cells appears to be due to overexpression of EPSPS, which in most cases is due to gene amplification. This mechanism may be peculiar to the tissue culture process itself as a result of epigenetic changes in somatic cells, since glyphosate resistance due to gene amplification was not heritable in chicory[24] and the process has not been confirmed in whole plant selection schemes.

The only exception to glyphosate resistance arising as a result of EPSPS overproduction in plant cell cultures was recently reported.[35] A Black Mexican Sweet maize cell line was obtained, without mutagenesis or selection of any kind, that was 30- to 60-fold more resistant to glyphosate than the control cell line. Cell extracts contained two chromatographically separable EPSPS activities, one of which was highly resistant to glyphosate inhibition, even at concentrations up to 30 mM. The resistant activity was tentatively assigned a cytosolic location, in contrast to other work demonstrating shikimate pathway localization in the chloroplast.[36] As seen in the mutated petunia EPSPS,[11] the $K_{m(app)}$ (PEP) of the partially purified insensitive maize EPSPS was increased about 12-fold over the susceptible maize activity, suggesting that a mutation in the EPSPS active site was responsible for resistance.[35] Growth rate of the selected line in the presence of glyphosate was about 80% of untreated control growth. It is perhaps surprising that an insensitive cytosolic EPSPS can confer such high levels of glyphosate resistance, since transgenic tobacco[37] and tomato[38] plants expressing a resistant *S. typhimurium aroA* gene in the cytosol were only 2- to 3-fold more resistant than the wild type, and growth rates were reduced after herbicide treatment.

Genetic Variability for Glyphosate Resistance

In contrast to the mainly *in vitro* screening described above, whole plant germplasm screening has demonstrated that variability in glyphosate sensitivity exists among various species. For example, fescue and perennial ryegrass cultivars,[42] soybean,[39,40] and *Convulvulus arvensis* L. lines[41] displayed varying degrees of resistance to glyphosate. Recurrent selection has been used to exploit this variability in the grass cultivars, resulting in lines exhibiting a 2- to 3-fold increase in glyphosate resistance over parental types.[42] In a similar approach, three germplasms of birdsfoot trefoil (*Lotus corniculatus* L.) were subjected to recurrent selection: after two cycles, selected populations showed an increase in mean resistance levels to glyphosate treatment.[43] Resistance was positively correlated with elevated EPSPS specific activities in three clones picked from selected populations, but selected population means were not different from parental means. The molecular mechanism of increased EPSPS activity has not been determined, and increased resistance levels may be due to simultaneous selection for multiple mechanisms.[44] In a field trial with no recurrent selection, 1457 cultivated tomato lines and 13 other *Lycopersicon* spp. were screened for glyphosate resistance, but none displayed useful levels of resistance.[45] In a similar approach, 173 accessions of seven *Glycine* spp. were screened in greenhouse trials for resistance to glyphosate.[40] After initial screening, substantial differences in glyphosate sensitivity were apparent among several species. Five accessions survived treatment with 1.1 kg ha^{-1} glyphosate, and one *G. tabacina* accession survived treatment with 2.2 kg ha^{-1}, while *G. max* was killed by 0.56 kg ha^{-1} glyphosate. However, resistance levels in cell suspension cultures of these accessions were not different from the sensitive *G. max* cultures, indicating that resistance was probably due to altered herbicide uptake and/or translocation at the whole plant level, and not an altered or overproduced EPSPS.[40] Although some of these selection programs are currently being pursued, the overall results have been generally disappointing. Relatively low levels of glyphosate resistance have been obtained, and in several cases excessive genotype by environment interactions have made further selection difficult.[44]

GLYPHOSATE METABOLISM

Metabolism in Bacteria

The primary means of glyphosate degradation in soil is through microbial catabolism.[46] Even though glyphosate half-lives vary widely in different soil types, apparently microorganisms able to degrade the molecule are nearly ubiquitous.[47] Soil bacteria degrade glyphosate by two general ways (Figure 1), leading to either glycine or aminomethylphosphonate (AMPA) as intermediate products.[48] The first step in the glycine pathway involves cleavage of the C-P bond of glyphosate to produce sarcosine, which is then converted by a sarcosine oxidase-dehydrogenase to glycine. Bacteria degrading glyphosate by this mechanism include *Pseudomonas* sp. strain PG2982[49] and *Arthrobacter* sp. strain GLP-1.[50]

The other degradative pathway proceeds by cleavage of the carboxymethyl carbon-nitrogen bond to produce AMPA, which can be further metabolized.

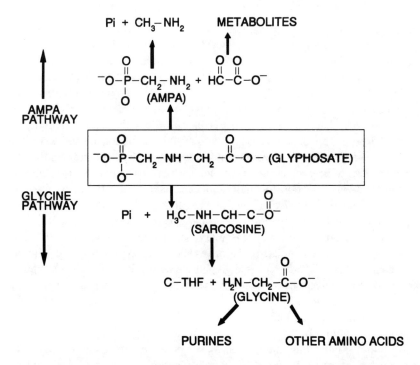

Figure 1. Pathways for glyphosate degradation in bacteria. THF, tetrahydrofolate.

Bacteria utilizing this pathway include some uncharacterized mixed cultures from soil[51] and a *Flavobacterium* sp. strain GDI.[52] A *Pseudomonas* sp. strain LBr has been described with a high capacity for glyphosate degradation, due to the presence of both metabolic pathways.[48] Both pathways are attractive candidates for use in introducing glyphosate resistance into crop plants, and research groups are undoubtedly pursuing this approach. However, few current results have been published in this area, probably because of patent considerations. The true potential of such an approach will depend ultimately on the nature of the bacterial glyphosate-degrading enzymes, and whether they will function appropriately in a plant cell.

Metabolism in Plants

For many years it was suggested that most plant species do not appreciably metabolize glyphosate,[53,54] and some[55] have even proposed that plants are unable to detoxify nonselective herbicides. Studies addressing this idea have demonstrated varying degrees of metabolism in several plant species, including Norway spruce,[56] soybean, maize, and cotton,[57] and *Equisetum arvense* L.[58] In the case of *E. arvense*, degradation was of sufficient magnitude for the authors to suggest that it was at least partially responsible for glyphosate's lack of performance on this weed.[58] However, in all of these studies, the unknown contribution to glyphosate metabolism by contaminating microbes in/on plant tissues prevented these data from being accepted as definitive. Recent studies using sterile cell suspension cultures of soybean, wheat, and maize have demonstrated unequivocally that plant cells metabolize glyphosate: soybean cells degraded 60% of applied [3-^{14}C]glyphosate to AMPA within 60 h.[59] Most AMPA was detected in the growth medium, although incorporation of AMPA into insoluble cell wall fractions was also shown. Proof of glyphosate metabolism in plant tissues and evidence of variability in metabolic levels among species demonstrates that it may be possible to select for metabolic resistance in weed species, if the selection pressure is stringent enough (see below).

ENGINEERING GLYPHOSATE RESISTANCE IN CROPS

Resistance to several herbicides has been introduced into crop plants, and the subject has been recently reviewed.[61] These advances have been made possible through the use of molecular biology and plant transformation techniques, and they represent some of the first and most highly publicized commercial applications of plant biotechnology.[60,62] The first report of introduced glyphosate resistance in transgenic plants was in 1985 by Comai et al.[37] A mutant *S. typhimurium aroA* gene encoding a resistant EPSPS was introduced into tobacco cells under control of the octopine synthase or mannopine synthase promoter. Transformed regenerated plants expressed the chimeric enzyme activity in leaves, and were two to three times more resistant to glyphosate although their growth rate was reduced compared to untreated controls.[37] This demonstration of glyphosate resistance was somewhat surprising, since the mutant enzyme activity was expressed in the cytosol, whereas almost all research indicated that the shikimate pathway was localized exclusively in the chloroplast (or plastid).[36] The same construction was subsequently introduced into tomato plants, with essentially the same results.[38]

Proper localization of a resistant EPSPS in the chloroplast was achieved by transforming plant cells with a chimeric gene combining a petunia transit peptide encoding sequence with a mutant *E. coli aroA* gene.[63] The chimeric preprotein was imported into petunia chloroplasts and regenerated plants showed reasonable levels of glyphosate resistance. During the course of these experiments, it was also demonstrated that glyphosate prevented import of the EPSPS preprotein into chloroplasts, suggesting an additional mechanism of plant inhibition.[64]

Glyphosate resistance has also been achieved in crop plants by overproduction of EPSPS. A wild-type petunia cDNA encoding EPSPS was linked to the constitutively expressed cauliflower mosaic virus 35S promoter and reintroduced into petunia.[32,65] The resulting transgenic plants were about 4-fold more resistant to glyphosate, but growth rates were once again reduced compared to controls. Similar results were obtained in transgenic *Arabidopsis thaliana* plants overexpressing EPSPS.[66]

Concerns over these relatively low levels of resistance and poor agronomic performance of the transgenic plants led to efforts to achieve higher levels of resistance. In order to identify mutant plant enzymes with high levels of resistance and normal kinetic efficiency, the petunia gene was subjected to site-specific mutagenesis and a number of mutant enzymes were recovered and tested.[67] In every case, the resistant EPSPS enzymes had reduced catalytic efficiencies. Some of these genes were overexpressed in transgenic tobacco: the plants were about 4-fold more resistant to glyphosate, and growth rates and agronomic performance were equal to controls. Similar constructions, including recently isolated mutant genes with no or very little loss of catalytic efficiency,[68] have now been introduced into several crop species. These mutant genes, some under the control of meristem-specific promoters, provide high levels of glyphosate resistance. However, there have been no published reports of yield trials documenting agronomic performance of the resistant cultivars. Another aspect yet to be fully addressed concerns the possibility of glyphosate residues in harvested portions of the crop. Some of these cultivars have been in field testing since 1987,[69] and resistant soybean[70] and canola cultivars may be marketed as early as 1995.[68]

Glyphosate Resistance in Weeds

Glyphosate has been used extensively since its commercial introduction in 1974. However, there are no confirmed reports of glyphosate-resistant weeds developing in field situations. Performance on some weedy species, such as *E. arvense* is erratic,[58] but the lack of control cannot be attributed to the presence of resistant types selected by the herbicide.

FUTURE LIKELY DEVELOPMENTS

The question now arises as to whether the use of glyphosate-resistant crops will increase the likelihood of selecting for resistant weeds. Several authors contend that this is highly unlikely, because 1) overall glyphosate usage is projected to increase only marginally in conjunction with resistant crops,[69] and 2) the complexity of engineering glyphosate resistance in crop plants indicates that naturally occurring mutations conferring resistance are extremely rare.[67,69] The suggestion was also made that development of resistance in weeds would require a combination of two mutations, one for mutation of EPSPS to confer resistance and another mutation for overproduction of the mutant enzyme.[67] However, the more recent identification of resistant EPSPS enzymes with essentially no loss of catalytic efficiency[68] now makes this contention less likely.

Clearly the use of glyphosate-resistant crops can be a very valuable addition to our current weed management strategies.[60] The ability to control weeds in crops using broadcast applications of a nonresidual, environmentally benign herbicide is certainly attractive, and should be pursued. When used in combination with conventional cultivars and in rotation with other crops and herbicides, glyphosate-resistant crops may become a fundamental part of sound weed management systems.

The key to making this situation a reality will depend on how glyphosate-resistant crops are perceived, marketed, and used. If, on one hand, resistant cultivars are viewed as a panacea for all weed management problems that farmers face in cropping systems, the tendency will be to use glyphosate exclusively for weed control in the crop. Given human nature as it is, farmers may be tempted to rely exclusively on one resistant cultivar for successive years, with the accompanying temptation to apply multiple glyphosate treatments during the growing season to control successive weed flushes. Although weed escapes may buffer the effect, multiple applications and few rotations would surely help set the stage for selecting resistant weeds, as has been clearly shown in the case of paraquat-resistant weeds.[60] Multiple applications during the growing season will effectively overcome glyphosate's lack of selection pressure due to nonpersistence in soil, and therefore selection for resistant weeds will depend more on the initial frequency of resistant individuals within populations. Although this frequency is unknown, the observations discussed above suggest that it may be very low.

Conversely, if glyphosate-resistant cultivars are marketed and viewed as an integral component of an integrated weed management system that includes other herbicides, nonresistant crops, and nonchemical control measures, these cultivars may well become incorporated into modern agricultural practices. Having made these statements, the degree to which glyphosate-resistant cultivars (and cultivars with engineered resistance to other herbicides) will be adopted remains unclear. Societal and regulatory pressures will undoubtedly influence the eventual market share to be captured by glyphosate-resistant cultivars, and perhaps control their destiny even more than the cultivars' agricultural and scientific worth.

REFERENCES

1. Baird, D. D., R. P. Upchurch, W. B. Homesley, and J. E. Franz. "Introduction of a New Broadspectrum Postemergence Herbicide Class with Utility for Herbaceous Perennial Weed Control," in *Proceedings of the 26th North Central Weed Control Conference* (Champaign, IL: Weed Science Society of America, 1971), pp. 64-68.
2. Franz, J. E. "Discovery, Development and Chemistry of Glyphosate," in *The Herbicide Glyphosate*, E. Grossbard and D. Atkinson, Eds. (London: Butterworths, 1985), pp. 3-17.
3. Cole, D. J. "Mode of Action of Glyphosate — A Literature Analysis," in *The Herbicide Glyphosate*, E. Grossbard and D. Atkinson, Eds. (London: Butterworths, 1985), pp. 48-74.
4. Duke, S. O. "Glyphosate," in *Herbicides: Chemistry, Degradation and Mode of Action,* Vol. 3, P. C. Kearney and D. D. Kaufman, Eds. (New York: Marcel Dekker, 1988), pp. 1-70.

5. Hoagland, R. E. and S. O. Duke. "Biochemical Effects of Glyphosate [N-(phosphonomethyl)glycine]," in *Biochemical Responses Induced by Herbicides*, D. E. Moreland, J. B. St. John, and F. D. Hess, Eds. (Washington, D.C.: American Chemical Society, 1982), pp. 175-205.
6. Jaworski, E. G. "Mode of Action of N-phosphonomethyl-glycine: Inhibition of Aromatic Amino Acid Biosynthesis," *J. Agric. Food Chem.* 20:1195-1198 (1982).
7. Gresshoff, P. M. "Growth Inhibition by Glyphosate and Reversal of its Action by Phenylalanine and Tyrosine," *Aust. J. Plant Physiol.* 6:177-185 (1979).
8. Steinrucken, J. and N. Amrhein. "The Herbicide Glyphosate is a Potent Inhibitor of 5-enolpyruvylshikimic Acid 3-phosphate Synthase," *Biochem. Biophys. Res. Commun.* 94:1207-1212 (1980).
9. Amrhein, N., D. Johänning, J. Schab, and A. Schulz. "Biochemical Basis for Glyphosate-tolerance in a Bacterium and a Plant Tissue Culture," *FEBS Lett.* 157:191-196 (1983).
10. Boocock, M. R. and J. R. Coggins. "Kinetics of 5-enolpyruvylshikimate-3-phosphate Synthase Inhibition by Glyphosate," *FEBS Lett.* 154:127-133 (1983).
11. Padgette, S. R., D. B. Re, C. S. Gasser, D. A. Eichholtz, R. B. Frazier, C. M. Hironaka, E. B. Levine, D. M. Shah, R. T. Fraley, and G. M. Kishore. "Site-directed Mutagenesis of a Conserved Region of the 5-enolpyruvylshikimate-3-phosphate Synthase Active Site," *J. Biol. Chem.* 266:22364-22369 (1991).
12. Shieh, W. J., D. R. Geiger, and J. C. Servaites. "Effect of N-(phosphonomethyl)glycine on Carbon Assimilation and Metabolism During a Simulated Natural Day," *Plant Physiol.* 97:1109-1114 (1991).
13. Comai, L., L. C. Sen, and D. M. Stalker. "An Altered *aroA* Gene Product Confers Resistance to the Herbicide Glyphosate," *Science* 221:370-371 (1983).
14. Stalker, D. M., W. R. Hiatt, and L. Comai. "A Single Amino Acid Substitution in the Enzyme 5-enolpyruvylshikimate-3-phosphate Synthase Confers Resistance to the Herbicide Glyphosate," *J. Biol. Chem.* 260:4724-4728 (1985).
15. Kishore, G. M., L. Brundage, K. Kolk, S. R. Padgette, D. Rochester, Q. K. Huynh, and G. della-Cioppa. "Isolation, Purification and Characterization of a Glyphosate Tolerant Mutant *E. coli* EPSP Synthase," *Fed. Proc.* 45:1506 (1986).
16. Sost, D. and N. Amrhein. "Substitution of Gly-96 to Ala in the 5-enolpyruvylshikimate 3-phosphate Synthase of *Klebsiella pneumoniae* Results in a Greatly Reduced Affinity for the Herbicide Glyphosate," *Arch. Biochem. Biophys.* 282:433-436 (1990).
17. Rogers, S. G., L. A. Brand, S. B. Holder, E. S. Sharps, and M. J. Brackin. "Amplification of the *aroA* Gene from *Escherichia coli* Results in Tolerance to the Herbicide Glyphosate," *Appl. Environ. Microbiol.* 46:37-43 (1983).
18. Reinbothe, S., A. Nelles, and B. Parthier. "N-(phosphonomethyl)glycine (Glyphosate) Tolerance in *Euglena gracilis* Acquired by Either Overproduced or Resistant 5-*enol*pyruvylshikimate-3-phosphate Synthase," *Eur. J. Biochem.* 198:365-373 (1991).
19. Nafziger, E. D., J. M. Widholm, H. C. Steinrücken, and J. L. Killmer. "Selection and Characterization of a Carrot Cell Line Tolerant to Glyphosate," *Plant Physiol.* 76:571-574 (1984).
20. Dyer, W. E., S. C. Weller, R. A. Bressan, and K. M. Herrmann. "Glyphosate Tolerance in Tobacco (*Nicotiana tabacum* L.)," *Plant Physiol.* 88:661-666 (1988).

21. Singer, S. R. and C. N. McDaniel. "Selection of Glyphosate-tolerant Tobacco Calli and the Expression of this Tolerance in Regenerated Plants," *Plant Physiol.* 78:411-416 (1985).
22. Smith, C. M., D. Pratt, and G. A. Thompson. "Increased 5-enolpyruvylshikimic Acid 3-phosphate Synthase Activity in a Glyphosate-tolerant Variant Strain of Tomato Cells," *Plant Cell Rep.* 5:298-301 (1986).
23. Steinrücken, H. C., A. Schulz, N. Amrhein, C. A. Porter, and R. T. Fraley. "Overproduction of 5-enolpyruvylshikimate 3-phosphate Synthase in a Glyphosate-tolerant *Petunia hybrida* Cell Line," *Arch. Biochem. Biophys.* 244:169-178 (1986).
24. Sellin, C., G. Forlani, J. Dubois, E. Nielsen, and J. Vasseur. "Glyphosate Tolerance in *Cichorium intybus* L. var. Magdebourg," *Plant Sci.* 85:223-231 (1992).
25. Ezhova, T. A., N. S. Tikhvinshaya, T. V. Petrova, A. M. Bagrova, I. R. Vasil'ev, D. N. Matorin, and S. A. Gostimskii. "Analysis of the Effects of Herbicides on Peas Seedlings and Calluses, and the Isolation of Herbicide-resistant Callus Lines and Regenerant Plants," *Sov. Genet.* 26:1317-1322 (1991).
26. Cresswell, R. C., M. W. Fowler, and A. H. Scragg. "Glyphosate-tolerance in *Catharanthus roseus*," *Plant Sci.* 54:55-63 (1988).
27. Smart, C. C., D. Johänning, G. Muller, and N. Amrhein. "Selective Overproduction of 5-enolpyruvylshikimic acid 3-phosphate Synthase in a Plant Cell Culture which Tolerates High Doses of the Herbicide Glyphosate," *J. Biol. Chem.* 260:16338-16346 (1985).
28. Hughes, K. "Selection for Herbicide Resistance," in *Handbook of Plant Cell Culture*, Vol. I, D. A. Evans, W. R. Sharp, P. V. Ammirato, and Y. Yamada, Eds. (New York: Macmillan, 1983), pp. 442-460.
29. Goldsbrough, P. B., E. M. Hatch, B. Huang, W. G. Kosinski, W. E. Dyer, K. M. Herrmann, and S. C. Weller. "Gene Amplification in Glyphosate Tolerant Tobacco Cells," *Plant Sci.* 72:53-62 (1990).
30. Wang, Y., J. D. Jones, S. C. Weller, and P. B. Goldsbrough. "Expression and Stability of Amplified Genes Encoding 5-enolpyruvylshikimate 3-phosphate Synthase in Glyphosate-tolerant Tobacco Cells," *Plant Mol. Biol.* 17:1127-1138 (1991).
31. Hauptmann, R. M., G. della-Cioppa, A. G. Smith, G. M. Kishore, and J. M. Widholm. "Expression of Glyphosate Resistance in Carrot Somatic Hybrid Cells Through the Transfer of an Amplified 5-enolpyruvylshikimic Acid-3-phosphate Synthase Gene," *Mol. Gen. Genet.* 211:357-363 (1988).
32. Shah, D. M., R. B. Horsch, H. J. Klee, G. M. Kishore, J. A. Winter, N. E. Tumer, C. M. Hironaka, P. R. Sanders, C. S. Gasser, S. A. Aykent, N. R. Siegel, S. G. Rogers, and R. T. Fraley. "Engineering Herbicide Tolerance in Transgenic Plants," *Science* 233:478-481 (1986).
33. Shyr, Y. J., A. G. Hepburn, and J. M. Widholm. "Glyphosate Selected Amplification of the 5-enolpyruvylshikimate-3-phosphate Synthase Gene in Cultured Carrot Cells," *Mol. Gen. Genet.* 232:377-382 (1992).
34. Holländer-Czytko, H., D. Johänning, H. E. Meyer, and N. Amrhein. "Molecular Basis for the Overproduction of 5-enolpyruvylshikimate 3-phosphate Synthase in a Glyphosate-tolerant Cell Suspension Culture of *Corydalis sempervirens*," *Plant Mol. Biol.* 11:215-220 (1988).
35. Forlani, G., E. Nielsen, and M. L. Racchi. "A Glyphosate-resistant 5-enol-pyruvylshikimate-3-phosphate Synthase Confers Tolerance to a Maize Cell Line," *Plant Sci.* 85:9-15 (1992).

36. Mousedale, P. M. and J. R. Coggins. "Subcellular Localization of the Common Shikimate Pathway Enzymes in *Pisum sativum* L.," *Planta* 163:241-249 (1985).
37. Comai, L., D. Facciotti, W. R. Hiatt, G. Thompson, R. E. Rose, and D. M. Stalker. "Expression in Plants of a Mutant *aroA* Gene from *Salmonella typhimurium* Confers Tolerance to Glyphosate," *Nature* 317:741-744 (1985).
38. Fillatti, J. J., J. Kiser, R. Rose, and L. Comai. "Efficient Transfer of a Glyphosate Tolerance Gene into Tomato Using a Binary *Agrobacterium tumefaciens* vector," *Bio/Technology* 5:726-730 (1987).
39. Hartwig, E. E. "Identification and Utilization of Variation in Herbicide Tolerance in Soybean (*Glycine max*) Breeding," *Weed Sci.* 35(Suppl. 1):4-8 (1987).
40. Loux, M. M., R. A. Liebl, and T. Hymowitz. "Examination of Wild Perennial *Glycine* Species for Glyphosate Tolerance," *Soybean Genet. Newsl.* 14:268-272 (1987).
41. Duncan, C. N. and S. C. Weller. "Heritability of Glyphosate Susceptibility Among Biotypes of Field Bindweed," *J. Hered.* 78:257-260 (1987).
42. Johnston, D. T. and J. S. Faulkner. "Herbicide Resistance in the Graminaceae — a Plant Breeder's View," in *Herbicide Resistance in Weeds and Crops*, J. C. Caseley, G. W. Cussans, and R. K. Atkin, Eds. (Oxford, UK: Butterworth-Heinemann, 1991), pp. 319-330.
43. Boerboom, C. M., N. J. Ehlke, D. L. Wyse, and D. A. Somers. "Recurrent Selection for Glyphosate Tolerance in Birdsfoot Trefoil," *Crop Sci.* 31:1124-1129 (1991).
44. Ehlke, N. Personal communication (1992).
45. Foy, C. L., R. Jacobsohn, and R. Jain. "Screening of *Lycopersicon* spp. for Glyphosate and/or *Orobanche aegyptiaca* Pers. Resistance," *Weed Res.* 28:383-391 (1988).
46. Torstensson, L. "Behaviour of Glyphosate in Soils and its Degradation," in *The Herbicide Glyphosate*, E. Grossbard and D. Atkinson, Eds. (London: Butterworths, 1985), pp. 137-150.
47. Liu, C. M., P. A. McLean, C. C. Sookdeo, and F. C. Cannon. "Degradation of the Herbicide Glyphosate by Members of the Family Rhizobiaceae," *Appl. Environ. Microbiol.* 57:1799-1804 (1991).
48. Jacob, G. S., J. R. Garbow, L. E. Hallas, N. M. Kimack, G. M. Kishore, and J. Schaefer. "Metabolism of Glyphosate in *Pseudomonas* sp. Strain LBr," *Appl. Environ. Microbiol.* 54:2953-2958. (1988).
49. Jacob, G. S., J. R. Garbow, J. Schaefer, and G. M. Kishore. "Solid-state NMR Studies of Regulation of Glyphosate and Glycine Metabolism in *Pseudomonas* sp. Strain PG2982," *J. Biol. Chem.* 262:1552-1557 (1987).
50. Pipke, R., N. Amrhein, G. S. Jacob, J. Schaefer, and G. M. Kishore. "Metabolism of Glyphosate in an *Arthrobacter* sp. GLP-1," *Eur. J. Biochem.* 165:267-273 (1987).
51. Rueppel, M. L., B. B. Brightwell, J. Schaefer, and J. T. Marvel. "Metabolism and Degradation of Glyphosate in Soil and Water," *J. Agric. Food Chem.* 25:517-528 (1977).
52. Balthazor, T. M. and L. E. Hallas. "Glyphosate-degrading Microorganisms from Industrial Activated Sludge," *Appl. Environ. Microbiol.* 51:432-434 (1986).
53. Coupland, D. "Metabolism of Glyphosate in Plants," in *The Herbicide Glyphosate*, E. Grossbard and D. Atkinson, Eds. (London: Butterworths, 1985), pp. 25-34.
54. Malik, J., G. Barry, and G. Kishore. "The Herbicide Glyphosate," *BioFactors* 2:17-25 (1989).

55. Schulz, A., F. Wengenmayer, and H. M. Goodman. "Genetic Engineering of Herbicide Resistance in Higher Plants," *Crit. Rev. Plant Sci.* 9:1-15 (1990).
56. Lund-Høie, K. "The Correlation Between the Tolerance of Norway Spruce (*Picea abies*) to Glyphosate (*N*-phosphonomethylglycine) and the Uptake, Distribution, and Metabolism of the Herbicide in the Spruce Plants," *Meld. Nor. Landbrukshoegsk.* 55:1-26 (1976).
57. Rueppel, M. L., J. T. Marvel, and L. A. Suba. "The Metabolism of *N*-phosphonomethylglycine in Corn, Soybeans, and Wheat," *Papers of the 170th American Chemical Society Meeting*, PEST 26 (1975).
58. Marshall, G., R. C. Kirkwood, and D. J. Martin. "Studies on the Mode of Action of Asulam, Aminotriazole and Glyphosate in *Equisetum arvense* L. (Field Horsetail). II. The Metabolism of [^{14}C]asulam, [^{14}C]aminotriazole and [^{14}C]glyphosate," *Pestic. Sci.* 18:65-77 (1987).
59. Komoßa, D., I. Gennity, and H. Sandermann. "Plant Metabolism of Herbicides with C-P bonds: Glyphosate," *Pestic. Biochem. Physiol.* 43:85-94 (1992).
60. Dyer, W. E. "Applications of Molecular Biology in Weed Science," *Weed Sci.* 39:482-488 (1991).
61. Mazur, B. J. and S. C. Falco. "The Development of Herbicide Resistant Crops," *Annu. Rev. Plant Physiol. Plant Mol. Biol.* 40:441-470 (1989).
62. Dyer, W. E., F. D. Hess, J. S. Holt, and S. O. Duke. "Potential Benefits and Risks of Herbicide-resistant Crops Produced by Biotechnology," *Hort. Rev.* 15:371-412 (1993).
63. della-Cioppa, G., S. C. Bauer, M. L. Taylor, D. E. Rochester, B. K. Klein, D. M. Shah, R. T. Fraley, and G. M. Kishore. "Targeting a Herbicide-resistant Enzyme from *Escherichia coli* to Chloroplasts of Higher Plants," *Bio/Technology* 5:579-584 (1987).
64. della-Cioppa, G. and G. M. Kishore. "Import of a Precursor Protein into Chloroplasts is Inhibited by the Herbicide Glyphosate," *EMBO J.* 7:1299-1305 (1988).
65. Kishore, G. M., D. Shah, S. R. Padgette, G. della-Cioppa, C. Gasser, D. Re, C. Hironaka, M. Taylor, J. Wibbenmeyer, D. Eichholtz, M. Hayford, N. Horrman, X. Delannay, R. Horsch, H. Klee, S. Rogers, D. Rochester, L. Brundage, P. Sanders, and R. T. Fraley. "5-enolpyruvylshikimate 3-phosphate Synthase: from Biochemistry to Genetic Engineering of Glyphosate Tolerance," *Am. Chem. Soc. Symp. Ser.* 379:37-48 (1988).
66. Klee, H. J., Y. M. Muskopf, and C. S. Gasser. "Cloning of *Arabidopsis thaliana* Gene Encoding 5-enolpyruvylshikimate-3-phosphate Synthase: Sequence Analysis and Manipulation to Obtain Glyphosate-tolerant Plants," *Mol. Gen. Genet.* 210:437-442 (1987).
67. Kishore, G. M. and D. Shah. "Amino Acid Biosynthesis Inhibitors as Herbicides," *Annu. Rev. Biochem.* 57:627-663 (1988).
68. Padgette, S. R. Personal communication (1992).
69. Waters, S. "Glyphosate-tolerant Crops for the Future: Development, Risks, and Benefits," in *Proceedings of the Brighton Crop Protection Conference — Weeds* (Surrey, U.K.: British Crop Protection Council, 1991), pp. 165-170.
70. Christou, P., D. E. McCabe, W. F. Swain, and D. R. Russel. "Legume Transformation," in *Control of Plant Gene Expression*, D. P. S. Verma, Ed. (Boca Raton, FL: CRC Press, 1993), pp. 547-564.

CHAPTER 9

Mechanisms Responsible for Cross Resistance and Multiple Resistance

Linda M. Hall, Joseph A. M. Holtum, and Stephen B. Powles

INTRODUCTION

In this chapter we address cross resistance and multiple resistance, best documented in the weed species *Lolium rigidum* Gaud. and *Alopecurus myosuroides* Huds. which exhibit resistance to many different herbicides. As has been described in the previous chapters, persistent herbicide application to a weed population selects for resistant individuals. Populations that result from this selection may be composed of individuals that differ in their complement of resistance genes. A herbicide is a selection agent for any heritable traits within a population that enable survival of individuals in the presence of herbicides. Therefore, survival mechanisms need only be sufficiently efficacious to confer survival at the rate of herbicide used. As herbicide treatments are nearly always at relatively low doses, this ensures survival of individuals expressing both strong resistance mechanisms (e.g., resistant target site enzyme) and relatively weak resistance mechanisms (e.g., enhanced rates of metabolism).[1] Plants with any genetically derived resistance mechanism will survive and contribute to the subsequent gene pool. When very large numbers of a highly variable weed species are treated, there can be survivors with quite different resistance mechanisms.[1]

The type of sexual reproduction in a weed species is crucial to the resistance profile a population may develop. In homogamous species, while more than one mechanism may be selected within a population, individuals carrying resistant genes seldom cross pollinate and their progeny are less likely to carry genes for more than one mechanism. In allogamous species, such as *L. rigidum* and *A. myosuroides*, resistant individuals carrying different resistance genes may cross pollinate and the progeny may then carry both resistance genes. Successive selection may thus accumulate resistance mechanisms within individuals and decrease the frequency of susceptible genes in the population. Individuals with more than one mechanism are more likely to survive applications of alternative herbicides. The allogamous nature of *L. rigidum* and *A. myosuroides*, along with their widespread distribution in cropping areas, has contributed to the development of multiple resistance (defined below).

There are no universally accepted definitions of cross and multiple resistance[2,3] and the two terms are often used interchangeably. In contrast to some former definitions which made reference to the selective agent, our definitions now are mechanistically based.

Cross resistance: Expression of a mechanism that endows the ability to withstand herbicides from different chemical classes. Cross resistance may be conferred either by a single gene or, in the case of quantitative inheritance, by two or more genes influencing a single mechanism (see Chapter 10). Two broad categories of mechanisms endowing cross resistance, are recognized:

- Target site cross resistance: The most common form of cross resistance occurs when a change at the site of action of one herbicide also confers resistance to herbicides from a different class that inhibits the same site of action. Examples include:

  1. Selection by aryloxyphenoxypropionic acid herbicides (APPs) for APP-resistant acetyl-CoA carboxylase (ACCase) that is also less sensitive to inhibition by cyclohexanedione herbicides (CHD) and vice versa;[4-7]

  2. Selection by sulfonylureas for sulfonylurea-resistant acetolactate synthase (ALS) that is also less sensitive to inhibition by imidazolinones and/or triazolopyrimidines[8] (see Chapter 4);

  3. Selection by triazines for triazine resistant D1 protein that is also less sensitive to triazinones[9] (see Chapter 2);

  4. Selection by dinitroanilines for dinitroaniline resistant microtubule assembly that also confers resistance to amiprophosmethyl[10] (see Chapter 7).

Target site cross resistance does not necessarily result in resistance to all herbicide classes with a similar mode of action or indeed all herbicides within a given herbicide class. Differences in the levels of resistance to herbicides with the same mode of action may result if a herbicide has more than one site of action or if the herbicide binding sites are not identical. For example, although triazines and substituted ureas bind to the D1 protein, the binding domains in higher plants overlap only partially. As a result, triazine selected mutations of the D1 protein observed in higher plants do not confer cross resistance to substituted ureas (see Chapter 2). Similarly, *Arabidopsis thaliana*, biotype GH90, resistant to imidazolinones, is not cross resistant to sulfonylureas or triazoloprymidines,[11] while biotype GH50 is resistant to sulfonylureas and cross resistant to triazolopyrimidines[12] and marginally cross resistant to imidazolinones[13] (see Chapter 4).

- Non-target site cross resistance: This is cross resistance conferred by a mechanism other than resistant enzyme target sites. Potential mechanisms include reduced herbicide uptake, reduced translocation, reduced herbicide activation, enhanced herbicide detoxification, changes in intra- or intercellular compartmentation and enhanced repair of herbicide-induced damage. Until recently documented for *L. rigidum* and *A. myosuroides*, non-target site cross resistance was unknown in plants but had been described in insects.[14,15]

Multiple resistance: Expression (within individuals or populations) of more than one resistance mechanism, endowing the ability to withstand herbicides from different chemical classes. Multiple resistant plants may possess two or many distinct resistance mechanisms.

The mechanisms of resistance in *L. rigidum* and *A. myosuroides*, two grass species which display cross and multiple resistance, are described below.

MECHANISMS OF RESISTANCE IN *LOLIUM RIGIDUM*

L. rigidum, a species native to the Middle East and Mediterranean regions, is a widespread, abundant weed within the cereal and grain-legume growing regions of southern Australia.[16] The first herbicide-resistant *L. rigidum* biotype was reported in 1982.[17] Since then the number of infestations has dramatically increased. A 1992 national survey, commissioned by the Agricultural and Veterinary Chemical Association of Australia, reported herbicide-resistant *L. rigidum* infested approximately 275,000 ha on more than 2,000 Australian farms in 1992.[18] What is striking about herbicide resistance in *L. rigidum* in Australia is the complex resistance patterns which develop across many different herbicide groups (Table 1). In some populations resistance exists to just one herbicide group, whereas in other populations resistance extends across many herbicide groups and modes of action. For example, of 242 resistant populations screened by Dr. G. Gill of the Western Australian Department of Agriculture, 39 and 11% exhibited resistance only to sulfonylureas and APPs, respectively. However, 32% were resistant to both sulfonylureas and APPs; 12% to sulfonylureas, APPs, and CHDs; and 6% were resistant to APPs, and CHDs.[19] Resistance to many herbicides makes *L. rigidum* control in Australia

Table 1. Herbicide groups to which biotypes of *L. rigidum* are resistant

PS II inhibitors	ALS inhibitors	ACCase inhibitors	Mitotic inhibitors	Other modes of action
s-triazines	Sulfonylureas	APPs	(Dinitroanilines)	(Chloroacetamides)
Substituted ureas	Imidazolinones	CHDs	(Carbamates)	(Isoxalolidinones)
Triazinones				(1-4 triazoles)

Parentheses indicate groups for which the resistance mechanism(s) have not been characterized.

a significant practical problem (see Chapter 12). Substantial research effort has been made to unravel the biochemical bases of multiple resistance in *L. rigidum* and the results of these studies are summarized below.

Target Site Cross Resistance

Resistant ACCase

ACCase resistant to inhibition by both APPs and CHDs has been documented in two biotypes of *L. rigidum*.[7,20] In one biotype (WLR 96),* resistance developed after 10 consecutive years of exposure to the APP diclofop-methyl, whereas the other biotype (SLR 3), received 3 consecutive years of exposure to the CHD sethoxydim. Both cases exhibit target site cross resistance to both the APP and CHD herbicides. Despite the dissimilar histories of herbicide exposure, resistance to APPs is higher than resistance to CHDs in both cases (Table 2). In a manner similar to that observed in maize,[23] resistant *L. rigidum* biotypes with resistant ACCase exhibit biotype-specific patterns of resistance at the whole plant level and in ACCase assays.[7] All herbicide-resistant plant ACCases examined to date exhibit inheritance patterns expected for single, nuclear encoded, dominant or partially dominant genes.[5,25,26]

The rates of metabolism of sethoxydim, fluazifop, or haloxyfop in *L. rigidum* biotypes with resistant ACCase are similar to those of susceptible plants with sensitive ACCase[7,21] (Figure 1). Even 190 h after exposure, when the resistant plants are healthy and have fresh weights four times those of susceptible plants, there are no differences in herbicide metabolism.

Resistant ALS

Biotypes of *L. rigidum* with resistant ALS are present in Australia.[27] These populations, typically resistant to a broad range of ALS inhibiting herbicides, have ALS with variable levels of resistance to sulfonylureas and imidazolinones (Table 3) (see Chapter 4). All populations known to possess herbicide-resistant

---

* Biotype designations for *L. rigidum* indicate the state of origin (S-South Australia; W-Western Australia; V-Victoria), the genus and species (LR-*Lolium rigidum*) followed by a number indicating the sequence of reporting. Biotypes of designations for *A. myosuroides* indicate the county of origin (Lincs.-Lincolnshire) followed by the paddock or field letter designation and a number indicating the sequence of reporting.

Table 2. Concentration of APPs and CHDs required to inhibit the activity of ACCase from susceptible and resistant *L. rigidum* by 50%

	Biotypes of *Lolium rigidum*				
	Susceptible		Resistant		
Herbicide	VLR 2	SLR 2	WLR 96	SLR 31	SLR 3
Aryloxyphenoxypropionates (APP)					
Diclofop acid (μM)	<0.2	0.3	25.4	0.6	7.3
Fluazifop acid (μM)	3.5	0.6	13.9	4.1	>>10.0
Haloxyfop acid (μM)	0.5	0.4	108	1.6	>10.0
Cyclohexanediones (CHD)					
Sethoxydim (μM)	1.0	2.7	5.1	2.8	13.3
Tralkoxydim (μM)	<0.2	0.3	1.1	0.6	1.9

Adapted from References 7, 21, and 22.

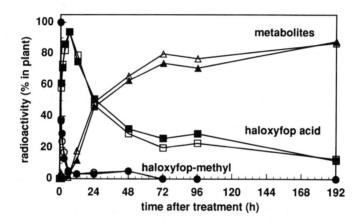

Figure 1. Metabolism of [^{14}C]haloxyfop-methyl (○,●) to haloxyfop acid (□, ■) and other metabolites (△, ▲) by two-leaf plants of a susceptible biotype of *L. rigidum* VLR 1 that contains haloxyfop-sensitive ACCase (○, □, △) and WLR 96, a resistant biotype that contains haloxyfop-insensitive ACCase (●, ■, ▲).[21]

ALS have a history of exposure to sulfonylurea herbicides, usually chlorsulfuron or triasulfuron. As expected, there is no evidence for the selection of resistant ALS in response to selection by the APP or CHD herbicides.

Non-Target Site Cross Resistance

Some *L. rigidum* biotypes have enhanced rates of herbicide metabolism. In such cases the degree of resistance at the whole plant level, while sufficient to provide resistance at the recommended rates, is much less than conferred by the target site cross resistance mechanisms.

Anecdotal evidence suggested selection of *L. rigidum* populations with the APP herbicide diclofop-methyl can result in APP resistance and non-target site cross resistance to some ALS inhibiting herbicides. Direct evidence of cross resistance was obtained by exposure of a susceptible population of *L. rigidum*

Table 3. Concentrations of sulfonylurea and imidazolinone herbicides required to inhibit the activity of ALS from susceptible and resistant *L. rigidum* by 50%

Herbicide	VLR 1 (S)	SLR 31 (R)	WLR 1 (R)
Sulfonylureas			
Chlorsulfuron (nM)	50 ± 15	38 ± 12	>1600
Sulfometuron-methyl (nM)	23 ± 10	16 ± 2	>1600
Imidazolinone			
Imazamethabenz-methyl (μM)	197 ± 30	230 ± 36	420 ± 20
Imazapyr (μM)	9 ± 2	9 ± 2	74 ± 1

Adapted from Reference 27.

to agricultural rates of the APP diclofop-methyl.[28] The plants that survived diclofop-methyl cross-pollinated among themselves, seed was collected and the selection with diclofop-methyl continued over the following two generations. After three generations of selection of an initially susceptible biotype with diclofop-methyl, APP resistance and cross resistance to chlorsulfuron were evident.[28] This study conclusively established that selection with an ACCase inhibiting herbicide can lead to a resistant population that displays non-target site cross resistance to ALS inhibiting herbicides without exposure to these herbicides.

Enhanced metabolism of ALS inhibiting herbicides — Some *L. rigidum* biotypes appear to have a similar ALS resistance mechanism to that of wheat. Wheat has a sensitive ALS, but can rapidly metabolize some ALS inhibitors by aryl-hydroxylation,[29] probably catalyzed by a Cyt P450 monooxygenase system, followed by glycosylation forming nontoxic metabolites. Similarly, the two chlorsulfuron-resistant *L. rigidum* biotypes studied in detail have sensitive ALS, a resistance profile similar to wheat, and can oxidatively metabolize chlorsulfuron more rapidly than the susceptible biotype.[30,31] The products of metabolism in *L. rigidum* and wheat are also similar.[30,31] Enhanced metabolism of ALS inhibiting herbicides in *L. rigidum* is probably due to a Cyt P450 enzyme(s). In SLR 31, chlorsulfuron resistance is largely overcome and metabolism of chlorsulfuron completely inhibited if malathion is combined with chlorsulfuron.[32] Malathion has been shown to inhibit the Cyt P450-dependent detoxification of sulfonylurea herbicides in microsome preparations from maize.[33] The reversal of resistance in SLR 31 confirms that detoxification plays a major role in chlorsulfuron resistance in this biotype. Definitive proof of the direct involvement of Cyt P450 enzymes in *L. rigidum* is still required.

Enhanced metabolism of PS II inhibiting herbicides — Metabolism-based resistance to triazine herbicides is another mechanism present in *L. rigidum* in Australia. Selection with atrazine in conjunction with the 1-4 triazole herbicide amitrole resulted in triazine resistance and cross resistance to substituted ureas in biotype WLR 2.[34] The proposed mechanism of simazine detoxification in resistant *L. rigidum* is enhanced capacity for N-dealkylation and increased de-ethyl simazine and di-de-ethyl simazine production.[35] Other metabolites are also detected, indicating that while N-dealkylation is a major pathway of metabolism, other pathways are also present.[35] For the substituted urea chlortoluron, resistant plants exhibit an enhanced capacity for N-demethylation as an initial metabolic step.[36] An enhanced capacity for N-demethylation would explain the broad cross resistance to many substituted ureas which have N-alkyl groups in common but differ in phenyl substituents. The pathways of subsequent herbicide metabolism are unresolved. Cyt P450 monooxygenase enzymes are likely to endow enhanced metabolism, as the Cyt P450 inhibitors ABT and PBO can antagonize resistance and inhibit metabolism of chlortoluron and simazine in this biotype.[35,36]

Enhanced metabolism of ACCase inhibiting herbicides — Some biotypes of *L. rigidum,* in which resistance is not due to resistant ACCase,[22,37] exhibit a marginal increase in the rate of diclofop-methyl metabolism.[37] The contribution of this increased metabolism to herbicide resistance is difficult to assess but is probably of minor importance. In addition to expression of the membrane recovery response (see Chapter 5), *L. rigidum* biotype SLR 31 exhibits a 10 to 15% increased capacity to metabolize diclofop-methyl.[37] Differences in diclofop-methyl metabolism between SLR 31 and susceptible biotypes are probably not due to secondary differences between herbicide-affected and -unaffected plants, as SLR 3 and WLR 96 plants, also unaffected by the herbicide due to a resistant ACCase, show no differences in metabolism rates.[7,21] At the whole plant level, diclofop-methyl resistance in SLR 31 is not altered in the presence of the cytochrome P450 inhibitors ABT, PBO, tetcyclasis, or malathion.

Membrane Recovery Response

A phenomenon associated with resistance to APP and possibly CHD herbicides has been detected in some biotypes where resistance is not due to resistant ACCase (see Chapter 5).[38,39] The biochemical bases of this so-called "membrane recovery response" are unclear, as is the causal connection between response at the membrane level and resistance at the whole plant level. In some biotypes, whole plant resistance correlates with the ability of coleoptiles and root tips to restore transmembrane potentials following exposure to low concentrations of APPs and CHDs that depolarize membranes (Figure 2). In etiolated coleoptiles of *L. rigidum,* 50% depolarization requires about 4 μM diclofop-acid. The recovery response is relatively rapid, requiring 10 to 20 min for full repolarization (in biotype SLR 31) following exposure to 50 μM diclofop-acid. Membrane polarity only recovers if the herbicide is removed from the vicinity of the tissue and there is resistance to the herbicide at the whole plant level. In the case of diclofop-acid, only resistant plants recover following depolarization in the presence of the herbicidal R-(+)-enantiomer, but both resistant and susceptible plants recover following depolarization in the presence of the nonherbicidal S-(–)-enantiomer, suggesting that the actions of the two enantiomers are not the same (Figure 3). The membrane recovery response is not restricted to *L. rigidum* in Australia, as similar responses have been observed in Canadian *Avena fatua* biotypes with APP[40] and CHD resistance.[41] Resistance in *A. fatua,* biotype UM

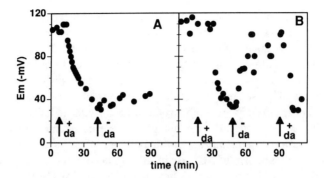

Figure 2. Effect of 50 μM diclofop acid on membrane potentials in etiolated coleoptiles of VLR 1, a susceptible biotype of *L. rigidum* (A) and in SLR 31, a resistant biotype (B). Arrows indicate the addition or removal of herbicide. During the experiments the coleoptiles were flushed continuously with 3 to 5 ml min^{-1} of 1 mM Pi buffer (pH 5.7) supplemented with 1 mM KCl, 1 mM Ca(NO$_3$)$_2$ and 0.25 mM MgSO$_4$.[38]

1, which exhibits the membrane response, is not due to enhanced herbicide metabolism or herbicide-resistant ACCase.

The mechanisms responsible for the membrane recovery response and the link between the response, resistance and cross resistance are unknown.[37] Speculations on this phenomenon are restricted by uncertainties about the intracellular concentrations of herbicide[42] and the extent and duration of membrane depolarization in meristems. Our current working hypothesis for the mechanism responsible for the regeneration of transmembrane potentials in resistant *L. rigidum*, and probably *A. fatua*, is that the concentrations of herbicide at the target sites in susceptible and resistant plants are not the same. This may be due to a different distribution of herbicide across the plasmalemma. Present evidence, based on enantiomeric specificity, K$^+$ responses and K$^+$/Na$^+$ ratios,[39] together with PCMBS (*p*-chloromercuribenzene sulfonic acid) responses[40] and the ability of root or coleoptile segments to acidify external media,[38,39,42,43] indicates the mechanism is mediated by a protein, probably membrane bound. Such a protein may be involved in the transport of the herbicide itself or may be involved in establishing transmembrane gradients that change the transmembrane partitioning of herbicide. For example, partitioning of diclofop-acid will be affected by the proportion that is dissociated, a ratio that will be sensitive to changes in pH or ionic concentration (see Chapter 5).

Differential compartmentation of herbicides between resistant and susceptible biotypes as a mechanism of resistance has not been adequately addressed. As the volume of the extracellular spaces is large in comparison to the cytoplasm, a small increase in extracellular herbicide concentration could result in a large decrease in the concentration in the cytoplasm and plastids. For example, if two compartments, one of a 50-fold greater volume than the other, contain a similar concentration of herbicide then a 2% increase in the concentration in the

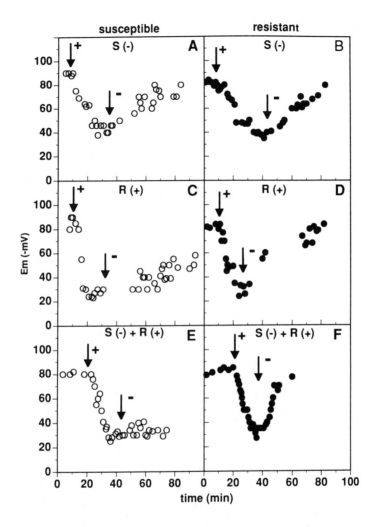

Figure 3. Effect of 25 µM S-(−)-diclofop acid (A, B), 25 µM R-(+)-diclofop acid (C, D), and a mixture of 25 µM S-(−)-diclofop acid and 25 µM R-(+)-diclofop acid (E, F) on membrane potentials in etiolated coleoptiles of VLR 1, a susceptible biotype of *L. rigidum* (○) and in SLR 31, a resistant biotype (●). Arrows indicate the addition or removal of herbicide. During the experiments the coleoptiles were flushed continuously with 3 to 5 ml min^{-1} of 1 mM Pi buffer (pH 5.7) supplemented with 1 mM KCl, 1 mM Ca(NO$_3$)$_2$ and 0.25 mM MgSO$_4$.[39]

larger compartment would account for all the herbicide in the smaller compartment. The observation that diclofop-methyl is continually degraded at similar rates in both susceptible and resistant *L. rigidum* and *A. fatua* indicates that either some herbicide leaks from the extracellular storage compartment or that the compartment exhibits an intrinsic capacity for detoxification.

Multiple Mechanisms of Resistance

Populations of *L. rigidum* may have more than one resistance mechanism and individuals within a population may differ in their component resistance genes.[1,44] Mechanisms found at varying frequencies within a biotype have been demonstrated in several populations, as illustrated in the following examples.

Biotype SLR 31 is the best-documented case of resistance conferred by multiple mechanisms. SLR 31 has a complex history of applications of many different herbicides and is resistant to herbicides from 9 different classes (Table 4). Several mechanisms, at various frequencies within the population, have been identified. A large proportion of the SLR 31 population demonstrates several mechanisms, including the "membrane recovery response", correlated with resistance to APP and (possibly) CHD herbicides;[38,39] a low level of enhanced herbicide metabolism of APP herbicide diclofop-methyl[37] which is not affected by all Cyt P450 inhibitors tested;[32] enhanced metabolism of chlorsulfuron,[27] which can be reversed by the Cyt P450 inhibitor malathion.[32] It has not yet been established whether enhanced metabolism of diclofop-methyl and chlorsulfuron are conferred by the same mechanism. Individuals comprising approximately 12% of the population are highly resistant to sethoxydim and possess an ACCase which is resistant to APP and CHD

Table 4. Herbicide history and resistance spectrum of the *L. rigidum* biotype SLR 31

Herbicide history			Resistance spectrum		
Year	Herbicide class	Herbicide	Herbicide class	Herbicide	Ref.
1969–1977	Dinitroaniline	Trifluralin	APP	Diclofop	22
				Haloxyflop	
1978	Dinitroaniline	Trifluralin		Quizalofop	
	Substituted urea	Linuron		Fluazifop	
				Chlorazifop	
1979–1982	Dinitroaniline	Trifluralin	CHD	Tralkoxydim	22
	APP	Diclofop		Sethoxydim	
1983	Carbamate	Chlorpropham	Sulfonylurea	Chlorsulfuron	27
1984	Sulfonylurea	Chlorsulfuron		Triasulfuron	
				Metsulfuron	
1985	Sulfonylurea	Metsulfuron	Imidazolinone	Imazamethabenz	27
1986	Sulfonylurea	Chlorsulfuron		Imazapyr	
1987	Sulfonylurea	Metsulfuron	Dinitroaniline	Trifluralin	45
	CHD	Sethoxydim		Ethalfluralin	
	Diphenyl ether	Oxyflurofen		Oryzalin	
				Pendimethalin	
1988	Bipyridyl	Paraquat		Isopropalin	
	Substituted urea	Linuron		Benfluralin	
	Carbamate	Carbetamide			
	CHD	Sethoxydim	Carbamate	Chlorpropham	45
			Thiocarbamate	Triallate	46
			Chloracetamide	Metolachlor	48
				Alachlor	
				Propachlor	
			Isoxalolidinone	Clomazone	47

Herbicide history information compiled from Reference 48.

herbicides.[44] ACCase from the balance of the population is sensitive to both types of herbicides, despite the plants being highly resistant to APPs.[22] Although considerable effort has been exerted towards characterizing the mechanisms present in this biotype, important questions remain unanswered, including whether herbicide compartmentation occurs and if so, its relative importance.[37] Also unresolved is the mechanistic link between the membrane recovery response and APP and CHD resistance.[39] However, this biotype does exemplify the complexity of resistance mechanisms which can be selected after many years of multiple herbicide use.

Another biotype, VLR 69 has also been exposed to many herbicides over 20 years and has similarly developed multiple resistance mechanisms. VLR 69 is resistant to the wheat selective sulfonylurea, chlorsulfuron, by virtue of an enhanced capacity for metabolism.[49] When ALS is extracted from the bulk of the population, herbicide inhibition kinetics for ALS are similar to that obtained from susceptible plants, indicating resistance is non-target site based. However, using a germination bioassay it is possible to select 2 to 3% of the population that contains a sulfonylurea-resistant ALS. This subset of the population, in addition to being resistant to chlorsulfuron, is resistant to the nonselective sulfonylurea, sulfometuron-methyl. Hence, there can be two mechanisms of resistance to the same herbicide within the same population, or individual.[49]

Populations with more than one mechanism are probably more common than is generally recognized (see Chapters 1 and 10). While *L. rigidum* in Australia is currently unique in the number and widespread distribution of resistant biotypes, herbicide resistance *A. myosuroides*, discussed below, illustrates that non-target site cross resistance and the accumulation of resistance mechanisms within populations is not unique to *L. rigidum*.

MECHANISMS OF RESISTANCE IN *ALOPECURUS MYOSUROIDES*

Resistant *A. myosuroides* biotypes have been reported in 19 counties in England,[50] and in France,[51] Spain,[52] and Germany.[53] However, the area affected has not been documented. *A. myosuroides*, like *L. rigidum*, exhibits many of the characteristics that favor the accumulation of herbicide resistance mechanisms. These characteristics include widespread distribution in cropping areas, high reproductive capacity, rapid seed bank turnover, allogamous reproduction, and genetic and phenotypic plasticity.[54] In general, the European cereal farming systems have been exposed to higher rates and a greater variety of herbicides than have the Australian systems. While it is not surprising that resistance in *A. myosuroides* has been reported to a variety of herbicide classes,[50,53,55] the number of resistant populations is currently a fraction of the number of resistant *L. rigidum* populations in Australia.[55] Resistance in *A. myosuroides* has been documented to one or more members of the substituted ureas, triazines, triazinones, sulfonylureas, imidazolinones, APPs, CHDs, thiocarbamates, carbamates, dinitroanilines, and chloroacetamides.[50]

Target Site Cross Resistance

Resistant ACCase

The biotype Lincs. E1 is resistant to the APP herbicides fenoxaprop-ethyl, diclofop-methyl, fluazifop-butyl, and quizalofop-ethyl, the CHD herbicide tralkoxydim, and is marginally resistant to sethoxydim.[56] ACCase extracted from this biotype showed consistent, low levels of resistance to APP and CHD herbicides. This level of resistance, if a property of the whole population, would be unlikely to be responsible for resistance. When the Lincs. E1 population was selected by germination on media containing fluazifop-ethyl, approximately 15% of the population appeared unaffected. This subset expresses an ACCase resistant to inhibition by diclofop-acid, compared to the balance of the resistant population and the susceptible biotype (Table 5).[56]

Non-Target Site Cross Resistance

Resistance to Substituted Ureas

In all examined biotypes of *A. myosuroides* resistant to substituted ureas, resistance is not due to a resistant PS II target site. A Spanish population studied by de Prado et al.[52] showed similar increase in leaf fluorescence emission and inhibition of the Hill reaction in R and S biotypes immediately following chlortoluron treatment. Fluorescence emission decreased after removal of chlortoluron in the R biotype only, suggesting either herbicide detoxification or sequestration confers resistance. This biotype is resistant to 13 other substituted ureas, but not to triazines or chloroacetamides.[52]

Similarly, PS II-dependent O_2 consumption measured in thylakoids isolated from the English biotype Peldon A1 and a susceptible biotype were inhibited equally by chlortoluron.[57] Preliminary evidence suggests Peldon A1 has an enhanced rate of metabolism of chlortoluron compared to the susceptible

Table 5. The concentration of APPs and CHDs required to inhibit by 50% the activity of ACCase from susceptible and resistant *A. myosuroides*

	I_{50} (μM)			
	Susceptible	Resistant		
Herbicide	Rothamsted	Lincs. E1	Lincs. E1 (subset)	Peldon A1
Diclofop	0.6	1.9	3.4	0.3
Fluazifop	1.7	13.5	>100	1.1
Sethoxydim	6.1	16.2	20.1	5.3

Adapted from Reference 55.

biotype.[58] The level of resistance can be reduced in the resistant biotype by the addition of the Cyt P450 inhibitors ABT and PBO.[57,59,60] Preliminary evidence indicates resistance to chlortoluron in Peldon A1 is quantitative, involving two nuclear genes with additive action[51] (see Chapter 10). The level of resistance to chlortoluron in English populations varies considerably.[50,55] It is not clear whether differences in resistance are due to differences in the relative frequency of the genes in the various populations, differences in the expression of herbicide detoxifying monooxygenases, or if resistance mechanisms vary.

Resistance to ACCase Inhibiting Herbicides

Many biotypes of *A. myosuroides* exhibit resistance to diclofop-methyl and fenoxaprop-ethyl but the level of resistance, and the extent of resistance to other APPs, varies considerably.[55] The two English biotypes which have been examined, Peldon A1 and Lincs. E1, are resistant to APPs diclofop-methyl, fenoxaprop-ethyl, and fluazifop-butyl and the CHD tralkoxydim, but only marginally to sethoxydim.[56,61] Both biotypes show enhanced metabolism of diclofop-methyl and fenoxaprop-ethyl.[56] Twenty-four hours after application of diclofop-methyl, the susceptible biotype contained 48% herbicidally active diclofop-acid, 3% diclofop-methyl, and 49% herbicide metabolites while the resistant biotypes, Lincs. E1 and Peldon A1, contained an average of 28% diclofop-acid, 3% diclofop-methyl, and 69% metabolites. Similar differences have been measured in fenoxaprop-ethyl metabolism.[57] Unlike *L. rigidum*, resistance to diclofop-methyl in whole plants can be partially reversed by the Cyt P450 inhibitor PBO,[57] suggesting the involvement of a monooxygenase system in conferring APP resistance.

Unidentified mechanism(s) — Resistance to fenoxaprop has also been deliberately selected from a susceptible population of *A. myosuroides* after four generations.[62] The mechanism conferring resistance in this population is unknown. The biotype expressed cross resistance to the aminopropionate herbicide flamprop but not to the APP quizalofop, CHD tralkoxydim, substituted urea chlortoluron or sulfonylurea chlorsulfuron. In this case resistance is nuclearly inherited and appears to involve a single gene.[62]

Enhanced Herbicide Metabolism

With resistant *A. myosuroides* biotype Peldon A1, there is a strong correlation between resistance and herbicide molecular structure, irrespective of chemical class. In general, resistance occurs to herbicides which are oxidatively metabolized by cereals (usually selective in small grain crops). This biotype can be controlled by herbicides which are not readily oxidatively metabolized. For example, Peldon A1 is resistant to the wheat selective sulfonylurea, chlorsulfuron, and the imidazolinone, imazamethabenz.[58] Peldon A1 is resistant to the dinitroaniline pendimethalin, but is susceptible to trifluralin, ethalfluralin, and isopropalin, which, unlike pendimethalin, do not have ring methyl groups susceptible to attack by monooxygenases.[55] Peldon A1 is controlled by the non-wheat selective carbamate, carbetamide, but not by the wheat selective barban.[55] There is no evidence of resistance to glyphosate or paraquat,[58] which are not readily metabolized.

Multiple Resistance

Two biotypes have been examined in detail for multiple mechanisms. One population, Lincs. E1, in addition to non-target site cross resistance conferred by enhanced herbicide metabolism of diclofop-methyl and fenoxaprop-ethyl, includes individuals (approximately 15% of the population) with resistant ACCase.[56] Peldon A1 does not contain a similar subset, confirming that multiple mechanisms may, or may not, occur in different populations. To date, no *A. myosuroides* biotypes have been examined for the "membrane recovery response".

While investigations of the mechanism(s) of resistance in *A. myosuroides* are limited, several conclusions can be tentatively drawn. Some biotypes, such as Peldon A1 and Lincs. E1, have developed a mechanism that confers a low level of resistance, sufficient to decrease mortality at field application rates, to a wide range of herbicides. This non-target site cross resistance mechanism is likely an enhanced, nonspecific, Cyt P450-mediated herbicide metabolism which appears to be quantitatively inherited. However, all biotypes do not exhibit the same pattern of resistance as Peldon A1 and Lincs. E1 and a single detoxification mechanism cannot be assumed.[50,56] Other *A. myosuroides* biotypes have developed changes at the site of action, including an altered PS II herbicide binding site which confers resistance to triazines[63] and ACCase less sensitive to inhibition by APPs.[52] Because the allogamous mating system favors mechanistic enrichment under continuing selection pressure, populations may contain both target site and non-target site cross resistance mechanisms at varying frequencies within populations.

CONCLUSIONS AND FUTURE DEVELOPMENTS

Resistance, multiple resistance, and cross resistance, endowed by a variety of physiological mechanisms, have appeared rapidly and repeatedly in *L. rigidum* and in *A. myosuroides* in response to local agronomic practices. Both species are fecund, annual, and allogamous. They infest high-technology, broad acre farming systems and many populations have been repeatedly exposed to herbicides from many different classes and modes of action. The possession of multiple resistance mechanisms in varying frequencies in populations is a direct result of the long-term use of herbicides and the subsequent concentration of resistance genes in survivors.

No evidence for differential fitness between biotypes of *L. rigidum* has been found to date.[64] Similarly, no fitness differential was found in the single resistant *A. myosuroides* biotype examined.[51] While generalizations cannot be extended to all the resistant biotypes of *L. rigidum* or *A. myosuroides*, reduced fitness does not appear to be universal.

While the biological attributes of both species contribute to the rapid development of multiple resistance, they are not particularly different from other weed species and, therefore, it should be expected that, with continuing herbicide use, the resistance responses observed in these two species will appear in other species. As has already occurred in insect species, cross resistance and multiple resistance will become less of a rarity in weed species. An inevitable consequence of the exposure of large weed populations to herbicides with a variety of chemistries and target mechanisms is the concentration in the survivors of genes that encode for both specific and general resistance mechanisms. Such weeds will become more difficult to control by chemical means. Reduced ability to control weeds will spur the development of non-chemical weed control technologies and techniques, and will foster research and adoption of integrated weed management strategies (see Chapter 12). Undoubtedly there will be pressure for a technological "fix", in particular for the development of effective herbicides with new modes of action. Technology will be used to widen the window of herbicide sensitivity between weeds and crops, probably by engineering resistance into crops and by the use of safeners, synergists, and other adjuvants. However, such technology will provide stop gap protection at best. In all probability, the emergence of resistance, multiple resistance, and cross resistance will dictate a reduction in herbicide use and the adoption of integrated weed management strategies.

REFERENCES

1. Powles, S. B., and J. M. Matthews. "Multiple Herbicide Resistance in Annual Ryegrass (*Lolium rigidum*), the Driving Force for the Adoption of Integrated Weed Management," in *Achievements and Developments in Combating Pest Resistance*, I. Denholm, A. Devonshire, and D. Holloman, Eds. (London: Elsevier, 1992), pp. 75-87.
2. Moss, S. R., and B. Rubin. "Herbicide-Resistant Weeds: A Worldwide Perspective," *J. Agric. Sci. Cambr.* 120:141-148 (1993).
3. Rubin, B. "Herbicide Resistance in Weeds and Crops, Progress and Perspectives," in *Herbicide Resistance in Weeds and Crops*, J. C. Caseley, G. W. Cussans, and R. K. Atkin, Eds. (Oxford: Butterworths-Heinemann, 1991), pp. 387-414.
4. Maneechote, C., J. A. M. Holtum, and S. B. Powles. "Resistant Acetyl-CoA Carboxylase is a Mechanism of Herbicide Resistance in a Biotype of *Avena sterilis* ssp. *ludoviciana*," *Plant Cell Physiol.* Submitted (1993).
5. Parker, W. B., L. C. Marshall, J. D. Burton,, D. A. Somers, D. L. Wyse, J. W. Gronwald, and B. G. Gengenbach. "Dominant Mutations Causing Alterations in Acetyl-Coenzyme A Carboxylase Confer Tolerance to Cyclohexanedione and Aryloxyphenoxypropionate Herbicides in Maize," *Proc. Natl. Acad. Sci. U.S.A.* 87:7175-7179 (1990).

6. Marles, M. A. S., M. D. Devine, and J. C. Hall. "Herbicide Resistance in *Setaria viridis* Conferred by a Less Sensitive Form of Acetyl Coenzyme A Carboxylase," *Pestic. Biochem. Physiol.* 46:7-14.
7. Tardif, F. J., J. A. M. Holtum, and S. B. Powles. "Occurrence of a Herbicide-Resistant Acetyl-Coenzyme A Carboxylase Mutant in Annual Ryegrass (*Lolium rigidum*) Selected by Sethoxydim," *Planta* 190:176-181 (1993).
8. Hall, L. M., and M. D. Devine. "Cross-resistance of a Chlorsulfuron-resistant Biotype of *Stellaria media* to a Triazolopyrimidine Herbicide," *Plant Physiol.* 93:962-966 (1992).
9. Pfister, K., and C. J. Arntzen. "The Mode of Action of Photosystem II–specific Inhibitors in Herbicide–resistant Weed Biotypes," *Z. Naturforsch.* 34c:996–1009 (1979).
10. Vaughn, K. C., and M. A. Vaughan. "Structural and Biochemical Characterisation of Dinitroaniline Resistant *Eleusine*," in *Managing Resistance to Agrochemicals: From Fundamental Research to Practical Strategies*, M. B. Green, H. M. LeBaron, and W. K. Moberg, Eds. (Washington, D.C.: American Chemical Society, 1990), pp. 364-375.
11. Haughn, G. W., and C. R. Somerville. "A Mutation Causing Imidazolinone Resistance Maps to the Csr1 Locus of *Arabidopsis thaliana*," *Plant Physiol.* 92:1081-1085 (1990).
12. Hall, L. M. "Chlorsulfuron Inhibition of Phloem Translocation in Chlorsulfuron-resistant and -susceptible *Arabidopsis thaliana*," Ph. D. Thesis, University of Saskatchewan, Saskatoon, Canada (1991).
13. Haughn, G. W., and C. Somerville. "Sulfonylurea-Resistant Mutants of *Arabidopsis thaliana*," *Mol. Gen. Genet.* 204:430-434 (1986).
14. Georghiou, G. P., and T. Saito. Eds. *Pest Resistance to Pesticides*, (New York: Plenum, 1983).
15. Georghiou, G. P. "The Magnitude of the Resistance Problem," in *Pesticide Resistance: Strategies and Tactics for Management*, Committee on Strategies for the Management of Pesticide Resistant Pest Populations, Board of Agriculture, National Research Council (Washington, D.C.: National Academy Press, 1986), pp. 14-43.
16. Kloot, P. M. "The Genus *Lolium* in Australia," *Aust. J. Bot.* 31:421-435 (1983).
17. Heap, J., and R. Knight. "A Population of Ryegrass Tolerant to the Herbicide Diclofop-Methyl," *J. Aust. Inst. Agric. Sci.* 48:156-157 (1982).
18. "The Incidence of Annual Ryegrass Herbicide Resistance," The AVCA Herbicide Resistance Action Committee (1992).
19. Gill, G. S. "Herbicide Resistance in Annual Ryegrass in Western Australia," Proceedings of the National Herbicide Resistance Extension Workshop, Glen Osmond, S. A., Australia (1992), pp. 8-11.
20. Holtum, J. A. M., and S. B. Powles. "Annual Ryegrass: an Abundance of Resistance, a Plethora of Mechanisms," *Brighton Crop Protection Conference — Weeds,* (Farnham, U.K.: The British Crop Protection Council, 1991), pp. 1071-1078.
21. Holtum, J. A. M., and S. B. Powles. Unpublished data (1992).
22. Matthews, J. M., J. A. M. Holtum, D. R. Liljegren, B. Furness, and S. B. Powles. "Cross-Resistance to Herbicides in Annual Ryegrass (*Lolium rigidum*). I. Properties of the Herbicide Target Enzymes Acetyl Coenzyme-A Carboxylase and Acetolactate Synthase" *Plant Physiol.* 94:1180-1186 (1990).

23. Marshall, L. C., D. A. Somers, P. D. Dotray, B. G. Gengenbach, D. L. Wyse, and J. W. Gronwald. "Allelic Mutations in Acetyl-Coenzyme A Carboxylase Confer Herbicide Tolerance in Maize," *Theor. Appl. Genet.* 83:435-442 (1992).
24. Betts, K. J., N. J. Ehlke, D. L. Wyse, J. W. Gronwald, and D. A. Somers. "Mechanism of Inheritance of Diclofop Resistance in Italian Ryegrass (*Lolium multiflorum*)," *Weed Sci.* 40:184-189 (1992).
25. Barr, A. R., A. M. Mansooji, J. A. M. Holtum, and S. B. Powles. "The Inheritance of Herbicide Resistance in *Avena sterilis* ssp. *ludoviciana*, Biotype SAS 1," *Proceedings of the First International Weed Control Congress*, (Melbourne: Weed Science Society of Victoria, 1992) pp. 70-72.
26. Tardif, F. J., M. W. Burnet, and S. B. Powles. "Inheritance of Resistance to Herbicides that Inhibit ACCase in Annual Ryegrass," Proceedings of the Genetics Society of Australia, 40th Annual Conference, Adelaide, (1993), p. 38.
27. Christopher, J. T., S. B. Powles, and J. A. M. Holtum. "Resistance to Acetolactate Synthase-Inhibiting Herbicides in Annual Ryegrass (*Lolium rigidum*) Involves at Least Two Mechanisms," *Plant Physiol.* 100:1909-1913 (1992).
28. Matthews, J. M., and S. B. Powles. "Aspects of population dynamics of selection for herbicide resistance in *Lolium rigidum* (Gaud.)," *Proceedings of the First International Weed Control Congress*, (Melbourne: Weed Science Society of Victoria, 1992) pp. 318-320.
29. Sweetser, P. B., G. S. Schow, and J. M. Hutchinson. "Metabolism of Chlorsulfuron by Plants. Biological Basis for Selectivity of a New Herbicide for Cereals," *Pestic. Biochem. Physiol.* 17:18-23 (1982).
30. Christopher, J. T., S. B. Powles, D. R. Liljegren, and J. A. M. Holtum. "Cross Resistance to Herbicides in Annual Ryegrass (*Lolium rigidum*). II. Chlorsulfuron Resistance Involves a Wheat-Like Detoxification System," *Plant Physiol.* 95:1036-1043 (1991).
31. Cotterman, J. C., and L. L. Saari. "Rapid Metabolic Inactivation is the Basis for Cross-Resistance to Chlorsulfuron in Diclofop-Methyl-Resistant Rigid Ryegrass (*Lolium rigidum*) SR4/84," *Pestic. Biochem. Physiol.* 43: 182-192 (1992).
32. Christopher, J. T., and S. B. Powles. Unpublished data (1993).
33. Kreuz, K., and R. Fonné-Pfister. "Herbicide-Insecticide Interaction in Maize: Malathion Inhibits Cytochrome P450-Dependent Primisulfuron Metabolism," *Pestic. Biochem. Physiol.* 43:232-240 (1992).
34. Burnet, M. W. M., O. B. Hildebrand, J. A. M. Holtum, and S. B. Powles. "Amitrole, Triazine, Substituted Urea, and Metribuzin Resistance in a Biotype of Rigid Ryegrass (*Lolium rigidum*)," *Weed Sci.* 39:317-323 (1992).
35. Burnet, M. W. M., B. R. Loveys, J. A. M. Holtum, and S. B. Powles. "Increased Detoxification is a Mechanism of Simazine Resistance in *Lolium rigidum*," *Pestic. Biochem. Physiol.* 46:207-218 (1993).
36. Burnet, M. W. M., B. R. Loveys, J. A. M. Holtum, and S. B. Powles. "A Mechanism of Chlortoluron Resistance in *Lolium rigidum*," *Planta* 190:182-189 (1993).
37. Holtum, J. A. M., J. M. Matthews, R. E. Häusler, and S. B. Powles. "Cross-Resistance to Herbicides in Annual Ryegrass (*Lolium rigidum*). III. On the Mechanism of Resistance to Diclofop-Methyl," *Plant Physiol.* 97:1026-1034 (1991).
38. Häusler, R. E., J. A. M. Holtum, and S. B. Powles. "Cross Resistance to Herbicides in Annual Ryegrass (*Lolium rigidum*). IV. Correlation Between Membrane Effects and Resistance to Graminicides," *Plant Physiol.* 97:1034-1043 (1991).

39. Holtum, J. A. M., R. E. Häusler, M. D. Devine, and S. B. Powles. "The Recovery of Transmembrane Potentials in Plants Resistant to Aryloxypropionate Herbicides: A Phenomenon Awaiting Explanation," *Weed Sci.* In press (1994).
40. Shimabukuro, R. H., and B. L. Hoffer. "Effect of Diclofop on the Membrane Potentials of Herbicide-Resistant and -Susceptible Annual Ryegrass Root Tips," *Plant Physiol.* 98:1415-1422 (1992).
41. Devine, M. D., S. A. MacIsaac, M. L. Romano, and J. C. Hall. "Investigation of the Mechanism of Diclofop Resistance in Two Biotypes of *Avena fatua*," *Pestic. Biochem. Physiol.* 42:88-96 (1992).
42. DiTomaso, J. M., P. H. Brown, A. E. Stowe, D. L. Linscott, and L. V. Kochian. "Effects of Diclofop and Diclofop-Methyl on Membrane Potentials in Roots of Intact Oat, Maize and Pea Seedlings," *Plant Physiol.* 95:1063-1069 (1991).
43. Devine, M. D., J. C. Hall, M. L. Romano, M. A. S. Marles, L. W. Thompson, and R. H. Shimabukuro. "Diclofop and Fenoxaprop Resistance in Wild Oat is Associated with an Altered Effect on the Plasma Membrane Electrogenic Potential," *Pestic. Biochem. Physiol.* 45:167-177 (1993).
44. Tardif, F. J., and S. B. Powles. "Multiple Herbicide Resistance in a *Lolium rigidum* Biotypes is Endowed by Multiple Mechanisms: Isolation of a Subset with Resistant Acetyl-Coenzyme A Carboxylase," *Plant Cell Physiol. U.S.A.* Submitted.
45. McAlister, F. M., J. A. M. Holtum, and S. B. Powles. "Dinitroaniline Resistance in Rigid Ryegrass (*Lolium rigidum*)," *Weed Sci.* Submitted (1993).
46. Tardif, F. J., and S. B. Powles. Unpublished data (1992).
47. Powles, S. B. Unpublished data (1993).
48. Burnet, M. W., A. R. Barr, and S. B. Powles. "Chloroacetamide Resistance in *Lolium rigidum*," *Weed Sci.* In press (1994).
49. Burnet, M. W. M., J. T. Christopher, J. A. M. Holtum, and S. B. Powles. "Identification of Two Mechanisms of Sulfonylurea Resistance Within One Population of Rigid Ryegrass (*Lolium rigidum*) Using a Selective Germination Medium," *Weed Sci.* In press (1994).
50. Moss, S. R., and G. W. Cussans. "The Development of Herbicide-Resistant Populations of *Alopecurus myosuroides* (Black-Grass) in England," in *Herbicide Resistance in Weeds and Crops*, J. C. Caseley, G. W. Cussans, and R. K. Atkin, Eds. (Oxford: Butterworths-Heinemann, 1991), pp. 45-56.
51. Chauvel, B. "Polymorphisme génétique et sélection de la resistance aux urées substituées chez *Alopecurus myosuroides* Huds.," Thèse pour le grade de Docteur en Science, Université de Paris-Sud, Centre d'Orsay (1991).
52. De Prado, R., J. Menendez, M. Tena, J. Caseley, and A. Taberner. "Response to Substituted Ureas, Triazines and Chloroacetanilides in a Biotype of *Alopecurus myosuroides* Resistant to Chlortoluron," *Brighton Crop Protection Conference — Weeds* (Farnham, U.K.: The British Crop Protection Council, 1991), 1065-1070.
53. Niemann, P., and W. Pestemer. "Resistenz verschiedener Herkunfte von Acker-Fuchsschwanz (*Alopecurus myosuroides*) gegenuber Herbizidbehandlungen," *Nachrichtenbl. Dtsch. Pflanzenschutzdienstes* 36:113-118 (1984).
54. Moss, S. R., and G. W. Cussans. "Variability in the Susceptibility of *Alopecurus myosuroides* (Black-Grass) to Chortoluron and Isoproturon," *Aspects Appl. Biol.* 9:91-98 (1985).

55. Moss, S. "Herbicide Resistance in the Weed *Alopecurus myosuroides* (Black-Grass): The Current Situation," in *Achievements and Developments in Combating Pesticide Resistance,* I. Denholm, A. Devonshire, and D. Holloman, Eds. (London: Elsevier, 1992), pp. 28-40.
56. Hall, L. M., S. R. Moss, and S. B. Powles. "Towards an Understanding of Resistance to APP Herbicides in *Alopecurus myosuroides*," Weed Management - Towards Tomorrow. 10th Council of Australian Weeds Science Societies and 14th Asian Pacific Weeds Science Society, Brisbane, Queensland (1993), pp. 299-301.
57. Hall, L. M. Unpublished data (1993).
58. Kemp, M. S. A., S. R. Moss, and T. H. Thomas. "Herbicide Resistance in *Alopecurus myosuroides*," in *Managing Resistance to Agrochemicals. From Fundamentals to Practical Strategies,* M. B. Green, H. LeBaron, and W. Moberg, Eds. (Washington, D.C.: American Chemical Society, 1991), pp. 376-393.
59. Kemp, M. S., and J. C. Caseley. "Synergistic Effect of 1-Aminobenzotriazole on the Phytotoxicity of Chlortoluron and Isoproturon in a Resistant Population of Black-Grass (*Alopecurus myosuroides*), *Brighton Crop Protection Conference — Weeds* (Farnham, U.K.: The British Crop Protection Council, 1987), pp. 895-898.
60. Kemp, M. S., and J. C. Caseley. "Synergists to Combat Herbicide Resistance," in *Herbicide Resistance in Weeds and Crops,* J. C. Caseley, G. W. Cussans, and R. K. Atkin, Eds. (Oxford: Butterworths-Heinemann, 1991), pp. 279-292.
61. Moss, S. R. "Herbicide Cross-Resistance in Slender Foxtail (*Alopecurus myosuroides*)," *Weed Sci.* 38:492-496 (1990).
62. Chauvel, B., J. Gasquez, M. A. Doucey, and F. Perreau. "Selection for Fenoxaprop-P-ethyl Resistance within Black-Grass (*Alopecurus myosuroides* Huds.) Populations," IXeme Colloque International sur la Biologie des Mauvaises Herbes, (Dijon: ANPP Annales, 1992), pp. 487-496.
63. Yaacoby, T., M. Schonfeld, and B. Rubin. "Characteristics of Atrazine Resistant Biotypes of Three Grass Weeds," *Weed Sci.* 34:181-184 (1986).
64. Matthews, J. M., and S. B. Powles, Unpublished data (1992).

CHAPTER 10

Genetics of Herbicide Resistance in Weeds and Crops

H. Darmency

INTRODUCTION

Weed research has only recently embraced genetics studies. Intraspecific variability was first described taxonomically and morphologically, giving the impression that different taxa have well separated and fixed traits. With such a paradigm, it was not easy for many individuals to imagine the potential of plants for rapid adaptation. By 1970, the morphological traits[1] used as taxonomic characters were used as genetic markers to investigate adaptive strategies in weeds.[2,3] Few subsequent studies have investigated the genetics of weeds and thus inheritance data for any trait in weeds is scarce. For some species, crossing is technically difficult and may have discouraged individuals from undertaking long and difficult genetic studies. Moreover, weed scientists frequently consider genetics too late in the development of their research program.

Herbicide resistance is a research area which would have benefited from good genetic data in studies of the physiology and epidemiology of herbicide resistance. Important questions such as predicting the occurrence and consequences of resistance on field management of resistant weeds remain unanswered. This is not due to a lack of understanding of plant genetics but to a paucity of specific data on the genetics of herbicide resistance. Plant breeders have utilized suitable theories and experimental tools to predict and study herbicide resistance in plants. In fact, herbicide resistance has been predicted on a mass selection basis since the 1950s.[4,5]

This chapter reports on the inheritance of herbicide resistance in weed species and demonstrates how genetic studies have provided, or could have provided, information on the appearance and mechanisms of resistance. A genetic approach to resistance involves three aspects. First, appropriate test conditions and a suitable scale for measuring variation in response to herbicides must be defined. As we will see, this is not a trivial pursuit. Second, the data analysis performed depends on whether resistance is controlled by cytoplasmic or nuclear factors, and, if nuclear factors are involved, whether the segregation response is continuous or discrete. Third, suitable material must be obtained for genetic analysis.

The first section considers a large amount of genetic data on resistance to triazine herbicides. In addition to describing chloroplastic inheritance of triazine resistance, inconsistent genetic results are scrutinized. Much data has been pooled, making an arbitrary division between resistance verified using a mortality/survival scoring system and resistance estimated by effects on growth parameters. The second section, quantitative inheritance, concerns herbicides which inhibit growth, rather than cause mortality. In such cases, the ability of resistant plants for regrowth and seed production are important traits and are best studied using quantitative genetics techniques. Examples of crop cultivars with differential response to herbicides are also considered. In the third section, cases of resistance that fit simple Mendelian inheritance are described for weeds and crops. Accuracy of methodology are also discussed. Conclusions, in the fourth section, concern the origin of resistance genes and their frequency and behavior in populations.

CHLOROPLAST ENDOWED TRIAZINE RESISTANCE

Fundamentals of Inheritance

In 1968, triazine resistance was reported in the U.S. in a biotype of *Senecio vulgaris* located in a plant nursery treated for 10 years with simazine. Until this report only a few cases of intraspecific variation of plant response to herbicides had been reported[6] and the genetics of variation had not been investigated in weeds. However, varietal differences in responses to triazine herbicides in crop species *Triticum aestivum*, *Sinapis alba*, and *Brassica napus* had been described and this showed that direct and rapid selection for resistance was possible.[7] It was also known that triazine resistance in maize was simply inherited and controlled by a single dominant nuclear gene.[8] However, quantitative estimates carried out with *Linum usitatissimum* showed a low heritability of triazine resistance, indicating selection for atrazine resistance in this species (and probably others) would be difficult.[9] This paucity of genetic information may explain why no inheritance studies were conducted on the resistant biotype of *S. vulgaris* at that time. Following the initial report of resistance in *S. vulgaris*,[6] triazine resistance has subsequently been reported in more than 55 species[10] and at numerous locations around the world (fully described in Chapter 2). However, for nearly 10 years, until 1976, the mechanism and inheritance of triazine resistance remained unknown. Had a cytoplasmic inheritance pattern been demonstrated through a simple crossing experiment, researchers may have focused their work earlier on the chloroplast.[11] In 1977 the first preliminary data on resistance inheritance in *Chenopodium album*[12] was reported following indications of the involvement of chloroplasts in resistance.[11] This data was inconclusive because of difficulties associated with crossing small flowers[13] and the lack of physical or morphological markers to distinguish parental plants. Uniparental inheritance through the female was first confirmed in *Brassica campestris*, a cross-pollinating species with large flowers suitable for hand emasculation. Cotyledon size was larger in the resistant biotype, providing a suitable morphological marker.[13] Subsequent research, either using parental markers or careful hand emasculation, confirmed triazine resistance was maternally inherited in many populations and species (Table 1).[14-28] Reciprocal crosses were made and reciprocal F_1 hybrids tested. Results on F_2 and F_3 generations of these crosses further established a clear picture of maternal inheritance. Several backcross generations, using pollen from susceptible plants, had no effect on the maternal transmission of resistance in *S. vulgaris*.[25] Finally, crossing resistant weed plants with closely related crop species, some containing a different chromosome number, did not alter the cytoplasmic inheritance in *Brassica napus*[29,30] and *Setaria italica*.[27]

Transfer of triazine resistance to crops has also been achieved by protoplast fusion after X-irradiation of nuclear DNA of resistant protoplasts. Regenerated hybrids were resistant due to the presence of mutant-type chloroplasts identified by restriction fragment length polymorphism analysis.[31] Molecular analysis of the resistance confirmed this trait was encoded by the chloroplast genome. It has been shown that, in resistant *Amaranthus hybridus*, a point mutation in the chloroplast DNA, at locus psbA codon 264, was responsible for a single amino acid change (Ser to Gly) in the 32,000-Da chloroplast protein.[32] This difference between R and S plants has been found repeatedly in at least ten other species.[33] Conclusive proof that the psbA gene mutation was the basis of triazine resistance was obtained when the chloroplast gene was introduced into tobacco nuclear genome via a plasmid transformation system and resulted in increased triazine resistance in the transformed plants.[34]

Inconsistent Features

As outlined above, while the chloroplast inheritance pattern of triazine resistance is firmly established, some inconsistent genetic data have revealed some discrepancies. In addition to chloroplastic inheritance, evidence of paternal cytoplasmic endowed inheritance has accumulated. Haploid plantlets of triazine-resistant species derived from anther culture expressed a resistance gene, providing evidence that pollen at an early stage of microsporogenesis contained the resistance trait.[35] Later, as pollen grains mature, the vegetative

Table 1. Inheritance studies of chloroplastic triazine resistance

Species	Crossing techniques	Generations studied	Nuclear markers	Resistance test	Ref.
Amaranthus bouchonii	Open pollination	F_1, F_2	Isozymes	4 kg ha^{-1}	14
Amaranthus retroflexus	Hand emasculation	F_1, F_2	Cotyledon size	3 kg ha^{-1}	15
Brassica campestris	Hand emasculation	F_1	Cotyledon size	3 kg ha^{-1}	13
Brassica campestris[a]	Hand emasculation	F_1, F_2, F_3, BC_1	Chlorotic cotyledon	3 kg ha^{-1}	16
Chenopodium album	Open pollination	F_1	None	3 kg ha^{-1}	12
Chenopodium album	Hand emasculation	F_1	None	Leaf response	17
Chenopodium album	Open pollination	F_1, F_2	Isozymes	0.5 kg ha^{-1} Leaf fluorescence	18
Chenopodium polyspermum	Open pollination	F_1, F_2	Isozymes	4 kg ha^{-1}	14
Echinochloa crus galli	Hand emasculation	F_1, F_2	None	10 kg ha^{-1}	19
Poa annua	Open pollination	F_1, F_2	Isozymes	Leaf fluorescence	20,21
Senecio vulgaris	Hand emasculation	F_1, F_2	Floret morphology	2.2 kg ha^{-1}	22,23
Senecio vulgaris	Hand emasculation	F_1, BC_6	None	10 kg ha^{-1}	24,25
Setaria viridis[a]	Partial emasculation	F_1, F_2, BC_4	Coleoptile morphology isozymes	2 kg ha^{-1}	26,27
Solanum nigrum	Hand emasculation	F_1, F_2	Fruit and leaf morphology	1 kg ha^{-1} Leaf fluorescence	28

[a]Cross between the weed species and a crop as pollen donor (*B. napus* and *S. italica*).

cell still has plastids, but not the generative cell.[36] The plastids are altered or destroyed in the development of the grain and are not transmitted. However, it was noticed in some studies (especially when a large number of susceptible females were crossed with resistant males) that the system that normally prevents plastid transmission through the pollen can sometimes fail (Table 2).[18-20,28] Hybrids having resistant paternal plastid DNA were found at a rate ranging from 0.2 to 2%. Such a low rate of paternal transmission had already been described for other traits in other species, although ultrastructural evidence is rare.[37] Segregation between triazine R and S phenotypes was recorded in the progeny of four of these F_1.[18,19] Such an elimination also occurred when resistant plastids were introduced into oilseed rape by means of protoplast fusion: in the absence of herbicide selective conditions during plant regeneration, the regenerated plants segregated for mutated chloroplasts.[31]

The progeny of hybrids of *C. album* that inherited resistance through the male parent maintained resistance to the F_5 generation. They expressed two different phenotypes (high and intermediate resistance) although the same mutation was present in the chloroplast genome.[38] The progeny of a highly

Table 2. Paternal transmission of chloroplastic triazine resistance in progeny of female susceptible plants crossed with male resistant plants

Species	Number of F_1		Ref.
	Susceptible	Resistant	
Chenopodium album	557	13	18
Echinochloa crus galli	156	3	19
Poa annua	> 5,000	13	20
Solanum nigrum	1,313	2	28

resistant plant segregated into intermediate and highly resistant plants; however, the proportion of highly resistant plants decreased with each generation. The F_1 generation comprised a large proportion of highly resistant plants, while the F_5 generation completely lacked such plants. Regression analysis of the proportion of resistant phenotypes as a function of the generation number showed the following linear equation: $y = 0.81 - 0.18 \times (r^2 = 0.70)$, where y is the percentage of highly resistant phenotypes and x is the generation number.[18] This was tentatively interpreted as the regular loss, in each generation, of a cytoplasmic factor encoding for high resistance. As this cannot be attributed to chloroplast loss, studies of mitochondrial involvement were initiated. In maize, a 420-bp segment homologous to the chloroplast psbA gene was found in the DNA of a mitochondrial plasmid.[39] A 270-bp DNA sequence with high psbA homology was also found in the mitochondrial genome of C. album.[40] This partial mitochondrial DNA copy of the chloroplast DNA gene included codon 264, but does not appear to be mutated at the same position. The mitochondrial sequences of both S and R plants showed no modification. However, the sequence was expressed as a mitochondrial transcript in R but not S plants. Whether this is related to the difference between the two resistant phenotypes has yet to be elucidated.

The occurrence of triazine-resistant phenotypes with different chloroplast characteristics to that described above has been documented in a few cases. In Poa annua, two resistant biotypes have been reported with a similar whole plant response to atrazine and with maternal inheritance of resistance, but with differences in chlorophyll fluorescence transients and atrazine inhibition of chloroplast activity.[21] As only one mutation has been detected in triazine-resistant higher plants, these differences should not be related to differences in the psbA chloroplast mutation. Rather, they indicate a possible interaction between variation in the genetic background of chloroplasts and the expression of resistance. This phenomenon may also be responsible for variation reported in the level of resistance among resistant populations of Solanum nigrum. While all populations had resistant chloroplasts and survived 6 kg ha^{-1} atrazine, one population was completely killed at 32 kg ha^{-1} while another showed 40% survival.[41] Since inheritance studies have seldom been carried out at rates of atrazine higher than 3 kg ha^{-1}, the presence of a second mechanism (functional at high herbicide rates) in addition to the chloroplast mutation cannot be excluded. Variation in the genetic background may affect some minor physiological processes that caused the variation in the whole plant response at high rates of atrazine.

Differences in plant response were also found at low atrazine rates. In greenhouse experiments, susceptible *C. album*, *C. polyspermum*, and *Amaranthus bouchonii* were killed by 0.15 kg ha^{-1}, intermediate plants were unaffected by 0.5 kg ha^{-1} but completely killed by 1 kg ha^{-1}, whereas resistant plants survived 3 kg ha^{-1}.[14,42] The intermediate phenotype showed the same biophysical and biological attributes as highly resistant plants carrying the chloroplast mutation, i.e., modified chlorophyll fluorescence transients and resistance to high atrazine rates.[14,38] Intermediate and high resistances were both maternally inherited in the F_1 generation. In the case of *C. album*, molecular DNA analysis showed both types have the same mutation in the *psbA* chloroplast gene.[38] Finally, segregation for intermediate and high resistance was found in the F_2 of reciprocal crosses between the two resistant types, indicating an extra chloroplastic mechanism was involved.[38] The link between intermediate and resistant phenotypes remains unknown. When intermediate plants are grown and treated with systemic pesticides or sublethal doses of atrazine or monolinuron, the seeds produced by self-pollination of the treated plants are all highly resistant. This phenotypic change is observed by leaf-fluorescence response as well as whole plant resistance.[38] It is inherited in subsequent generations and segregation of phenotypes is no longer observed. The origin of such a phenotypic change and the additional mechanism involved are yet to be elucidated.

Clearly, expression of chloroplast encoded resistance may interact with other mechanisms, resulting in plants with various degrees of resistance. The explanation for this variation may be found in studying the difference between phenotype and genotype. The presence of the very efficient chloroplast *psbA* mutation endowing high-level resistance may obscure other, less effective, resistance mechanisms. Even in the absence of genes that encode for some other resistance mechanisms (as described in the next section), variation in the genetic background may affect the level of resistance among species and populations, and between plants of the same population. Therefore, it can be difficult to determine which individual represents the correct control susceptible plant or population, and which is the genotype of a resistant phenotype.

Several methodological concerns are evident in selecting triazine-resistant and -susceptible material for study. To determine if there are pleiotropic effects associated with resistance genes, isogenic R and S lines must be used in order to avoid excessive variation. Too much research has been conducted with plant material of undefined parentage, often from different origins! Fortunately, several accurate analyses have been made on growth and physiological parameters of weeds using reciprocal F_1 hybrids,[24,43,44] F_2 individuals of reciprocal F_1,[45] back-crossed generations,[25,46] and "newly" mutated resistant individuals along with their susceptible sibling plants.[47,48] Several studies have also been conducted with back-crossed generations after hybridization with oilseed rape and foxtail millet crops.[27,48-51] These studies, as discussed in Chapters 2 and 11, generally confirm the reduced fitness of triazine R versus S plants. The fitness of R plants growing among a S population is difficult to measure. To correctly assess if fitness changes are associated with resistance, intrapopulation genetic variability must be studied. For instance, the chlorophyll a/b ratio decreased by 10% in R versus S isogenic *C. album*, but also varies naturally by 10% among plants within a uniformly susceptible population.[48] Therefore, the mutation did not bring any change to the population with respect to the existing range of variation observed for this trait. It may be assumed that this holds true for the fitness of resistant plants. Because of intrapopulation variability, it is likely that some S plants may have a fitness lower than that of the R mutants. Since these plants persist within populations, resistant mutants may also be expected to survive in the field in the absence of herbicide selection pressure. Although the comparison of fitness of resistant plants to the whole population is the only way to predict the population dynamics of resistant weeds in the field, no work has been done to date on this topic.

These inconsistent data and possible interpretations indicate there are deficits in our understanding of chloroplast endowed triazine resistance. The final section of this chapter, on the origin and dynamics of resistance genes, provides additional information on how resistance has appeared in plants and populations. The genetic basis of resistance is more complex than commonly assumed and thus numerous questions remain unresolved.

QUANTITATIVE INHERITANCE

Some herbicides are used at rates that do not kill all target weeds but rather inhibit growth sufficiently to enable the crop species to have a competitive advantage. As a consequence, scoring survival versus mortality at high herbicide rates is not appropriate and growth rather than mortality should be measured. Since growth is a function of many genetic factors, and variability within populations is known for this trait, the genetic factors involved in resistance must be studied using quantitative genetics techniques. Herbicide efficacy is usually assessed with a subjective visual injury rating based on leaf necrosis and growth reduction (on a 1–9 scale or a similar scoring system) instead of objective measurements like leaf area, plant height, or weight.

Research on the quantitative inheritance of herbicide resistance in crop species is aimed primarily at improving the resistance level of crop cultivars or preventing phytotoxicity from herbicide soil residues affecting rotational crop species. Quantitative estimates of genetic gains resulting in higher yields has been emphasized in these studies and little account taken of the mortality of plants. However, research on crop species is relevant as it is certain that similar quantitative herbicide resistance genes can be selected in weed species (Table 3).

Table 3. Quantitative inheritance of herbicide resistance in weeds and wild populations

Herbicide	Species	Plant material	Resistance test	Scoring	Heritability	Ref.
Barban	Avena fatua	Selfed families, 7 populations (100 families each)	0.15 kg ha^{-1} at 2- to 3-leaf stage	Visual injury rating	0–0.63	58
Chlortoluron	Triticum dicoccoides	208 families from 11 populations	5.5 kg ha^{-1} at 4- to 5-leaf stage	Visual injury rating	0.72	54
Glyphosate	Convolvulus arvensis	Diallele (20 F$_1$)	1.1 kg ha^{-1} on 4- to 5-month-old plant	Visual injury rating	Additive	60
Simazine	Senecio vulgaris	Selfed families (5 families)	0.7 kg ha^{-1} by soil incorporation	Seedling mortality	0.22	52

Weed Species

Resistance to Photosystem II Inhibitors

Plants may become resistant to the triazine herbicides due to mechanisms other than target site mutation (see Chapter 2). In resistant biotypes of *S. vulgaris* originating from fields with a long history of atrazine treatment, a positive linear relationship was found between the percentage survival of a population at 0.7 kg ha^{-1} and the number of years of triazine application.[52] This indicated effective selection due to repeated use of the same herbicide. Genetic factors were investigated through intrapopulation studies. In *S. vulgaris*, genetic and environmental components of variances were obtained using a selfing series procedure. Broad sense heritability was low, 0.22, which indicated that increased triazine resistance was inherited in a polygenic fashion.[52] Similarly, selection of *Alopecurus myosuroides* with a low rate of simazine for 4 years was sufficient to select for plants showing 50% or more root biomass than control plants. This was interpreted as selection for larger plants with root systems that could penetrate through herbicide-treated soil layer.[53] Variation in resistance to chlortoluron, another PS II inhibitor, has been reported in wheat and in wild populations of *Triticum dicoccoides*.[54] Although chlortoluron resistance in wheat is under the control of a single major gene located on chromosome 6B,[55] the phenotypic variance of the wild species due to other genetic or environmental means implied that quantitative techniques must be used for analysis.

Field experiments were carried out to examine the response to chlortoluron of individual families from 11 populations of the wild *T. dicoccoides*.[54] The majority of the 11 wild populations of *T. dicoccoides* were polymorphic for chlortoluron response. Individual families were generally homogeneous for either resistance or susceptibility, although segregation of resistance was found in a few families. The susceptible phenotype appeared to be at a slightly higher frequency than the R one. The analysis of variance of the visual injury rating confirmed highly significant differences among families, within and between populations. Considering all families taken at random within a unique large population, broad sense heritability was estimated at 0.72. Hence, if these wild populations were placed under herbicide selection pressure then resistance would rapidly evolve.

A similar analysis has been performed with variability in response to difenzoquat.[54] Cultivated wheats are polymorphic for difenzoquat resistance, which is controlled by a major gene located on chromosome 2B.[56] None of the wild families tested were susceptible to 2 kg ha^{-1} of difenzoquat.[54] This suggests resistance is the wild-type state and that genes for susceptibility to difenzoquat evolved with selection for cultivated wheats.

Resistance to Thiocarbamate Herbicides

Several studies have dealt with variability in response to the thiocarbamate herbicides within and between populations of *Avena fatua*. Populations with a long history of exposure to the herbicide produced significantly higher frequencies of resistant plants.[57] Variation in response to a carbamate herbicide, barban, was also found in *A. fatua* populations with no exposure history. Genetic variation for response to barban was quantified using calculations of genotypic and phenotypic variances for unexposed populations.[58] After 2 years, differences between *A. fatua* families were significant in six of seven populations. One population showed no genetic potential for increased herbicide resistance, while in the six other families, observed heritability values ranged from 0.12 to 0.63, suggesting a high potential for the development of barban resistance. Inheritance of the response to barban was also studied in the F_1 generation. The level of resistance in one generation was highly correlated to the level of resistance in the next generation.[59]

Resistance to Glyphosate

Resistance to glyphosate (a nonselective herbicide) has not yet evolved in weed populations (see Chapter 8). However, screening experiments have revealed variability to glyphosate in previously unexposed biotypes of the perennial *Convolvulus arvensis*, which indicated that there is significant variation in glyphosate susceptibility in this species. A diallel cross between five of the biotypes (including reciprocal crosses) was carried out to determine the heritability of glyphosate tolerance.[60] As this species is perennial and produces rhizomes, clonal propagation of the hybrids was possible, providing relatively uniform plant material. Visual injury rating confirmed large differences in herbicide response. Analysis of variance components showed additive gene action, indicating the response to glyphosate is a quantitative characteristic controlled by multiple loci or by multiple alleles at a single locus. A broad heritability estimate could not be drawn from this experiment because the biotypes used were not collected at random and it is not possible to extrapolate the genetic gain found in this experiment to that which may occur in wild populations. However, the high significance of the general combining ability predicts that the presence of a few families is sufficient to build up a resistant population. It is possible that glyphosate resistance in a *C. arvensis* population could be enhanced by selection pressure of repeated sublethal herbicide spraying. As *C. arvensis* is a self-incompatible species, seed formation on the surviving plants would be the result of gene exchange between resistant biotypes. This may result in the accumulation of genes enhancing herbicide resistance in a simple, additive way.

Crop Species

Additional information on the quantitative inheritance of herbicide resistance is evident from classical crop breeding literature (Table 4). Such studies provide detailed reports on methods and experimental designs to obtain heritability estimates and to determine the number of genes involved in each

Table 4. Quantitative inheritance of herbicide resistance in crops

Herbicide	Species	Plant material	Resistance test	Scoring	Heritability	Ref.
Atrazine	Linum usitatissimum	F_1, F_3, BC	0.8 kg ha^{-1} soil incorporation	Dry matter reduction	0.29–0.34	9
Atrazine	Zea mays	Diallel (90 F_1)	6 ppm in hydroponics	Leaf mass	0.43–0.70	61
Bensulfuron/ Pretilachlor	Oryza sativa	Diallel (15 F_1)	0.36–2 kg ha-1 1- to 2-leaf stage	Plant height	0.52	62
Chloramben	Cucumis sativus	Diallel (4 F_1) F_2, BC	3 ppm in hydroponics	Visual injury rating, height and weight	0.36–0.87	63
Clomazone	Zea mays	Parent/offspring	0.02 to 1.02 kg ha^{-1} soil incorporation	Bleaching rate	0.84	64
Diclofop	Zea mays	F_1, F_2	0.8 kg/ha at 6- to 7-leaf stage	Visual injury rating	0.90–0.99	65
MCPA	Linum usitatissimum	F_3, F_4, BC	1.1 kg/ha on 1-month-old plant	Visual injury rating, height, maturity	0.16–0.36	66
Metribuzin	Ipomea batatas	Parent/offspring	0.6 ppm in hydroponics	Visual injury rating, weight	0.84–1.0	67
Metribuzin	Triticum aestivum	F_1, F_2, BC	Fluorescence	Photosynthetic inhibition	Cytoplasm + nuclear genes	68
Paraquat	Lolium perenne	Population/ offspring	0.4 kg ha^{-1}	Visual injury rating, yields	0.51–0.72	69
Propanil	Glycine max	F_1, F_2	0.8 kg ha^{-1} at 8-nodes stage	Visual injury rating	0.38–0.50	70
Simazine	Brassica napus	Diallel (42 F_1)	50 ppm in hydroponics	Visual injury rating	0.62	71
Simazine	Brassica campestris	Cross-composite population	50 ppm in hydroponics	Visual injury rating	0.57	72

case.[9,61-72] Progeny analysis (F_1 to F_4 and back-cross generations), parent/offspring regression, and diallel studies have been used. The principal scoring methods used were visual injury rating and effect on dry weight. Heritability values have been found to be high in most cases, indicating a high potential for rapid selection. However, very few crop cultivars with improved resistance to herbicides have been released using these techniques, which suggests natural variation among modern crop cultivars is insufficient to easily provide new resistant germplasm.

Prospects

As outlined above, quantitative inheritance of certain traits may lead to practical weed control failures due to herbicide resistance. With the *C. arvensis* and *T. dicoccoides* studies,[54,60] significant variation for resistance within and between populations was revealed. With *S. vulgaris* and *A. fatua*,[52,58] an increasing frequency of the resistant phenotypes was documented in weeds infesting fields with long histories of herbicide treatment. From studies with crop species, it can be concluded that moderate herbicide selection pressure on initially susceptible crop species may result in improved tolerance in progeny. One may extrapolate to conclude the same for weed species if the selection conditions are favorable. To date, such polygenic resistance mechanisms have not been documented in the field. This could be due to some, or all, of at least five different factors. (1) The accumulation of resistance genes is more prevalent with self-incompatible species but not for autogamous and preferentially autogamous species. (2) The rate of increase of resistance genes within weed populations can be limited, even if herbicide rates are high, if susceptible individuals survive and hybridize with resistant individuals. (3) A buffering effect of the soil seed bank results from discontinuous germination of seeds produced within and between years. (4) In a rotation system, the herbicide applied may change each year, resulting in the dilution of resistance genes in a susceptible genetic background. (5) Quantitative characters do not provide more than a two- to threefold resistance which may not allow a clear-cut discrimination between S and R plants. Despite these factors which can limit the development of polygenic resistance, it is virtually certain that quantitative shifts in response to herbicide exposure has changed the response of plant species towards resistance. Anecdotal statements from weed control practitioners in many countries imply that these quantitative shifts have occurred widely; however, they have not been diagnosed or documented.

MENDELIAN INHERITANCE OF RESISTANCE GENES

If herbicide effects on R and S plants are distinct, i.e., involve clear-cut plant mortality, then inheritance studies can be interpreted in a Mendelian fashion. In such cases, inheritance is likely due to a single resistance gene, although hypotheses involving more genes cannot be excluded. Genetic data available for resistance in weed species to various herbicides are summarized in Table 5. Results are also reported for resistance genes naturally occurring in the germplasm of crop species (Table 6) as well as results from artificial selection (Table 7).

Weed Species

Resistance to Photosystem II Inhibitors

Resistance to triazine herbicides can be endowed by mechanisms other than the modification of the *psbA* gene (as discussed in the preceding section and Chapter 2). In a resistant biotype of *Abutilon theophrasti*, resistance is due to a fourfold increase in glutathione-*S*-transferase activity leading to a more rapid detoxification of the herbicide.[73] Hand emasculation and pollination of this species is difficult to achieve, but a few successful crosses were obtained. When grown in hydroponics in the presence of 1 ppm atrazine, reciprocal F_1 populations showed an intermediate response. The segregation ratio in the F_2 generations was 1:2:1 (resistant:intermediate:susceptible), indicating that resistance is controlled by a single incompletely dominant gene.[74] The study of several F_3 families confirmed this hypothesis. The F_2 plants expressing an intermediate phenotype produced F_3 families segregating for resistance in a 1:2:1 ratio, while no segregation occurred in F_3 families from either the resistant or susceptible phenotypes.

Another case of simple Mendelian inheritance of resistance to a PS II inhibitor has been reported in a weed species, albeit not as a result of field selection. The potential to develop resistance to siduron has been studied among a number of random collections of *Hordeum jubatum*. The length of the seedling radicle was chosen as the resistance criterion, although (unfortunately) a correlation between this criterion and the field response has not been clarified, nor has the link with the mode of action of siduron (normally considered a PS II inhibitor). Resistant seedlings were affected by the herbicide since their radicles were five times shorter in the presence of siduron compared to the untreated control. F_1 families, raised between R and S accessions, demonstrated that 8 out of 12 different families agreed with a hypothesis of resistance being conferred by three complementary dominant genes.[75] Segregation from the four remaining crosses agreed with a hypothesis of resistance endowed by two complementary genes.

Selection pressure with substituted urea herbicides has led to resistant populations of the grass weed species *Alopecurus myosuroides* in the U.K., following repeated field treatments with chlortoluron.[76] The genes involved in resistance in *A. myosuroides* are not the same as those found in siduron-resistant *H. jubatum*, since various patterns of non-target site cross resistance are evident in *A. myosuroides*.[76] Preliminary data only are available on the genetics of chlortoluron resistance in *A. myosuroides*. Three classes of phenotypes,

Table 5. Mendelian inheritance of herbicide resistance in weed biotypes

Herbicide	Species	Plant material	Resistance test	Scoring	Number of genes	Ref.
Atrazine	Abutilon theophrasti	F_1, F_2, F_3	1 ppm in hydroponics	Dead/alive + intermediate	1 semidominant	74
Chlortoluron	Alopecurus myosuroides	Bulk F_1	Leaf fluorescence	Inhibition recovery	2 additive	77
Diclofop	Lolium multiflorum	F_1, F_2	7.5 kg ha^{-1} in vitro test	Dry weight and visual injury rating	1 semidominant	92
Fenaxoprop	Avena sterilis	F_1, F_2	0.6 kg ha^{-1} at 2-leaf stage	Visual injury rating	1 semidominant	94
Fluazifop	Avena sterilis	F_2	0.55 kg ha^{-1}	Visual injury rating	1 semidominant	94
Haloxyfop	Lolium rigidum	F_1, F_2	Up to 0.2 kg ha^{-1}	Dead/alive	1 semidominant	93
Metsulfuron	Lactuca serriola	F_1, F_2, F_3	0.5 ppm	Dead/alive + intermediate	1 semidominant	86
Paraquat	Arctotheca calendula	F_1, F_2, BC	0.8 kg ha^{-1} at 5- to 6-leaf stage	Visual injury rating	1 semidominant	84
Paraquat	Conyza philadelphicus	F_1, F_2, BC	0.5 kg ha^{-1} on seedlings	Dead/alive	1 dominant	79
Paraquat	Conyza bonariensis	F_1, F_2	10^{-5} M at rosette stage or in vitro test	Dead/alive	1 dominant	80
Paraquat	Erigeron canadensis	F_1, F_2	10^{-5} M at germination and at cotyledon stages	Dead/alive	1 dominant	81
Paraquat	Hordeum glaucum	F_2, F_3	0.1 kg ha^{-1} at tillering	Dead/alive + intermediate	1 semidominant	83
Paraquat	Hordeum leporinum	F_1, F_2	0.2 kg ha^{-1}	Visual injury rating	1 semidominant	84
Siduron	Hozdeum jubatum	F_2	2.2 kg ha^{-1} at germination	Radicle length	3 complimentary	75
Trifluralin	Setaria viridis	F_2	0.6 ppm at germination	Radicle length	1 recessive	100

Table 6. Mendelian inheritance of herbicide resistance in crops

Herbicide	Species	Plant	Resistance material	Scoring test	Number of genes	Ref.
Atrazine	Zea mays	F_1, F_2, BC	Recommended rate	Dead/alive	1 dominant	8
Barban	Hordeum vulgare	F_2, F_3	0.8 kg ha^{-1}	Leaf chlorosis or apical inhibition	1 recessive or quantitative	101
Bensulfuron	Oryza sativa	F_1, F_2	Germination on 10^{-5} M agar	Visual rating	1 or 2 recessive	103
Bentazon	Capsicum annuum	F_1, F_2, BC	4.5 kg ha^{-1} on 25-day-old seedling	Visual rating	1 dominant	104
Butachlor	Oryza sativa	F_1, F_2	30 kg/ha 3- to 4-leaf seedling	Visual rating	1 recessive	105
Chlortoluron	Triticum aestivum	Substitution lines F_1, F_4	4.8 kg ha^{-1}	Visual rating	1 dominant	55
2,4-D	Triticum aestivum	F_1, F_2	0.6 kg ha^{-1} on 45-day-old plant.	Dead/alive	1 dominant	106
Diclofop	Avena sativa	F_2, BC	0.7 kg ha^{-1} at 3-leaf stage	Visual rating	2 semidominant	107
Difenzoquat	Triticum aestivum	F_2, monosomic lines, F_4	1.7 kg ha^{-1} 3- to 4-leaf stage	Visual rating	1 dominant	56,108
Metoxuron	Triticum aestivum	F_3, F_4, F_5	13.5 kg ha^{-1} at 2-leaf stage	Visual rating	1 recessive	109
Metribuzin	Glycine max	F_1, F_2, BC	0.12 ppm in hydroponics	Dead/alive	1 dominant	110
Metribuzin	Lycopersicon esculentum	Diallel (F_1) F_2, BC	0.25 ppm in hydroponics	Dead/alive, height and weight	1 recessive	102
Metribuzin	Solanum tuberosum	F_1, F_2, BC	0.12 ppm in hydroponics	Visual rating	1 dominant	111
Oxyfluorfen	Oryza sativa	F_1, F_2	2 kg ha^{-1} on 10-day-old seedling	Visual rating	1 recessive	103
Trifluralin	Cucurbita moschata	F_1, F_2, BC	1.1 kg ha^{-1} soil-incorporated	Visual rating	1 dominant and 1 epistatic	112

Table 7. Inheritance of herbicide resistance in laboratory-generated resistant plants

Herbicide	Species	Inheritance	Ref.
Bensulfuron methyl	Oryza sativa	1 dominant + interaction	113
Bentazon	Nicotiana tabacum	1 or 2 recessive (2 loci)	114
Chlorsulfuron	Arabidopsis thaliana	1 dominant	115
Chlorsulfuron	Glycine max	1 dominant	116
Chlorsulfuron	Glycine max	1 recessive (3 loci)	117
Chlorsulfuron	Nicotiana tabacum	1 dominant (2 loci)	118,119
2,4-D	Arabidopsis thaliana	1 almost recessive	120
Haloxyfop	Zea mays	1 dominant (2 loci)	121
Imazapyr	Arabidopsis thaliana	1 dominant	122
Imazapyr	Zea mays	1 semidominant (2 loci)	123
Imazethapyr	Brassica napus	1 or 2 semidominant (2 loci) additive	124
Imazethapyr	Triticum aestivum	1 semidominant	125
Isoxaben	Arabidopsis thaliana	1 semidominant (2 loci)	126,127
Paraquat	Ceratopteris richardii	1 recessive	128
Picloram	Nicotiana tabacum	1 dominant or semidominant (2 loci)	129, 130
Primisulfuron	Nicotiana tabacum	1 dominant	131
Primisulfuron	Zea mays	1 dominant	132
Phenmedipham	Nicotiana tabacum	1 recessive	114
Sethoxydim	Zea mays	1 semidominant	121,133

resistant, intermediate and susceptible, were distinguished according to the ability of plants to metabolize herbicide, as indirectly measured by PS II fluorescence response of isolated leaves. As this species is preferentially allogamous, bulk F_1 generation seed were obtained by crossing groups of parent plants. Progeny of crosses between R plants showed one third intermediate plants, indicating parental R plants were not homozygous. Several nuclear loci were certainly involved. The segregation ratio in progeny of crosses between R and S plants was shown to be approximately 4:11:1 (resistant:intermediate:susceptible). These results agree with the hypothesis of two nuclear genes having additive actions: 3 and 4 alleles in the R, 1 and 2 in the intermediate, and 0 in the S plants.[77] However, this remains to be confirmed using a quantitative genetic approach with known homozygous progenitors.

Resistance to Photosystem I Disruptors

As discussed in Chapter 3, paraquat resistance has evolved in several weed species and different mechanisms are evident.[78] In three species, resistance to paraquat is conferred by a single gene. In *Erigeron philadephicus*, crossing experiments provided evidence that a single dominant gene is responsible for paraquat resistance.[79] As there is a low rate of self-pollination in this species, neither emasculation nor nuclear parental markers was used to check the F_1; however, the segregation ratio in F_2 and BC generations confirmed the single-gene hypothesis. In resistant *Conyza bonariensis,* inheritance studies were combined with a physiological study of enzymes involved in superoxide detoxification. Reciprocal crosses were made and dose-response curves of the parental and F_1 families using leaf disks floating on paraquat were compared and indicated nuclear dominance of resistance.[80] F_2 plants at the rosette stage were sprayed with 10^{-4} M paraquat and 75% survived, suggesting inheritance due to a single dominant gene. Enzyme activity assays on non-treated plants showed the same high activity in the F_1 and the R parent, clearly different from that of the susceptible parent. Levels of superoxide dismutase and glutathione reductase were measured immunologically in single parents and F_2 plants. The R F_2 plants had the same elevated levels of these two enzymes as the R parent, and the S F_2 plants had the same low levels as the S parent. These results indicate either a very tight linkage between resistance and the levels of the two enzymes, or that one locus controls resistance by pleiotropically enhancing the levels of detoxification enzymes.[80] If the hypothesis of pleiotropy is true, the gene conferring resistance is likely to have a regulatory role. Such a gene remains to be isolated and characterized. In a third species, *Erigeron canadensis,* reciprocal crosses between R and S plants were made after removal of hermaphrodite flowers from the capitulum.[81] To confirm that hybridization was successful, individual progeny were tested for paraquat resistance. At 10^{-4} M paraquat, leaf disks of the reciprocal F_1 plants were injured while complete necrosis was observed in S plants. A concentration of 10^{-3} M paraquat was necessary to affect resistant plants. The F_2 generation segregated in a 3:1 (alive/dead) ratio, indicating that resistance was controlled by a single nuclear gene. When seed was germinated on 10^{-5} mM paraquat, the individual reciprocal F_2 families segregated in a 3:1 ratio, again indicating that one gene is involved.

In three paraquat-resistant species, resistance has been shown to be inherited as a semi-dominant gene. In *Hordeum glaucum,* it is likely that paraquat resistance is due to reduced efficacy of herbicide movement across the cell wall and sequestration of the herbicide within the leaf apoplast.[82] Despite the difficulty of obtaining crosses in this self-pollinating species, some F_1 seeds were bulked using the S plant as the female parent. F_2 seeds were obtained from F_1 plants and three classes of plants were scored after paraquat treatment (survivors, intermediates, and deaths). A subclass of intermediate plants had to be considered because one third of the F_2 showed severe bleaching of leaves a week after treatment. The F_2 segregation ratio was 1:2:1, with this segregation ratio confirming the hybrid status of the F_1 plant. F_3 progeny from unaffected F_2 plants bred true for resistance, whereas those from intermediate F_2 plants showed a 1:2:1 segregation 6 days after treatment. Paraquat resistance in *H. glaucum,* therefore, is controlled by a single, incompletely dominant gene.[83] A similar study carried out with R and S plants of the closely related *H. leporinum* led to the same conclusion of resistance endowed by a single, semidominant gene.[84] In a resistant biotype of *Arctotheca calendula*, an obligate allogamous species, a large number of F_1 individuals were obtained and their survival and growth at various herbicide concentrations determined. The LD_{50} of reciprocal F_1 plants was 0.08 kg ha^{-1}, while it was 0.03 kg ha^{-1} and greater than 0.2 kg ha^{-1} for S and R biotypes, respectively. F_2 and back-cross plants showed the segregation ratios expected in the case of a nuclear, semidominant gene, i.e., 1:2:1 (alive:intermediate:dead) in the F_2, and 1:1 (intermediate:dead) in BC generations.[84]

Resistance to Acetolactate Synthase Inhibiting Herbicides

As discussed in Chapter 4, resistance to ALS inhibiting herbicides has appeared in several weed species.[85] Two mechanisms are responsible for this resistance: metabolic inactivation of the herbicide, and ALS target site resistance.[85] Except for the (as yet) unpublished studies with *Kochia scoparia* (see Chapter 4), only one study of the genetic control of sulfonylurea resistance in weeds has been published.[86] In *Lactuca serriola*, resistance is due to an alteration of the target enzyme ALS. Reciprocal crosses between the R biotype and a S biotype of *L. sativa* (cv. Bibb) were made and F_1 seedlings from a S female parent were sprayed with 13 g ha^{-1} metsulfuron to determine successful crosses (no other genetic markers were available). Some F_1 seedlings from the reciprocal crosses showed only slight herbicidal symptoms and were not included in the analysis. The F_2 generation was sprayed with metsulfuron and was scored as resistant (unaffected), intermediate (with chlorosis symptoms), or susceptible (dead). Segregation of the F_2 generation fitted a 1:2:1 ratio, indicating resistance is controlled by a single nuclear gene with incomplete dominance. Further evidence was obtained with F_3 seedlings in which virtually all the F_3 progeny from resistant F_2 were resistant. However, the F_3 seedlings from the intermediate F_2 segregated in a 3:1 ratio (resistant + intermediate/susceptible). Therefore, resistance in *L. serriola* is inherited as a dominant or semidominant nuclear gene.[86]

The genetics and molecular origins of resistance to ALS inhibiting herbicides has been studied, as fully discussed in Chapter 4. Experiments with transgenic plants derived from the transfer of an ALS mutated gene showed resistance is due to amino acid substitutions on the ALS protein.[87] Most DNA sequence changes associated with resistance to ALS inhibitors in crops and yeast are located in a highly conserved domain of 13 amino acids.[85] The sequence of that DNA region of the ALS gene has been determined in *L. serriola*.[88] Only one nucleotide substitution has been detected, the replacement of proline at amino acid 173 in the susceptible biotype with a histidine residue in the resistant biotype.[88] Such a single point mutation fits well with single-gene inheritance. Similarly, a single amino acid change at the same codon has been observed in a resistant *K. scoparia* biotype, but in this case there is a proline to threonine replacement. This again suggests single-gene inheritance. However, one collection of resistant *K. scoparia* lacked evidence of that particular mutation of the ALS gene, and hence some alternate basis for resistance is likely.[88] Variation in the source of resistance may also be inferred from DNA studies of *Salsola iberica* in which R and S biotypes showed no differences in that DNA sequence.[88] These results show a fundamental difference to the situation observed in the chloroplast encoded triazine resistance, in which only a single nucleotide change and a single amino acid change have been consistently found (see Chapter 2). Therefore, it is not possible to extrapolate the presence of a single constant point mutation at the proline codon to all resistant weed plants having altered ALS sensitivity. In addition, recent studies have shown that different resistance mechanisms may occur in the same individual,[89] and within the same field population[90] (see Chapters 9 and 12). Further knowledge on the resistance to ALS inhibitors can be drawn from crops and plants derived through mutagenesis, especially for cross resistance to a range of sulfonylurea and imidazolinone herbicides (see crop section).

Resistance to Acetyl-Coenzyme A Carboxylase Inhibiting Herbicides

As discussed in Chapter 5, two classes of herbicides inhibit the plastidic enzyme acetyl-coenzyme A carboxylase (ACCase).[91] Biotypes of *Lolium multiflorum* from Oregon and the very closely related *L. rigidum* from Australia are resistant due to a modification of the enzyme target site.[92,93] Since these species are almost totally self-incompatible, F_1 individuals were obtained by spontaneous crosses between R and S inflorescences isolated at anthesis. F_2 plants were generated by sib-mating of F_1 plants, and test-cross families obtained following crossing with S individuals.[92] In the *L. multiflorum* study, dry weight 15 days after treatment and ACCase activity were used to evaluate the response to diclofop. Plants were scored as resistant (no injury), intermediate (severe reduction in growth rate without chlorosis), and susceptible (dead), 3 weeks after treatment. The dry weight of reciprocal F_1 individuals after herbicide treatment was intermediate between the two parents, indicating nuclear inheritance with partial dominance. Characteristics of ACCase inhibition paralleled the response at the whole plant level. The diclofop concentration required to inhibit 50% of ACCase activity was not different between reciprocal F_1 plants, and was intermediate between the values of the S and R parents. At a rate of 7.5 kg ha^{-1}, the F_2 families segregated in a 1:2:1 ratio (resistant/intermediate/susceptible) and segregation in test-cross families fitted a ratio of 1:1 (intermediate/susceptible). The ACCase and whole plant responses to herbicide in F_2 families were also consistent, demonstrating ACCase-herbicide resistance in biotypes of these two species is due to the presence of an altered form of the ACCase under the control of a single, partially dominant gene.[92,93] Similar studies have been conducted with a resistant biotype of *Avena sterilis* which is resistant to a range of ACCase inhibiting herbicides due to a resistant ACCase enzyme.[94] Intermediate responses to fluazifop and fenoxaprop were observed in F_1 individuals. Their F_2 families segregated as 1:2:1 (resistant:intermediate:susceptible). As with *Lolium* spp., inheritance is due to a single, partially dominant gene.[94]

Resistance to ACCase inhibiting herbicides in weeds is not always due to modifications of the enzyme target site. As discussed in Chapters 5 and 9, in many biotypes of *L. rigidum* in Australia, resistance is not related to any changes in ACCase, but rather is correlated with the ability of the resistant biotypes to recover from herbicide-induced membrane depolarization.[95] Some biotypes exhibit cross resistance to a range of ALS inhibitors.[96] Cross resistance and multiple herbicide resistance is a complex phenomenon.[97,98] Its biochemical and genetic bases are not yet fully understood (see Chapter 9, but it is possible that different mutations affecting the same membrane protein can confer various patterns of herbicide detoxification[95]).

Resistance to Dinitroaniline Herbicides

Resistance to dinitroaniline herbicides, a class of mitotic inhibitors, is documented in a number of weed species (see Chapter 7). In *Eleusine indica*, two resistant biotypes show different resistance levels, likely due to different mechanisms. Biochemical investigations indicate the most resistant biotype has altered tubulin proteins, the site of action of the herbicide.[99] Hybrids between R and S plants exhibit resistance, but not at the same level as the resistant parent. The F_2 generation shows a complex inheritance pattern, probably because tubulins are encoded by a multicopy gene family.[99] An inheritance study has also been conducted with *Setaria viridis* biotypes selected in the field following repeated uses of trifluralin.[100] No data are available to date on the physiological basis of resistance. F_1 individuals from crosses between R and S plants were identified using morphological markers. F_2 families were scored for resistance using radicle length and shoot growth of seedlings in petri dishes wetted with trifluralin solutions. The data fit a 1:3 ratio (resistant:susceptible), indicating that resistance is controlled by a single nuclear recessive gene. This is apparently the first report of resistance endowed by a recessive gene in a weed species.

Crop Species

Herbicide resistance traits have also been found to be inherited in a Mendelian fashion in several studies with crop species. One of the first reports of single-gene inheritance was the discovery of triazine resistance due to detoxification in *Zea mays*, conferred by a single dominant gene.[8] Further work on germplasm and cultivar variability showed single dominant genes to be responsible for resistance in many crops, but recessive gene controls have also been found in some cases (Table 6). Apart from progeny analysis, a diversity of genetic methods has been used in these studies. In a pioneering study of the inheritance of resistance to barban in *Hordeum vulgare*, leaf chlorosis symptoms were first used to identify phenotypes.[101] Using this criterion, the F_2 segregation fitted a 1:3 ratio (resistant:susceptible), indicating the action of a recessive nuclear gene. However, using another criterion (apical inhibiting effect), resistance appeared to be quantitatively inherited.[101] Also, the proportion of apparently R and S F_2 progeny varied according to the amount of barban applied, suggesting the action of more than one gene. In *Lycopersicon esculentum*, a visual rating of injury in an F_1 diallel, F_2 and BC generations suggested the action of one recessive gene involved in metribuzin resistance.[102] However, frequency distributions of seedling weight and height in the F_2 generation were not fully consistent with discrete segregation as expected with the single-gene hypothesis. A quantitative analysis showed high heritability values ranging from 0.58 to 0.72 and, assuming gene additivity and implementing the variance components associated with the various generations, the number of genes was estimated to be between 0.9 and 1.1. This confirmed that one major gene controls metribuzin resistance in *L. esculentum*.[102] In wheat, F_2 monosomic families were used to determine on which chromosome a major gene encoding resistance is located. Such an accurate analysis was possible, due to the availability of aneuploid and single chromosome substitution lines. Thus, single dominant genes conferring resistance to difenzoquat and chlortoluron were located on chromosomes 2B and 6B, respectively.[55,56] Some minor genes also appeared to influence the resistance of F_2 plants in wheat. Other studies on this topic are summarized in Table 6.[101-113]

Data on the inheritance of herbicide resistance in plants have also been obtained from characterization of laboratory-generated mutants. This is worthwhile reporting, since laboratory mutagenesis may be considered an evolutionary shortcut towards what may occur more slowly in fields. Single-gene mutation has been found in most cases (Table 7).[113-133] As discussed in Chapter 4, numerous mutants have been found to be resistant to ALS inhibiting herbicides due to mutant ALS which is expressed as a dominant or semidominant gene. Mutations at a proline codon of the ALS gene were found in several mutants, but other codons were also found to be altered and two loci were identified in some cases.[134] Resistance across a range of ALS inhibitors was found to depend on which codon of the gene was altered.[87] Mutants selected for imidazolinone resistance and exhibiting a semidominant gene did not always show target site cross resistance to sulfonylurea herbicides.[123,125] As discussed in Chapter 4, the degree of resistance may differ according to where the mutation occurs and the number of mutations in the ALS encoding genes.[119] In addition, the number of ALS genes varies from species to species: 1 in *Arabidopsis thaliana*,[134] 2 in *Nicotiana tabacum*,[87] and up to 5 in *B. napus*.[135] Multiplicity of ALS encoding genes may account for the complexity of inheritance in some cases.[124] When resistance to ALS inhibitors is not conferred by a modification of the ALS target enzyme, as in some *Glycine max*,[117] *B. napus*,[114] and *Zea mays*[132] mutants, inheritance of alternate resistance mechanisms may be either dominant or recessive. The inheritance of resistance to other herbicides in other mutant derived plants shows recessive control.[114,120,128] In some cases, attempts to regenerate resistant plantlets from callus have failed, especially when the basis of resistance is gene-amplification or over-expression of the target enzyme.[136-138] In other cases, sterility of regenerated plants or loss of resistance in the absence of herbicide selection precluded the possibility of any inheritance study, although sexual transfer was shown to be possible.[139,140] Finally, transfer of genes from foreign species (different plant species and yeast) through genetic engineering may also confer herbicide resistance and in such cases, resistance is expected to be inherited as a single dominant or semidominant gene.[141-143]

Prospects

Clearly a single resistance gene can ultimately change the herbicide response of an entire population, leading to large infestations of resistant weeds. For crop species, the presence or absence of such genes can determine selectivity. Comparing inheritance studies conducted with weed versus crop species highlights the weaknesses of much of the weed genetic data. The first concern is with the identification of F_1 hybrids. Since in a majority of weed species, crosses are difficult to make, and no genetic markers are available, it is often assumed that F_1 plants showing an intermediate to high level of resistance following herbicide treatment are true hybrids. This can be misleading, especially when the number of putative hybrids is very low, making it difficult to check contamination from different origins. Problems are apparent in four situations:

1. When resistance is encoded by the cytoplasm, it is not clear whether crosses have failed or not.

2. When nuclear resistance is highly dominant, the progeny of heterozygous resistant parents can be confounded with that of homozygous parents.

3. When resistance is semidominant, hybrids can be confounded with the progeny of heterozygous resistant parents. Indeed, spontaneous heterozygous resistant plants have seldom been found in the studies reported above and this is certainly not a reflection of what really happens in the field but rather a result of the limitations of the experimental procedure.

4. When resistance is recessive, hybrids cannot be distinguished from progeny of susceptible parents and will be therefore discarded in the breeding process. Should several resistance genes exist within a resistant weed population, hybrid individuals having recessive resistance genes will not be retained for further studies and it would be concluded that only semidominant and/or dominant genes endow resistance in that weed population.

Evaluation of plant response to herbicides can often be clear-cut (i.e., survival vs. death), although several minor genes may also influence the degree of resistance observed, as is apparent when studies of herbicide resistance in crop species have combined a quantitative genetic approach with classical Mendelian methodology. In crop breeding experiments, quantitative variance components of the parents and F_1 plants are often used to determine management of the F_2 generation intermediate class. However, in the discipline of weed science, most researchers assume evaluation of herbicide effect can be reduced to a "dead or alive" alternative and a class of plants expressing symptoms intermediate between the two parents is "created" to account for the degree of resistance of heterozygotes. This contradicts the dead/alive scoring system. Methodological problems include the subjective or objective basis for determining the intermediate phenotypes, the environmental conditions under which experiments are conducted, and the rate of herbicides used to discriminate between phenotypes. Because only one intermediate phenotype is usually defined from visual estimates, this increases the likelihood of finding a segregation ratio in the F_2 generation that fits a single-gene inheritance hypothesis. In addition to this arbitrary simplification, analysis of the F_3 generation is seldom conducted and if it is, the number of families is often too low to give reliable results. The ultimate result of methodological imprecision is that inheritance of herbicide resistance is interpreted as a single-gene trait and (unfortunately) any influence of minor or recessive genes is overlooked. It is therefore clear that studies on the genetics of herbicide resistance will benefit from improvements in methodology. While simple dominant mutations may produce highly selectable resistant mutants (as in triazine and ALS herbicide-resistant plants), studies with crop species illustrate other genetic events are possible and comprehensive studies with weeds must consider these possibilities.

ORIGIN AND DYNAMICS OF RESISTANCE GENES

It is generally assumed that repeated use of herbicides has favored selection of existing resistance genes and that herbicides do not generate the mutations endowing herbicide resistance. However, the frequency of resistance genes in weed populations before herbicide treatment has seldom been measured. It is widely believed that mutations for resistance occur spontaneously at a low rate with an underlying hypothesis of uniform or random distribution throughout the world. Most cases of herbicide resistance have not been sufficiently studied to confirm or deny such assumptions. However, there are some interesting observations on the chloroplastic resistance to triazine herbicides worth mentioning.

In the triazine-resistant weed species for which the resistance gene has been sequenced, the chloroplast DNA contains a mutation at the same position on the *psbA* gene, i.e., a serine to glycine substitution.[33] However, when mutations in higher plants were obtained after *in vitro* cell culture and mutagenesis, the same position was involved, but the amino acid substitution was different: serine to asparagine[144] or threonine.[145] Moreover, other mutations at the *psbA* gene are known to confer atrazine resistance in algae and cyanobacteria.[33] The repeated occurrence of only one mutation in weed populations while a diversity of mutants could have been expected, indicates either a site-directed mutation or the removal of other mutants by natural selection. A plastome mutator system has been suggested to account for the high rate of mutation in a resistant accession of *S. nigrum*.[146] A mutator would increase the probability of chloroplast mutations occurring naturally and thus the rate of appearance of triazine-resistant mutants following triazine herbicide application. There is evidence that other mutations are induced by this system since a high number of plants were found with variegated leaves.[146] This capacity to mutate is transmitted by a nuclear gene with chloroplast mutations being stable and cytoplasmically inherited.

Another indication of a high mutation rate was found in *C. album* in which *psbA* resistant mutants appeared spontaneously from a few susceptible plants collected from sources with no history of triazine herbicides. Up to 8% of the plants from a population could produce *psbA* mutants and mutants were found to represent up to 12% of all the seedlings produced by a parent plant. This resulted in mutation frequencies within populations, ranging from 10^{-4} to 3×10^{-3}, which is much higher than expected through random mutation alone.[147] When analyzed for isozyme polymorphisms, all the plants that produced mutants had the same electrophoretic pattern while plants which did not produce mutants had a variety of patterns.[148] Therefore, the potential to produce a high rate of mutants seemed to be restricted to certain genotypes within the populations. This agrees with a nuclear mutator genome hypothesis, but also raises the question of what influence the genetic background has on the likelihood of a mutant appearing. Indeed, the resistant *C. album* found in farmers' fields in a given region had the same isozyme pattern as the spontaneous mutants found in gardens of the same region. This phenomenon certainly applied for most resistant populations of *C. album* that were sampled in France and in Canada as each one showed only one isozyme pattern.[149,150] A different isozyme pattern was found in each region, indicating no direct relationship between the mutation and a specific isozyme pattern. The triazine resistance mutation is not distributed over all the possible nuclear genomes in a population. However, after herbicide treatment, all the S plants are removed and a founder effect commences from the (initially) few R plants, i.e., the subsequent resistant population is monomorphic. Resistance is further maintained in a species such as *C. album* due to low allogamy and isolation from outside pollen. When a species is predominantly allogamous, as for *A. retroflexus*, or when favorable conditions promote cross pollination and heterozygote advantage, as for *P. annua*, resistant populations are polymorphic, but to a lesser degree than susceptible ones. This lack of polymorphism is attributed to the founder effect.[151,152] As a consequence, the genetic structure of a triazine-resistant weed population can be different from the previously unselected population.

The association of a given isozyme pattern with the ability to produce a higher frequency of mutants in each region is likely to involve other traits apart from triazine-resistance. It is noteworthy that most reports of triazine-resistant weeds in Europe and America have come from the northern, cooler regions of these continents. Reports from southern regions are rare. As discussed in Chapter 11, this could be the result of stronger selection against resistant plants in warmer climates. This view is supported by work with isogenic lines of *S. nigrum* that showed a decreased tolerance to high temperature due to the triazine resistance mutation.[44] This is likely to be a significant factor contributing to the lower fitness of triazine-resistant plants in the absence of herbicide selection pressure. However, experiments on the growth of these two biotypes and their reciprocal F_1 did not support the hypothesis that reduced growth in the southern regions was due to an inability to withstand higher temperature.[153] No evidence was found of a differential effect on the biomass production of triazine R and S isogenic lines in *S. vulgaris*.[46] Additional studies on this topic are needed, since these results apply only to conditions in which leaf temperature remained below 35°C and therefore high-temperature inhibition of photosynthesis[44] is likely to be of little importance. It is possible that low-temperature hardiness may be an alternate and more appropriate criterion. More subtle changes were recently detected between photosynthetic performances of R and S isogenic *B. napus*.[51] These data indicate the superiority of one biotype relative to the other was a function of time of day and the plant age and suggest that resistant plants may have some adaptive advantage in certain unfavorable ecological niches such as cool, low-light environments early and late in the day.[51] Also, it is noteworthy that studies on fitness seldom encompass the whole life cycle of the plant, from seed germination to germination of the next generation. Seed production, and subsequent seedling survival are important steps of the life cycle and should not be overlooked (see Chapter 11). Finally, if we accept the hypothesis of directed mutation and the involvement of a peculiar genetic background we must reconsider also the need for isogenic lines in the studies of triazine resistance effect on fitness. Association of mutability of resistance with nuclear genome traits may have occurred at random in each region, so that the relative fitness of a given resistant plant within a population cannot be predicted. Indeed, nuclear controlled variation has been demonstrated to be extensive in plants and may be more important than that controlled by the *psbA* chloroplastic mutation.[46,48] Alternatively, it is possible the association between triazine resistance mutation and a specific nuclear genome does not occur at random but rather from a set of stable genotypes. Thus, specific associations between nuclear and chloroplast genomes that co-evolved would be lost in the hybrids and back-crosses as used in genetics studies and the use of isogenic lines would not be appropriate.

Many questions still remain unanswered. Models (see Chapter 1) that predict the rate of development of herbicide resistance under given selection pressure have been proposed,[154,155] but the accuracy of these predictions is highly dependent on the initial frequency of the genes controlling herbicide resistance and the fitness of resistant individuals (see Chapter 1). The genetic investigations reported above suggest that a diversity of inheritance mechanisms plays a role in resistance: chloroplast encoded, nuclear dominant, recessive or additive genes at one or more loci. The same mutation may arise independently at different locations, but it is not always easy to distinguish these from simple seed migration from a distant resistant source. In addition, several mechanisms may be found within the same population or individuals.[97,98] Some traits may confer resistance to only a single herbicide whereas others may confer resistance to several herbicides. To date, clear multiple resistance has evolved in obligate cross-pollinated species with high fecundity. Exposure of large and highly variable populations to herbicides certainly increases the risk of appearance of resistant genotypes in the field. To account for this, the extent of genetic variation for herbicide resistance and its prediction through association with the degree of polymorphism for enzymic and morphological traits has to be studied.[156-158] Adaptability may be due to the presence of peculiar mutable genotypes, as in triazine-resistant *C. album*,[147] or high gene exchange among plants within and between populations, as in *L. rigidum*[98] and *A. myosuroides*.[158] The experience with *L. rigidum* provides two contrasting examples of weed response to two different weed control strategies: herbicide rotation or mixtures to avoid resistance. In one instance multiple resistance and non-target site cross resistance arose after sequential selection with four herbicides from four different chemical classes: trifluralin, diclofop, chlorsulfuron, and sethoxydim.[98] In another occurrence, non-target site cross resistance arose after repeated applications of a mixture of atrazine and amitrole.[159] Every new case of resistance is likely to originate from a different genetic situation. Biological attributes that make a plant a weed are numerous and depend on a multiplicity of ecological conditions. The question of the appearance of resistant plants must be studied on a case by case basis. According to its own biology, each weed species will find a different way to escape herbicides. This includes novel biochemical resistance mechanisms, as well as crop mimicry and introgression of resistance genes from closely related crops.[160] The complex interactions between herbicide(s), weed species and environmental constraints makes resistance mechanisms and the genetic mutations which confer them difficult to predict. What can be predicted is that novel and multiple mechanisms will develop and generalizations should be made with extreme caution.[97]

REFERENCES

1. Prante, G. "Ein Beitrag zur Systematik des Flughafers (*Avena fatua* L.)," *Z. Pflanzenkr.* 78: 675-694 (1971).
2. Imam, A. G., and R. W. Allard. "Population Studies in Predominantly Self-pollinated Species. VI. Genetic Variability Between and Within Natural Populations of Wild Oats from Differing Habitats in California," *Genetics* 51: 49-62 (1966).

3. Barrett, S. C. H. "Genetics and Evolution of Agricultural Weeds," in *Weed Management in Agroecosystems: Ecological Approaches*, M. A. Altieri, and M. Liebman, Eds. (Boca Raton, CRC Press, Inc., 1988), pp. 57-75.
4. Blackman, G. E. "Selective Toxicity and the Development of Selective Weedkillers," *J. R. Soc. Arts* 98: 499-517 (1950).
5. Harper, J. L. "The Evolution of Weeds in Relation to Resistance to Herbicides," in *Proceedings of the 3rd British Weed Control Conference* (Brighton, 1956), pp. 179-188.
6. Ryan, G. F. "Resistance of Common Groundsel to Simazine and Atrazine," *Weed Sci.* 18: 614-616 (1970).
7. Karim, A., and D. Bradshaw. "Genetic Variation in Simazine Resistance in Wheat, Rape and Mustard," *Weed Res.* 8: 283-291 (1968).
8. Grogan, C. O., E. F. Eastin, and R. D. Palmer. "Inheritance of Susceptibility of a Line of Maize to Simazine and Atrazine," *Crop Sci.* 3: 451 (1963).
9. Comstock, V. E., and R. N. Andersen. "An Inheritance Study of Tolerance to Atrazine in a Cross of Flax (*Linum usitatissimum* L.)," *Crop Sci.* 8: 508-509 (1968).
10. LeBaron, H. M., and J. McFarland. "Overview and Prognosis of Herbicide Resistance in Weeds and Crops, " in *Managing Resistance to Agrochemicals*, M. B. Green, W. K. Moberg, and H. M. LeBaron, Eds. (Washington, D.C., American Chemical Society Symposium Series 421, 1990), pp. 336-352.
11. Radosevich, S. R., and O. T. Devilliers. "Studies on the Mechanism of s-Triazine Resistance in Common Groundsel," *Weed Sci.* 24: 229-232 (1976).
12. Souza-Machado, V., and J. D. Bandeen. "Cross-Pollination and F_1 Segregation of Atrazine Tolerant and Susceptible Biotypes of Lamb's-quarters, " in *Research Report of the Canada Weed Committee* (East Sect. Fredericton N 22, 1977), p. 305.
13. Souza-Machado, V., J. D. Bandeen, G. R. Stephenson, and P. Lavigne. "Uniparental Inheritance of Chloroplast Atrazine Tolerance in *Brassica campestris*," *Can. J. Plant Sci.* 58: 977-981 (1978).
14. Solymosi, P., Z. Kostyal, and E. Lehoczki. "Characterization of Intermediate Biotypes in Atrazine-susceptible Populations of *Chenopodium polyspermum* L. and *Amaranthus bouchonii* Thell. in Hungary," *Plant Sci.* 47: 173-179 (1986).
15. Solymosi, P. "Az *Amaranthus retroflexus* L. Triazin-rezisztenciajanak rkldese," *Nvnytermeles* 30: 57-60 (1981).
16. Souza-Machado, V., and J. D. Bandeen. "Genetic Analysis of Chloroplast Atrazine Resistance in *Brassica campestris* — Cytoplasmic Inheritance," *Weed Sci.* 30: 281-285 (1982).
17. Warwick, S. I., and L. Black. "Uniparental Inheritance of Atrazine Resistance in *Chenopodium album*," *Can. J. Plant Sci.* 60: 751-753 (1980).
18. Gasquez, J., A. Al Mouemar, and H. Darmency. "Quels Gènes pour la Résistance Chloroplastique aux Triazines chez *Chenopodium album* L.?," in *Proceeding of the 7th Colloque International sur l'Ecologie, la Biologie et la Systématique des Mauvaises Herbes* (Paris-Columa, 1984), pp. 281-286.
19. Gawronski, S. W. "Inheritance of Resistance to Triazine Herbicides by *Echinochloa crus galli* (L.)," in *Proceedings of the 5th International Congress Society for the Advancement of Breeding Research in Asia and Oceania* (Bangkok, 1985), pp. 797-801.
20. Darmency, H., and J. Gasquez. "Inheritance of Triazine Resistance in *Poa annua*: Consequences for Population Dynamics," *New Phytol.* 89: 487-493 (1981).

21. Gasquez, J., and H. Darmency. "Variation for Chloroplast Properties Between two Triazine Resistant Biotypes of *Poa annua* L.," *Plant Sci. Lett.* 30:99-106 (1983).
22. Scott, K. R., and P. D. Putwain. "Maternal Inheritance of Simazine Resistance in a Population of *Senecio vulgaris*," *Weed Res.* 21: 137-140 (1981).
23. Scott, K. R., and P. D. Putwain. "Maternal Inheritance of Simazine Resistance in Four Populations of *Senecio vulgaris* L.," *Prot. Ecol.* 5: 359-367 (1983).
24. Stowe, A. M., and J. S. Holt. "Comparison of Triazine-resistant and Susceptible Biotypes of *Senecio vulgaris* and their F_1 Hybrids," *Plant Physiol.* 87: 183-189 (1988).
25. McCloskey, W. B., and J. S. Holt. "Triazine Resistance in *Senecio vulgaris* Parental and Nearly Isonuclear Backcrossed Biotypes is Correlated with Reduced Productivity," *Plant Physiol.* 92: 954-962 (1990).
26. Darmency, H., and J. Pernès. "Use of Wild *Setaria viridis* (L.) Beauv. to Improve Triazine Resistance in Cultivated *S. italica* (L.) Beauv. by Hybridization," *Weed Res.* 25: 175-179 (1985).
27. Darmency, H., and J. Pernès. "Agronomic Performance of a Triazine Resistant Foxtail Millet (*Setaria italica* (L.) Beauv.)," *Weed Res.* 29: 147-150 (1989).
28. Gasquez, J., H. Darmency, and J. P. Compoint. "Etude de la Transmission de la Résistance Chloroplastique aux Triazines chez *Solanum nigrum* L. C.R. Acad. Sci. Paris 292: 847-849 (1981).
29. Beversdorf, W. D., J. Weiss-Lerman, L. R. Erickson, and V. Souza-Machado. "Transfer of Cytoplasmically-Inherited Triazine Resistance from Bird's Rape to Cultivated Oilseed Rape (*Brassica campestris* and *B. napus*)," *Can. J. Genet. Cytol.* 22: 167-172 (1980).
30. Ali, A., E. P. Fuerst, C. J. Arntzen, and V. Souza-Machado. "Stability of Chloroplastic Triazine Resistance in Rutabaga Backcross Generations," *Plant Physiol.* 80: 511-514 (1986).
31. Thomzik, K. E., and R. Hain. "Transfer and Segregation of Triazine Tolerant Chloroplasts in *Brassica napus* L.," *Theor. Appl. Genet.* 76: 165-171 (1988).
32. Hirschberg, J., and L. McIntosh. "Molecular Basis of Herbicide Resistance in *Amaranthus hybridus*," *Science* 222: 1346-1349 (1983).
33. Darmency, H., and J. Gasquez. "Fate of Herbicide Resistance Genes in Weeds," in *Managing Resistance to Agrochemicals*, M. B. Green, W. K. Moberg, and H. M. LeBaron, Eds. (Washington, D.C., American Chemical Society Symposium Series 421, 1990), pp. 353-363.
34. Cheung, A.Y., L. Bogorad, M. Van Montagu, and J. Schell. "Relocating a Gene for Herbicide Tolerance: A Chloroplast Gene is Converted into a Nuclear Gene," *Proc. Natl. Acad. Sci. U.S.A.* 85: 391-395 (1988).
35. Souza-Machado, V., J. Shupe, and W. A. Keller. "Cytoplasmic-Inherited Atrazine Resistance Transmitted Through Anther Culture in Rutabaga," *Z. Pflanzenzuecht.* 95: 179-184 (1985).
36. Vaughn, K. C. "Physical Basis for the Maternal Inheritance of Triazine Resistance in *Amaranthus hybridus*," *Weed Res.* 25: 15-19 (1985).
37. Corriveau, J. L., and A. W. Coleman. "Rapid Screening Method to Detect Potential Biparental Inheritance of Plastid DNA and Results for Over 200 Angiosperm Species," *Am. J. Bot.* 75: 1443-1458 (1988).
38. Bettini, P., S. McNally, M. Savignac, H. Darmency, J. Gasquez, and M. Dron. "Atrazine Resistance in *Chenopodium album*: Low and High Levels of Resistance to the Herbicide Are Related to the Same Chloroplast *psbA* Mutation," *Plant Physiol.* 84: 1442-1446 (1987).

39. Sedoroff, R. R., P. Ronald, P. Bedinger, C. Rivin, V. Walbot, M. Bland, and C. S. Levings. "Maize Mitochondrial Plasmid S-1 Sequences Share Homology with Chloroplast Gene *psbA*," *Genetics* 113: 469-482 (1986).
40. Bettini, P., S. McNally, M. Savignac, and M. Dron. "A Mitochondrial Transcript with Homology to the 3' End of the Chloroplast *psbA* Gene is Present only in the Atrazine-resistant Biotype of *Chenopodium album*," *Theor. Appl. Genet.* 75: 291-297 (1988).
41. Zanin, G., V. Vecchio, and J. Gasquez. "Indagini Sperimentali su Popolazioni di Dicotiledoni Resistenti all'Atrazina," *Riv. Agron.* 15: 196-207 (1981).
42. Gasquez, J., A. Al Mouemar, and H. Darmency. "Triazine Herbicide Resistance in *Chenopodium album* L.: Occurrence and Characteristics of an Intermediate Biotype," *Pestic. Sci.* 16: 392-396 (1985).
43. Jacobs, B. F., J. H. Duesing, J. Antonovics, and D.T. Patterson. "Growth Performance of Triazine-resistant and Susceptible Biotypes of *Solanum nigrum* over a Range of Temperatures," *Can. J. Bot.* 66: 847-850 (1988).
44. Ducruet, J. M., and D. R. Ort. "Enhanced Susceptibility of Photosynthesis to High Leaf Temperature in Triazine-resistant *Solanum nigrum* L. Evidence for Photosystem II D1 Protein Site of Action," *Plant Sci.* 56: 38-48 (1988).
45. Touraud, G., M. T. Leydecker, and H. Darmency. "Abscissic Acid in Triazine-resistant and Susceptible *Poa annua*," *Plant Sci.* 49: 81-83 (1987).
46. McCloskey, W. B., and J. S. Holt. "Effect of Growth Temperature on Biomass Production of Nearly Isonuclear Triazine-resistant and Susceptible Common Groundsel (*Senecio vulgaris* L.)," *Plant Cell Environ.* 14: 699-705 (1991).
47. Trémolières, A., H. Darmency, J. Gasquez, M. Dron, and A. Connan. "Variation of Transhexadecenoic Acid Content in two Triazine-resistant Mutants of *Chenopodium album* and their Susceptible Progenitor," *Plant Physiol.* 86: 967-970 (1988).
48. Darmency, H., B. Chauvel, J. Gasquez, and A. Matejicek. "Variation of Chlorophyll a/b Ratio in Relation to Population Polymorphism and Mutation of Triazine Resistance," *Plant Physiol. Biochem.* 30: 57-63 (1992).
49. Gressel, J., and G. Ben-Sinai. "Low Intraspecific Competition Fitness in a Triazine-resistant Nearly Nuclear-Isogenic Line of *Brassica napus*," *Plant Sci.* 38: 29-32 (1985).
50. Hart, J. J., S. R. Radosevich, and A. Stemler. "Influence of Light Intensity on Growth of Triazine-resistant Rapeseed (*Brassica napus*)," *Weed Res.* 32: 349-356 (1992).
51. Dekker, J. H., and R. G. Burmester. "Pleiotropy in Triazine Resistant *Brassica napus*: Ontogenetic and Diurnal Influences on Photosynthesis," *Plant Physiol.* 100: 2052-2058 (1992).
52. Holliday, R. J., and P. D. Putwain. "Evolution of Herbicide Resistance in *Senecio vulgaris*: Variation in Susceptibility to Simazine Between and Within Populations," *J. Appl. Ecol.* 17: 779-791 (1980).
53. Darmency, H. "Some Effects of Herbicide-selection on *Alopecurus myosuroides* Huds.," *Plant Soil* 59: 491-494 (1981).
54. Snape, J. W., E. Nevo, B. B. Parker, D. Leckie, and A. Morgunov. "Herbicide Response Polymorphism in Wild Populations of Emmer Wheat," *Heredity* 66: 251-257 (1991).
55. Snape, J. W., and R. B. Parker. "Chemical Response Polymorphisms: an Additional Source of Genetic Markers in Wheat," in *Proceedings Seventh International Wheat Genetics Symposium*, T.E. Miller, and R.D. Koebner, Eds. (Cambridge, Institute of Plant Science Research, 1988), pp. 651-656.

56. Snape, J. W., W. J. Angus, B. Parker, and D. Leckie. "The Chromosomal Locations in Wheat of Genes Conferring Differential Response to the Wild Oat Herbicide, Difenzoquat," *J. Agric. Sci. Cambridge* 108: 543-548 (1987).
57. Jana, S., and J. M. Naylor. "Adaptation for Herbicide Tolerance in Populations of *Avena fatua*," *Can. J. Bot.* 60: 1611-1617 (1982).
58. Price, S. C., J. E. Hill, and R. W. Allard. "Genetic Variability for Herbicide Reaction in Plant Populations," *Weed Sci.* 31: 652-657 (1983).
59. Thai, K. M., S. Jana, and J. M. Naylor. "Variability for Response to Herbicides in Wild Oat (*Avena fatua*) Populations," *Weed Sci.* 33: 829-835 (1985).
60. Duncan, C. N., and S. C. Weller. "Heritability of Glyphosate Susceptibility Among Biotypes of Field Bindweed," *J. Hered.* 78: 257-260 (1987).
61. Le Court De Billot, M. R., P. C. Nel, and H. O. Gevers. "Inheritance of Atrazine Tolerance in South Africa Maize," *S. Afr. J. Plant Soil* 7: 81-86 (1990).
62. Roh, S. E., Y. M. Lee, and J. O. Guh. "Test of Resistance to Herbicides and Genetic Analysis by Diallel Cross in Rice," in *Proceedings Twelfth Conference Asian-Pacific Weed Science Society* (Taipei, 1989), pp. 261-265.
63. Miller, J. C., L. R. Baker, and D. Penner. "Inheritance of Tolerance to Chloramben Methyl Ester in Cucumber," *J. Am. Hortic.* Sci. 98: 386-389 (1973).
64. Keifer, D. W. "Tolerance of Corn (*Zea mays*) Lines to Clomazone," *Weed Sci.* 37: 622-628 (1989).
65. Geadelmann, J. L., and R. M. Andersen. "Inheritance of Tolerance to Hoe 23408 in Corn," *Crop Sci.* 17: 601-603 (1977).
66. Stafford, R. E., V. E. Comstock, and J. H. Ford. "Inheritance of Tolerance in Flax (*Linum usitatissimum* L.) Treated with MCPA," *Crop Sci.* 8: 423-426 (1968).
67. Harrison, H. F., A. Jones, and P. D. Dukes. "Heritability of Metribuzin Tolerance in Sweet Potatoes (*Ipomea batatas*)," *Weed Sci.* 35: 715-719 (1987).
68. Ratliff, R. L., B. F. Carver, and T. F. Peeper. "Expression of Metribuzin Sensitivity in Winter Wheat (*Triticum aestivum*) populations," *Weed Sci.* 39: 130-133 (1991).
69. Faulkner, J. S. "Heritability of Paraquat Tolerance in *Lolium perenne* L.," *Euphytica* 23: 281-288 (1974).
70. Karazawa, M., and C. E. Caviness. "Genetic Variability for Resistance to Propanil Injury in Soybeans," *Crop Sci.* 19: 739-740 (1979).
71. McGuire, G. M., and N. Thurling. "Nuclear Genetic Control of Variation in Simazine Tolerance in Oilseed Brassicas. I. *Brassica napus*," *Euphytica* 59: 221-229 (1992).
72. McGuire, G. M., and N. Thurling. "Nuclear Genetic Control of Variation in Simazine Tolerance in Oilseed Brassicas. II. Selection for Simazine Tolerance in a *Brassica campestris* Population," *Euphytica* 61: 153-160 (1992).
73. Gronwald, J. W., R. N. Andersen, and E. C. Ye. "Atrazine-resistance in Velvetleaf (*Abutilon theophrasti*) due to Enhanced Atrazine Detoxification," *Pestic. Biochem. Physiol.* 34: 149-163 (1989).
74. Andersen, R. N., and J. W. Gronwald. "Noncytoplasmic Inheritance of Atrazine Tolerance in Velvetleaf (*Abutilon theophrasti*)," *Weed Sci.* 35: 496-498 (1987).
75. Shooler, A. B., A. R. Bell, and J. D. Nalewaja. "Inheritance of Siduron Tolerance in Foxtail Barley," *Weed Sci.* 20: 167-169 (1972).
76. Moss, S. R. "Herbicide Cross-resistance in Slender Foxtail (*Alopecurus myosuroides*)," *Weed Sci.* 38: 492-496 (1990).
77. Chauvel, B. "Polymorphisme génétique et sélection de la résistance aux urées substituées chez *Alopecurus myosuroides* Huds.," Ph. D. Thesis, University of Paris-Orsay (1991).

78. Fuerst, E. P., and K. C. Vaughn. "Mechanism of Paraquat Resistance," *Weed Technol.* 4: 150-156 (1990).
79. Itoh, K., and M. Miyahara. "Inheritance of Paraquat Resistance in *Erigeron philadelphicus* L.," *Weed Res.* (Japan) 29: 301-307 (1984).
80. Shaaltiel, Y., N. H. Chua, S. Gepstein, and J. Gressel. "Dominant Pleitropy Controls Enzymes Co-segregating with Paraquat Resistance in *Conyza bonariensis*," *Theor. Appl. Genet.* 75: 850-856 (1988).
81. Yamasue, Y., K. Kamiyama, Y. Hanoika, and T. Kusanagi. "Paraquat Resistance and its Inheritance in Seed Germination of the Foliar-resistant Biotypes of *Erigeron canadensis* L. and *E. sumatrensis* Retz.," *Pestic. Biochem. Physiol.* 44: 21-27 (1992).
82. Preston, C., J. A. M. Holtum, and S. B. Powles. "On the Mechanism of Resistance to Paraquat in *Hordeum glaucum* and *H. leporinum*: Delayed Inhibition of Photosynthetic O_2 Evolution after Paraquat Application," *Plant Physiol.* 100: 630-636 (1992).
83. Islam, A. K. M. R., and S. B. Powles. "Inheritance of Resistance to Paraquat in Barley Grass *Hordeum glaucum* Stend.," *Weed Res.* 28: 393-397 (1988).
84. Purba, E., C. Preston, and S. B. Powles. "Inheritance of Bipyridyl Herbicide Resistance in *Arctotheca calendula* and *Hordeum leporinum*," *Theor. Appl. Genet.* 87: 598-602 (1993).
85. Holt, J., J. A. M. Holtum, and S. B. Powles. "Mechanisms and Agronomic Aspects of Herbicide Resistance," *Annu. Rev. Plant Physiol. Plant Mol. Biol.* 44: 203-229 (1993).
86. Mallory-Smith, C. A., D. C. Thill, M. J. Dial, and R. S. Zemetra. "Inheritance of Sulfonylurea Herbicide Resistance in *Lactuca* sp.," *Weed Technol.* 4: 787-790 (1990).
87. Lee, K. Y., J. Townsend, J. Tepperman, M. Black, C. F. Chui, B. Mazur, P. Dunsmuir, and J. Bedbrook. "The Molecular Basis of Sulfonylurea Herbicide Resistance," *EMBO J.* 7: 1241-1248 (1988).
88. Guttieri, M. J., C. V. Eberlein, C. A. Mallory-Smith, D. C. Thill, and D. L. Hoffman. "DNA Sequence Variation in Domain A of the Acetolactate Synthase Genes of Herbicide Resistant and Susceptible Weed Biotypes," *Weed Sci.* 40: 670-676 (1992).
89. Christopher, J. T., S. B. Powles, and J. A. M. Holtum. "Resistance to Acetolactate Synthase-Inhibiting Herbicides in Annual Ryegrass (*Lolium rigidum*) Involves at Least Two Mechanisms," *Plant Physiol.* 100: 1909-1913 (1992).
90. Burnet, M. W. M., J. T. Christopher, J. A. M. Holtum, and S. B. Powles. "Identification of Two Mechanisms of Sulfonylurea Resistance within one Population of Rigid Ryegrass (*Lolium rigidum*) Using a Selective Germination Medium," *Weed Sci.* (in press).
91. Gronwald, J. W. "Lipid Biosynthesis Inhibitors," *Weed Sci.* 39: 435-449 (1991).
92. Betts, K. J., N. J. Ehlke, D. L. Wyse, J. W. Gronwald, and D. A. Somers. "Mechanism of Inheritance of Diclofop Resistance in Italian Ryegrass (*Lolium multiflorum*)," *Weed Sci.* 40: 184-189 (1992).
93. Tardif, F. J., and S. B. Powles. "Inheritance of Resistance to ACCase inhibiting herbicides in a biotype of *Lolium rigidum*," *Theor. Appl. Genet.* (submitted).
94. Barr, A. R., A. L. Mansooji, J. A. M. Holtum, and S. B. Powles. "The Inheritance of Herbicide Resistance in *Avena sterilis* ssp. *ludoviciana*, Biotype SAS 1," in *Proceedings First International Weed Control Congress* (Melbourne, Weed Science Society of Victoria [WSSV], 1992), pp. 70-72.

95. Häusler, R. E., J. A. M. Holtum, and S. B. Powles. "Cross-resistance to Herbicides in Annual Ryegrass (*Lolium rigidum*). IV. Correlation Between Membrane Effect and Resistance to Graminicides," *Plant Physiol.* 97: 1035-1043 (1991).
96. Heap, I. M., and R. Knight. "Variation in Herbicide Cross-resistance Among Populations of Annual Ryegrass (*Lolium rigidum*) Resistant to Diclofop-methyl," *Aust. J. Agric. Res.* 41: 121-128 (1990).
97. Powles, S. B., and J. M. Matthews. "Multiple Herbicide Resistance in Annual Ryegrass (*Lolium rigidum*): a Driving Force for the Adoption of Integrated Weed Management," in *Achievements and Developments in Combating Pest Resistance*, I. Denholm, A. Devonshire, and D. Hollomon, Eds. (Elsevier, London, 1992), pp. 75-87.
98. Holtum, J. A. M. and S. B. Powles. "Annual Ryegrass: an Abundance of Resistance, a Plethora of Mechanisms," in *Proceedings British Crop Protection Conference — Weeds* (Brighton, British Crop Protection Council, 1991), pp. 1071-1077.
99. Vaughn, K. C., and S. O. Duke. "Biochemical Basis of Herbicide Resistance," in Chemistry of Plant Protection: 7, W. E. Bing, Ed. (Berlin, Springer-Verlag, 1991), pp. 141-169.
100. Jasieniuk, M., A. L. Brûlé-Babel, and I. N. Morrison. "Inheritance of Trifluralin Resistance in Green Foxtail (*Setaria viridis*)," in *Weed Sci.* (in press).
101. Hayes, J. D., R. K. Pfeiffer, and M. S. Rana. "The Genetic Response of Barley to DDT and Barban and its Significance in Crop Protection," *Weed Res.* 5: 191-206 (1965).
102. Souza-Machado, V., S. C. Phatak, and I. L. Nonnecke. "Inheritance of Tolerance of the Tomato (*Lycopersicon esculentum* Mill.) to Metribuzin Herbicide," *Euphytica* 31: 129-138 (1982).
103. Kim, C. S., Y. M. Lee, and J. O. Guh. "Genetic Analysis of Resistance to Oxyfluorfen and Bensulfuron in Rice," in *Proceedings Twelth Conference Asian-Pacific Weed Science Society* (Taipei, 1989), pp. 267-270.
104. Fery, R. L., and H. F. Harrison, Jr. "Inheritance and Assessment of Bentazon Herbicide Tolerance in "Santaka" Pepper," *J. Am. Soc. Hortic. Sci.* 115: 854-847 (1990).
105. Park, H. H., Y. M. Lee, J. O. Guh, and K. H. Lee. "Screening for Varietal Resistance to Butachlor and its Inheritance in Rice," in *Proceedings Eleventh Conference Asian-Pacific Weed Science Society* (Taipei, 1987), pp. 277-282.
106. Randhama, A. S., H. S. Dhaliwal, S. K. Sharma, and D. S. Multani. "Inheritance of 2,4-D Tolerance in Wheat," *Curr. Sci.* 56: 191-192 (1987).
107. Warkentin, T. D., G. Marshall, R. I. H. McKenzie, and I. N. Morrison. "Diclofop-methyl Tolerance in Cultivated Oats (*Avena sativa* L.)," *Weed Res.* 28: 27-35 (1988).
108. Bush, R., R. Behrens, A. Ageez, and M. Elakkad. "Inheritance of Tolerance to, and Agronomic Effects of, Difenzoquat Herbicide in Spring Wheat," *Crop Sci.* 29:47-50 (1989).
109. Lupton, F. G. H., and R. H. Oliver. "The Inheritance of Metoxuron Susceptibility in Winter Wheat," in *Proceedings British Crop Protection Conference — Weeds* (Brighton, British Crop Protection Council, 1976), pp. 473-478.
110. Edwards, C. J., W. L. Barrentine, and T. C. Kilen. "Inheritance of Sensitivity to Metribuzin in Soybeans," *Crop Sci.* 16: 119-120 (1976).
111. Dejong, H. "Inheritance of Sensitivity to the Herbicide Metribuzin in Cultivated Diploid Potatoes," *Euphytica* 32: 41-48 (1983).

112. Adeniji, A. A., and D. P. Coyne. "Inheritance of Resistance to Trifluralin Toxicity in *Cucurbita moschata* Poir," *Hortscience* 16: 774-775 (1981).
113. Terakawa, T., and K. Wakasa. "Rice Mutant to the Herbicide Bensulfuron Methyl (BSA) by in Vitro Selection," *Jpn. J. Breed.* 42: 267-275 (1992).
114. Radin, D. N., and P. S. Carlson. "Herbicide Resistant Tobacco Mutant Selected in situ Recovered via Regeneration from Cell Culture," *Genet. Res.* 32: 85-89 (1978).
115. Haughn, G. W., and C. Somerville. "Sulfonylurea-resistant Mutants of *Arabidopsis thaliana*," *Mol. Gen. Genet.* 204: 430-434 (1986).
116. Sebastian, S. A., G. M. Fader, J. F. Ulrich, D. R. Forney, and R. S. Chaleff. "Semidominant Soybean Mutation for Resistance to Sulfonylurea Herbicides," *Crop Sci.* 29: 1403-1408 (1989).
117. Sebastian, S. A., and R. S. Chaleff. "Soybean Mutants with Increased Tolerance for Sulfonylurea Herbicides," *Crop Sci.* 27: 948-952 (1987).
118. Chaleff, R. S., and T. B. Ray. "Herbicide-resistant Mutants from Tobacco Cell Cultures," *Science* 223: 1148-1151 (1984).
119. Creason, G. L., and R. S. Chaleff. "A Second Mutation Enhances Resistance to a Tobacco Mutant to Sulfonylurea Herbicides," *Theor. Appl. Genet.* 76: 177-182 (1988).
120. Maher, E. P., and S. J. B. Martindale. "Mutants of *Arabidopsis thaliana* with Altered Responses to Auxins and Gravity," *Biochem. Genet.* 18: 1041-1053 (1980).
121. Marshall, L. C., D. A. Somers, P. D. Dotray, B. G. Gengenbach, D. L. Wyse, and J. W. Gronwald. "Allelic Mutations in Acetyl Coenzyme A Carboxylase Confer Herbicide Tolerance in Maize," *Theor. Appl. Genet.* 83: 435-442 (1992).
122. Haughn, G. W., and C. R. Somerville. "A Mutation Causing Imidazolinone Resistance Maps to the Csr 1 Locus of *Arabidopsis thaliana*," *Plant Physiol.* 92: 1081-1085 (1990).
123. Anderson, P. C., and M. Georgeson. "Herbicide-Tolerant Mutants of Corn," *Genome* 31: 994-999 (1989).
124. Swanson, E. B., M. J. Herrgesell, M. Amoldo, D. W. Sipell, and R. S. C. Wong. "Microspore Mutagenesis and Selection: Canola Plants with Field Tolerance to Imidazolinones," *Theor. Appl. Genet.* 78: 525-530 (1989).
125. Newhouse, K. E., W. A. Smith, M. A. Starrett, T. J. Schaefer, and B. K. Singh. "Tolerance to Imidazolinone Herbicides in Wheat," *Plant Physiol.* 100: 882-886 (1992).
126. Heim, D. R., J. L. Roberts, P. D. Pike, and I. M. Larrinua. "Mutation of a Locus of *Arabidopsis thaliana* Confers Resistance to the Herbicide Isoxaben," *Plant Physiol.* 90: 146-150 (1989).
127. Heim, D. R., J. L. Pike, and I. M. Larrinua. "A second Locus, *Ixr B1* in *Arabidopsis thaliana*, that Confers Resistance to the Herbicide Isoxaben," *Plant Physiol.* 92: 858-861 (1990).
128. Hickok, L. G., and O. J. Schwarz. "An in Vitro Whole Plant Selection System: Paraquat Tolerant Mutants in the Fern *Ceratopteris*," *Theor. Appl. Genet.* 72: 302-306 (1986).
129. Chaleff, R. S., and M. F. Parsons. "Direct Selection in Vitro for Herbicide-resistant Mutants of *Nicotiana tabacum*," *Proc. Natl. Acad. Sci. U.S.A.* 75: 5104-5107 (1978).
130. Chaleff, R. S. "Further Characterization of Picloram-Tolerant Mutants of *Nicotiana tabacum*," *Theor. Appl. Genet.* 58: 91-95 (1980).

131. Harms, C. T., J. J. Di Maio, S. M. Jayne, L. A. Middlesteadt, D. V. Negrotto, H. Thompson-Taylor, and A. L. Montoya. "Primisulfuron Herbicide-resistant Tobacco Plants Mutant Selection in Vitro by Adventitious Shoot Formation from Cultured Leaf Discs," *Plant Sci.* 79: 77-85 (1991).
132. Harms, C. T., A. L. Montoya, L. S. Privalle, and R. W. Briggs. "Genetic and Biochemical Characterization of Corn Inbred Lines Tolerant to the Sulfonylurea Herbicide Primisulfuron," *Theor. Appl. Genet.* 80: 353-358 (1990).
133. Parker, W. B., L. C. Marshall, J. D. Burton, D. A. Somers, D. L. Wyse, J. W. Gronwald, and B. G. Gengenbach. "Dominant Mutations Causing Alterations in Acetyl-Coenzyme A Carboxylase Confer Tolerance to Cyclohexanedione and Aryloxyphenoxypropionate Herbicides in Maize," *Proc. Natl. Acad. Sci. U.S.A.* 87: 7175-7179 (1990).
134. Mazur, B. J., and S. C. Falco. "The Development of Herbicide Resistant Crops," *Annu. Rev. Plant Physiol. Plant Mol. Biol.* 40: 441-470 (1989).
135. Hattori, J., R. G. Rutledge, and B. L. Miki. "DNA Sequence Relationships and Origins of Acetohydroxy Acid Synthase Genes of *Brassica napus*," *Can. J. Bot.* 70:1957-1963 (1992).
136. Deak, M., G. Donn, A. Feher, and D. Dudits. "Dominant Expression of a Gene-Amplification-related Herbicide Resistance in *Medicago* Cell Hybrids," *Plant Cell Rep.* 7: 158-161 (1988).
137. Smith, C. M., D. Pratt, and G. A. Thompson. "Increased 5-enolpyruvylshikimic acid 3 Phosphate Synthase Activity in a Glyphosate Tolerant Variant Strain of Tomato Cells," *Plant Cell Rep.* 5: 298-301 (1986).
138. Parker, W. B., D. A. Somers, D. L. Wyse, R. A. Keith, J. D. Burton, J. W. Gronwald, and B. G. Gengenbach. "Selection and Characterization of Sethoxydim-Tolerant Maize Tissue Culture," *Plant Physiol.* 92: 1220-1225 (1990).
139. Thomas, B. R., and D. Pratt. "Isolation of Paraquat-Tolerant Mutants from Tomato Cell Cultures," *Theor. Appl. Genet.* 63: 169-176 (1982).
140. Singer, S. R., and C. N. McDaniel. "Selection of Amitrole Tolerant Tobacco Calli and the Expression of this Tolerance in Regenerated Plants and Progeny," *Theor. Appl. Genet.* 67: 427-432 (1984).
141. Gabard, J. M., P. Charest, V. N. Lyer, and B. Miki. "Cross-resistance to Short Residual Sulfonylurea Herbicides in Transgenic Tobacco Plants," *Plant Physiol.* 91: 574-580 (1989).
142. DeBlock, M., J. Botterman, M. Vandewiele, J. Dockx, C. Thoen, V. Gossel, N. Rao Movva, C. Thompson, M. Van Montagu, and J. Leemans. "Engineering Herbicide Resistance in Plants by Expression of a Detoxifying Enzyme," *EMBO J.* 6: 2513-2518 (1987).
143. Bayley, C., N. Trolinder, C. Ray, M. Morgan, J. E. Quisenberry, and D.W. Ow. "Engineering 2,4-D Resistance in Cotton," *Theor. Appl. Genet.* 84: 645-649 (1992).
144. Pay, A., M. A. Smith, F. Nagy, and L. Marton. "Sequence of the *psbA* Gene from Wild Type and Triazine-resistant *Nicotiana plumbaginifolia*," *Nucleic Acids Res.* 16: 8176 (1988).
145. Sato, F., Y. Shigematsu, and Y. Yamada. "Selection of an Atrazine-resistant Tobacco Cell Line Having a Mutant *psbA* Gene," *Mol. Gen. Genet.* 214: 358-360 (1988).

146. Arntzen, C. J., and J. H. Duesing. "Chloroplast-Encoded Herbicide Resistance," in *Advances in Gene Technology: Molecular Genetics of Plants and Animals*, K. Downey, R. W. Voellmy, F. Ahmad, and J. Schultz, Eds. (New York, Academic Press, Inc., 1983), pp. 273-294.
147. Darmency, H., and J. Gasquez. "Appearance and Spread of Triazine-resistance in Common Lambsquarters (*Chenopodium album*)," *Weed Technol.* 4: 173-177 (1990).
148. Al Mouemar, A., and J. Gasquez. "Environmental Conditions and Isozyme Polymorphism in *Chenopodium album* L.," *Weed Res.* 23: 141-149 (1983).
149. Gasquez, J., and J. P. Compoint. "Isoenzymatic Variations in Populations of *Chenopodium album* L. Resistant and Susceptible to Triazines," *Agro-Ecosystem* 7: 1-10 (1981).
150. Warwick, S. I., and P. B. Marriage. "Geographical Variation in Populations of *Chenopodium album* Resistant and Susceptible to Atrazine. I. Between and Within Population Variation in Growth Response to Atrazine," *Can. J. Bot.* 60: 483-493 (1982).
151. Warwick, S. I., and L. D. Black. "Electrophoretic Variation in Triazine-resistant and Susceptible Populations of *Amaranthus retroflexus* L.," *New Phytol.* 104: 661-670 (1986).
152. Darmency, H., and J. Gasquez. "Interpreting the Evolution of a Triazine Resistant Population of *Poa annua* L," *New Phytol.* 95: 299-304 (1983).
153. Jacobs, B. F., J. H. Duesing, J. Antonovics, and D. T. Patterson. "Growth Performance of Triazine Resistant and Susceptible Biotypes of *Solanum nigrum* Over a Range of Temperatures," *Can. J. Bot.* 66: 847-850 (1988).
154. Gressel, J., and L. A. Segel. "Modelling the Effectiveness of Herbicide Rotations and Mixtures as Strategies to Delay or Preclude Resistance," *Weed Technol.* 4: 186-198 (1990).
155. Maxwell, B. D., M. L. Roush, and S. R. Radosevich. "Predicting the Evolution and Dynamics of Herbicide Resistance in Weed Populations," *Weed Technol.* 4: 2-13 (1990).
156. Price, S. C., R. W. Allard, J. E. Hill, and J. Naylor. "Associations Between Discrete Genetic Loci and Genetic Variability for Herbicide Reaction in Plant Populations," *Weed Sci.* 33: 650-653 (1985).
157. Nevo, E., J.W. Snape, B. Lavie, and A. Beiles. "Herbicide Response Polymorphism in Wild Emmer Wheat: Ecological and Isozyme Correlations," *Theor. Appl. Genet.* 84: 209-216 (1992).
158. Chauvel, B., and J. Gasquez. "Relationships Between Genetic Polymorphism and Herbicide Resistance Within *Alopecurus myosuroides* Huds.," *Heredity* 71: (in press).
159. Burnet, M. W. M., O. B. Hildebrand, J. A. M. Holtum, and S. B. Powles. "Amitrole, Triazine, Substituted Urea, and Metribuzin Resistance in a Biotype of Rigid Ryegrass (*Lolium rigidum*)," *Weed Sci.* 39: 317-323 (1991).
160. Darmency, H. "The Impact of Hybrids Between Genetically Modified Crop Plants and their Related Species: Introgression and Weediness," *Molecular Ecology* 3: 37-40 (1994).

CHAPTER 11

Growth and Productivity of Resistant Plants

J. S. Holt and D. C. Thill

AGROECOLOGY OF HERBICIDE RESISTANCE

Long-term prevention and management of herbicide resistance will require adopting practices that reduce selection pressures that favor resistant plants (see Chapter 12). Achieving these goals requires an understanding of the evolution and dynamics of resistant populations. As considered in Chapter 1, evolutionary change occurs through natural or artificial selection acting on populations that possess heritable phenotypic variation. When one phenotype leaves more offspring than another, because of its superior ability to survive and reproduce, that phenotype has an evolutionary advantage relative to other phenotypes in that environment. Survival and reproductive success combine to determine the relative fitness of a phenotype in a particular environment.[1]

A gene that replaces another in a population may be less adaptive initially if it has some physiological disadvantages relative to the original gene.[2] For example, alteration of the normal regulation of an enzyme may render it resistant to a pesticide but may interfere with its original function. The genotype with this mutation may suffer a disadvantage relative to the wild type in the absence of artificial selection by the pesticide.[3] This disadvantage would keep the new gene at a very low frequency in the population. The new gene may become more adaptive when changes in the environment occur that favor it. In an environment where a herbicide is used, the fitness of individuals resistant to that herbicide is increased relative to that of susceptible individuals. Theoretically, when the herbicide is removed, a recently selected R genotype could suffer some cost in fitness relative to the original S genotype. Over time, the physiological disadvantages of the new gene may be gradually compensated by selection of modifying traits.[2] Until established in the population, however, the new gene may cause increased mortality or reduced fertility of the organism. Thus, the pleiotropic cost that accompanies a herbicide resistance mutation can be decreased fitness. Since selection of modifying traits is an evolutionary process that occurs over a long time, in the absence of the herbicide, a more fit S genotype would replace the less fit R genotype.[4]

Studies of growth, physiological performance, and competition of biotypes R or S to triazine herbicides, the oldest documented case of evolved herbicide resistance,[5] have shown repeatedly that triazine resistance has negative physiological consequences.[6,7] Because triazine R biotypes are less fit than their S counterparts, it has been often stated that plants resistant to other herbicide classes would also be less fit than S biotypes.[4] However, comparative studies of biotypes R or S to other herbicide classes, and some studies with triazine R and S biotypes, have not always supported this hypothesis. In some cases, lack of data on genetics and mechanisms of resistance precludes an understanding of the consequences of resistance to the organism (see Chapter 10). In other cases, investigations have not measured fitness appropriately. With some of the more recently discovered cases of resistance (e.g., ALS inhibitor resistance, see Chapter 4), evidence suggests that R biotypes are not less fit than S biotypes. In order to predict and manage herbicide resistance, it is imperative that we reach a better understanding of relative fitness of herbicide R and S biotypes in the absence of herbicide selection pressure.

Population Models

As has also been discussed in Chapter 1, models to describe and simulate weed population dynamics have been developed specifically to predict the evolution and dynamics of herbicide resistance in weed populations.[8-10] The recent model by Gressel and Segel[8] and an earlier version of the same,[4] stress the importance of herbicide selection pressure, seedbank dynamics, and fitness (described by a single term) in regulating changes in abundance of R individuals in mixed populations of R and S plants. Putwain and Mortimer[10] include parameters in their model that account for effects of density and competition on R and S weed population dynamics. The model by Maxwell et al.[9] expands upon these other models by including gene flow, via immigration, as an important factor in the evolution and dynamics of herbicide-resistant weed populations (see Chapter 1). In addition, Maxwell et al.[9] define fitness in terms of numerous component life history processes that more accurately describe success of an individual phenotype. Using data derived from field experiments, these models can be used for mathematical simulations to predict potential outcomes of current weed management practices and to suggest alternative management strategies that could decrease selection for resistance.[8,11,12]

Components of Fitness

The relative allocation of resources to survival and reproduction throughout the life cycle of an organism determines its fitness. Survivorship may be further partitioned into success of seed, seedling, and mature plant, which in turn are functions of processes such as germination, dormancy, establishment, and growth.[1,6,9] Reproductive success is determined by pollen and seed production throughout the life cycle of the plant. Fitness of a particular phenotype is also determined in the context of prevailing environmental conditions and other phenotypes in that environment. Therefore, both adaptation and competitiveness will define fitness. This chapter reviews research on ecological and physiological characteristics of R and S weeds to assess the impact of resistance on relative fitness in the absence of the herbicide selector.

CHARACTERISTICS OF HERBICIDE-RESISTANT WEEDS

Resistance to Photosystem II Inhibiting Herbicides

The three major classes of photosystem II (PS II) inhibitors are the triazine, urea, and uracil herbicides, all of which inhibit electron transport at the QB site in PS II.[13] As discussed in Chapter 2, evolved resistance in weeds has been documented to all of these herbicide classes. By far the most prevalent and well studied of these cases is triazine resistance, which has been documented in 57 species in over 1000 sites worldwide.[5,14] Resistance to substituted ureas has been documented in 7 weed species, while uracil resistance has been reported in only 2.[14] A significant body of research has been conducted to evaluate effects of triazine resistance on physiology, ecology, and fitness of plants. By virtue of its status as the first documented case of evolved herbicide resistance in weeds, triazine resistance has served as the model system upon which many theories and principles about herbicide resistance have been based. This is despite the fact that triazine resistance is atypical of cases of xenobiotic resistance reported in other organisms and in subsequently reported cases of herbicide resistance (Chapters 3 to 7).

In nearly all species studied, triazine resistance is caused by a mutation in the chloroplast *psbA* gene that encodes the 32-kDa herbicide binding (called QB or D1) protein of PS II.[15] Since the QB protein is a chloroplast gene product, triazine resistance is maternally inherited.[16] This mutation not only decreases affinity of QB for triazines[17] but also reduces the rate of electron transfer between PS II acceptors QA and QB.[18-20] As this effect occurs in a process that is fundamentally important for plant survival and growth, it is not surprising that R biotypes generally have reduced photosynthetic rates, quantum yield, biomass production, fecundity, and competitiveness relative to S biotypes.[6,7]

Photosynthesis, Growth, and Competition

Photosynthesis and productivity have been studied in a myriad of triazine R and S weed biotypes. Lower rates of CO_2 fixation have been measured in R biotypes of *Amaranthus* spp.,[21,22] *Brassica* spp.,[23] and *Senecio vulgaris*.[19,24,25] Reduced rates of PS II electron transport and lower quantum yields are also characteristic of triazine R biotypes.[18-21,24-27] R biotypes of several species possess "shade"-type leaf anatomy and chloroplast ultrastructure, including a higher degree of grana stacking, higher chlorophyll *b/a* ratios, and more light-harvesting proteins.[26,28,29] These adaptations appear to be compensatory responses to reduced PS II efficiency in triazine R plants.

As expected from reduced photosynthetic potential of triazine R biotypes, in the absence of triazine herbicides, R biotypes are generally less productive than S biotypes. Biomass and seed production were lower in S biotypes of *Chenopodium album*,[30] *S. vulgaris*,[31-33] and *Solanum nigrum*.[34] R biotypes were less competitive than S in studies with *Amaranthus powellii*,[35] *A. retroflexus*,[35,36] *C. album*,[37] and *S. vulgaris*.[31,36] All of these reports confirm that the triazine resistance trait is correlated with reduced fitness.

Not all the literature on fitness of triazine R weeds is unequivocal in support of the contention that R biotypes are less fit. Several cases have been reported where triazine R and S biotypes had similar rates of photosynthesis and productivity or where R biotypes outperformed S.[22,38,39] In addition, comparative studies of several R and S populations of *A. powellii* and *A. retroflexus*[40] and *C. album*[41,42] showed that variation among populations from disparate locations was often greater than variation between R and S populations (see also Chapter 10). Since triazine resistance is encoded on the chloroplast genome, even R and S biotypes collected in the same field are likely to differ in nuclear genome controlled traits that may compensate to some extent for detrimental effects of triazine resistance.[43] Results of research using isonuclear biotypes, which differ only in the trait of triazine resistance, have confirmed that impaired chloroplast function in R biotypes limits photosynthesis, growth, and productivity at the whole plant level.[20,43-48] However, since these biotypes do not occur naturally in the field, meaningful evaluations of fitness that will be relevant for weed management should be made on field-selected triazine R and S biotypes under appropriate environmental conditions.

Despite extensive research on triazine resistance using non- and isonuclear biotypes, questions remain about the mechanism whereby the mutation affects whole plant performance. Many triazine R weeds first appeared in cool temperate regions, suggesting that productivity of R biotypes may be limited by high temperatures. Greater high-temperature inhibition of PS II in R plants relative to S has been observed in several field-collected biotype pairs.[49-53] In studies with isonuclear biotypes, PS II electron transport was more sensitive to extremes of high temperature (greater than 38°C) in R than in S *S. nigrum*, indicating that high-temperature sensitivity is a direct consequence of the QB mutation.[54] However, growth of isonuclear R and S biotypes of *S. nigrum*[55] and *S. vulgaris*[56] and photosynthetic rates of *S. vulgaris*[56] were not affected differently by growth or measurement temperature. Comparisons of temperature responses of growth of non-isonuclear R and S biotypes have proved inconclusive,[57] suggesting that temperature responses under control of the nuclear genome may be greater than those due to triazine resistance. Thus, responses of PS II to growth temperatures normally encountered in temperate climates are not responsible for reduced growth and limited distribution of triazine R biotypes.[56]

Studies of isonuclear R and S biotypes of *Brassica napus* grown in different light regimes have indicated that increased sensitivity to photoinhibition may be responsible for reduced productivity of triazine R plants in the field. Biotypes grown under low light had similar quantum yields and productivity, while R plants grown in high light had lower quantum yield and biomass production than S plants.[58,59] Exposure of low-light-grown plants to high light decreased PS II activity more in R than in S plants.[60] Enhanced sensitivity to photoinhibition is most likely a direct result of reduced electron transfer in PS II caused by the QB mutation conferring triazine resistance.[60]

Seed Germination

Several studies have been conducted to evaluate seed germination characteristics of non-isonuclear triazine R and S biotypes. No consistent differences between R and S biotypes of *A. retroflexus, C. album, Polygonum lapathifolium*, or *S. nigrum* were found to suggest that resistance has a detrimental effect on germination.[61] In contrast, S biotypes of *Brassica campestris* had earlier seed germination and seedling emergence than R biotypes.[62] When biotypes of two species of *Amaranthus* were compared, differences between R and S plants were not consistent between the species and in some germination parameters, differences between species were greater than those between biotypes.[63] Similarly, seedbank dynamics of R and S biotypes of *S. vulgaris* were more affected by weed management practices than by differences in germination characteristics between biotypes.[64]

In summary, the considerable research that has been conducted with triazine R and S biotypes of numerous weed species provides unequivocal evidence that the target site mutation conferring resistance has a detrimental effect on overall fitness of plants possessing this trait. Research using isonuclear biotypes has allowed investigation and better understanding of the mechanisms responsible for this negative pleiotropic effect. However, in the field environment, selection imposed by environmental factors and cultural practices leads to biotypic differences in nuclear genome controlled traits that may compensate for the effects of triazine resistance. It is the combination of chloroplast and nuclear genome effects that determine overall fitness of plants in the field.

Resistance to Photosystem I Inhibiting Herbicides

Evolved resistance to paraquat (and diquat) has been documented in 18 weed species and found in numerous locations.[14] Ecological studies to assess relative fitness of R biotypes have been conducted on only a few of these species. Since paraquat resistance may be conferred by more than one mechanism,[65] it is likely that fitness of R biotypes could differ among species possessing this trait (see Chapter 3).

Seed and Seedbank Characteristics

Experiments were conducted to quantify seed dormancy characteristics and seedbank longevity of R and S biotypes of *Hordeum glaucum* in Australia.[66] Seeds of both biotypes had strong innate dormancy and failed to germinate when fresh. Following a short period of after-ripening, nearly all seeds of both biotypes germinated, indicating that seedbank life is short in this species. When new seed production in the field was prevented, the seedbank of both biotypes was eliminated after three years.[66] While no differences between R and S biotypes were detected in this study, lack of a residual seedbank of *H. glaucum* indicates that control of both biotypes could be achieved by prevention of seed production. In contrast, in similar experiments with R and S *Arctotheca calendula*, viable seeds remained in the seedbank even after six years of prevention of seed production.[67] For these two weed species, seed and seedbank characteristics are more important factors determining population dynamics than are any differences between paraquat R and S biotypes.

Physiology, Growth, and Competition

Field studies conducted in Australia with *H. glaucum* biotypes planted in a replacement series design showed that the S biotype was marginally more competitive than the R.[68] Similar experiments using *H. leporinum* biotypes showed no differences between S and R biotypes in competitiveness.[69] In experiments conducted in Japan with *Conyza canadensis*, growth of the paraquat S biotype was more vigorous than that of the R.[70] While it appears that some reduction in productivity may accompany paraquat resistance, it is not possible from these limited studies to generalize about effects of paraquat resistance on fitness.

Photosynthetic characteristics were examined in paraquat R, paraquat/atrazine co-R, and S biotypes of *C. canadensis* in Hungary.[71] Photosynthetic electron transport from plastoquinone to the cytochrome b6/f complex was similar in the three biotypes, and intersystem pool size in paraquat R and S plants was the same.[71] In contrast, intersystem pool size in the paraquat/atrazine co-R biotype was twofold larger than that in R and S plants and similar to that found in triazine R biotypes.[72] These studies detected no effects associated with the paraquat R mutation that could account for any growth differential between biotypes. Similar studies with Australian biotypes R or S to paraquat revealed no differences in light-limited or light-saturated photosynthetic performance.[73]

Resistance to Acetyl CoA Carboxylase Inhibiting Herbicides

As detailed in Chapter 5, evolved resistance to aryloxyphenoxypropionate (APP) and cyclohexanedione (CHD) herbicides was discovered in 1982 and has since been documented in eight grass species in many sites.[14] The few abstracts published on fitness of R and S weed biotypes have focused on germination, phenology, and gene flow.[74-76] These reports are all from experiments conducted in the field and provide valuable information on population dynamics and implications for resistance management. However, because little information is available on differences between biotypes in relative growth, competitiveness, and fecundity, few conclusions can be drawn regarding effects of resistance to acetyl CoA carboxylase (ACCase) inhibiting herbicides on fitness.

Seed Germination

Ghersa et al.[74] examined germination phenology of diclofop-methyl R and S *Lolium multiflorum* biotypes in a wheat field. They hypothesized that seedling populations emerging during the cropping season would be subjected more often to herbicide selection pressure than would populations emerging before wheat sowing, since early germinators would likely be removed by cultivation during seedbed preparation.[74] Results of field experiments conducted over two years supported this hypothesis. Prior to the typical wheat sowing date, four times more S seedlings emerged than R, while later in the cropping season more R seedlings emerged than S. Differences between R and S biotypes in this case were not due to pleiotropic effects of the resistance mutation. In other field experiments with *L. multiflorum*, the R biotype emerged over a narrower range of temperatures than the S, suggesting that the R soil seedbank was more dormant than the S. These studies suggest that current agricultural practices to control *L. multiflorum* might be simultaneously selecting for herbicide resistance as well as seed dormancy and germination characteristics.[77]

Phenology and Growth

In the same experiment described above, Roush et al.[75] examined phenology of diclofop-methyl R and S *L. multiflorum* biotypes. In addition to germinating later than S, the R plants were developmentally delayed through to flowering and seed set relative to S plants. Growth rate, leaf area, and vegetative and reproductive biomass were similar in the two biotypes.[75] Growth chamber experiments with graminicide R and S biotypes of *Eleusine indica* yielded different results. The R biotype produced 46% less shoot biomass than the S, but flowered earlier and allocated relatively more biomass to reproductive structures than the S.[78]

In field experiments with *L. multiflorum*, pollen was dispersed one week earlier in the S than the R biotype and S plants produced greater volumes of pollen for most of the growing season.[76] As a result, R plants had a higher probability of being fertilized by pollen from S plants, while S plants had a very low probability of being fertilized by R pollen.[76] Phenological disparity in pollen flow in these biotypes tends to reduce the level of resistance in a mixed population, and thus to decrease seedling survival after herbicide use.[79]

Although the literature on diclofop-methyl R and S biotypes does not address relative fitness, per se, this work has demonstrated the important effect of agricultural practices on population dynamics of herbicide resistance. It is clear that cultural, as well as chemical, practices used in crop production impose significant selection pressure on weed populations. The effects of such selection pressure may be more significant to weed evolution than intrinsic differences in relative fitness between herbicide R and S biotypes.

Resistance to Acetolactate Synthase Inhibiting Herbicides

Weed biotypes resistant to sulfonylurea herbicides were detected in fields within five years of introduction of this herbicide class (Chapter 4). Sulfonylurea resistance has now been documented in biotypes of 14 weed species and found in many sites, primarily in North America and Australia.[14] In 10 of these species, the mechanism of resistance involved mutations in ALS that confer different spectra of resistance to sulfonylurea, imidazolinone, and triazolopyrimidine herbicides,[80,81] all of which share the same acetolactate synthase (ALS) binding site. While the effect of these mutations on ALS structure and function are mostly unknown, rapid selection of resistance to ALS inhibitors in the field suggests that the effect of the mutations on fitness of R biotypes may be insignificant.

Studies with transgenic crops indicate that plant vigor may be unaffected by site-of-action resistance to ALS inhibitors. Seed yield, maturity date, and disease tolerance of *Brassica napus* (oilseed rape) selected for resistance to imidazolinone herbicides were similar to those of the S biotype.[82] Likewise, no deleterious effects attributed to the imidazolinone resistance allele engineered in corn have been observed.[83] Seedling shoot heights of sulfonylurea and imidazolinone R corn mutants were similar to those of the S biotype.[84]

Components of fitness have been compared in sulfonylurea R and S biotypes of the weeds *Lactuca serriola* and *Kochia scoparia* from several locations in North America. Plant characteristics that have been examined include seed germination,[85-88] growth and competitiveness,[89-92] seed production,[85,90] and seed longevity in soil.[85] Reports of these studies indicate that differences between R and S biotypes exist, but the overall effect of the resistance mutation on fitness is not pronounced.

Seed Germination

Seeds of sulfonylurea R and S biotypes of *L. serriola* collected in Idaho were incubated at 24°C (12 h light) and rate and percentage of germination were evaluated.[85] All R and S seed lots reached nearly 100% germination within 7 days. However, seeds from R plants always germinated as fast or faster than seeds from S plants. Germination of seeds of nearly isonuclear R and S lines of Bibb lettuce, containing the ALS resistance gene from *L. serriola*, was compared at 8, 18, and 28°C. Results were similar to those for *L. serriola*. In all cases, the R line germinated faster than the S line.[86]

Germination data for sulfonylurea R and S biotypes of *K. scoparia* collected in Kansas are also similar to those of *L. serriola*. When incubated at 8°C, R and S seeds attained 50% germination by 111 and 180 h, respectively. At 18°C, 50% germination was attained at 50 h for the R and 62 h for the S biotype. Both biotypes germinated at the same rate at 28°C.[93] Similarly, R *K. scoparia* biotypes from Montana germinated faster than S biotypes at 4.6 and 7.2°C, but germination rate was equal in the two biotypes at 10.5 and 16.8°C.[87] The seeds of the R biotypes from Montana contained higher levels of free valine, leucine, and isoleucine compared to S biotypes. It is not known how elevated concentrations of these amino acids affect germination.[88] It appears from these studies that sulfonylurea resistance in these particular biotypes is associated with more rapid seed germination, particularly at low temperatures. However, it is not clear whether this association is a pleiotropic effect or the result of other selection factors.

Growth and Competition

The relative growth rate of sulfonylurea R and S biotypes of *L. serriola* was determined in greenhouse experiments, in which individual plants were harvested every 3 to 7 days from 10 days after germination until bud stage.[89] The S biotype produced 24% more biomass and accumulated biomass 52% faster than the R biotype. In similar studies, R and S *K. scoparia* biotypes had similar growth rates.[93]

The relative competitiveness of R and S *L. serriola* biotypes was also determined in greenhouse experiments using plants grown in 75-l pots until the floral bud stage.[89] When data were averaged over all densities and proportions, the S biotype produced 31% more biomass than the R biotype. The ratio of R to S biomass ranged from 0.7 in lower-density plantings to 0.97 in the highest densities, where yield of both biotypes was severely affected by competition. The calculated relative competitive ability of the R and S biotypes was similar, 0.94 and 0.91, respectively.[89] Similar to data from *L. serriola* plants grown without competition, these data indicate that S plants are more productive than R.

Data from sulfonylurea R and S biotypes of *K. scoparia* are much less conclusive. In greenhouse experiments, R biotypes of *K. scoparia* from North Dakota and Kansas tended to have equal or greater leaf and stem dry weight, shoot height, and stem and shoot diameter than their S biotype counterparts 13 weeks after emergence.[90] Greenhouse studies conducted with *K. scoparia* biotypes from Colorado produced contradictory results. One report suggested that S plants were more productive than R both in competition and when grown alone,[91] while a second report by the same authors indicated that no differences in biomass, leaf area production, or competitiveness were found between these biotypes the following year.[92] As these reports are abstracts that provide no data, definitive conclusions cannot yet be drawn about *K. scoparia*.

Seed Production

Seeds from R and S biotypes of *L. serriola* were sown in the field during autumn, then beginning at bud stage the following year, ripe seeds from each flower head on a subsample of plants were collected by hand and counted for a 53-day period.[85] Plants of both biotypes began flowering at the same time. During the sampling period, R and S biotypes produced 4870 and 4160 seeds per plant, respectively. Numbers of flower heads and seeds per plant were somewhat greater for R plants, but values were within the range of other reports for *L. serriola* and biotype differences were not significant.[85] Seed production was also measured on greenhouse-grown *K. scoparia* biotypes. At maturity (13 weeks after establishment) R and S plants had produced similar numbers of seeds per plant.[90] In these two species, therefore, resistance to ALS inhibitors has no apparent detrimental effects on fecundity.

Seed Longevity

While seed longevity is a major factor regulating weed seedbank dynamics, only one study has been reported that examined the effect of sulfonylurea resistance on seed longevity.[85] Seeds from sulfonylurea R and S *L. serriola* were collected from field-grown plants, and used in field burial studies. Seeds left on the soil surface and buried 7.5 and 15 cm deep were exhumed every 6 months over 3 years and tested for viability. As expected, longevity was increased in buried seeds of both biotypes compared to seeds placed on the soil surface. No differences in seed viability were found between R and S biotypes, regardless of burial position or study location.[85]

Results from this array of seed and plant ecological studies indicate that resistance to ALS inhibitors confers no consistent disadvantage on biotypes possessing the trait. In fact, the data show a striking level of variability and inconsistency. The mutation in ALS conferring resistance is not the same in *L. serriola* and *K. scoparia*, and this may cause fitness to differ between these two species. In addition, there is a high degree of genetic and phenotypic variability in these species which is unrelated to sulfonylurea resistance. Based on this data it appears that if selection pressure by ALS inhibitors was removed from a field, S plants might only slowly (if at all) replace R ones in a mixed population.

Resistance to Mitotic Disrupting Herbicides

As considered in Chapter 7 weed biotypes within a number of species exhibit resistance to the dinitroaniline herbicide class.[14] Extensive work with dinitroaniline R and S biotypes of *Eleusine indica* showed that the R biotype contains altered microtubules which are hyperstabilized relative to those in the S biotype (see Chapter 7). As a result of this hyperstabilization, microtubule functioning during mitosis in the R biotype is abnormal, particularly during cell plate formation.[94] Because mitosis is an essential component of normal growth and development, it seems likely that this mutation would have a detrimental effect. Two studies that have addressed the effect of this mutation at the whole plant level present evidence that fitness is somewhat reduced in R biotypes of one species.

Growth and Competition

Growth and development of several populations of dinitroaniline R and S biotypes of *E. indica* from the southeastern United States were examined in the field.[95] The only consistent difference detected between R and S biotypes was greater total inflorescence dry weight in S plants. In all other characteristics measured, variation within and among R and S populations was greater than variation between R and S populations.[95] Other work using three dinitroaniline R and three S *E. indica* biotypes from a wider region in the southeastern United States confirmed these findings.[96] Additionally, R biotypes were less competitive than S and responded to competition with reduced reproductive output relative to S biotypes.[96] Since these data were obtained from a total of 9 R and 7 S populations of *E. indica*,[95,96] it seems clear that dinitroaniline resistance in this species is correlated with reduced fitness. The mechanism of this effect and its relationship to impaired functioning of tubulin remains unknown. As discussed in Chapter 7, preliminary studies with R and S *Amaranthus* indicate that there may not be a fitness penalty associated with dinitroaniline resistance in these biotypes.

Multiple Herbicide Resistance

As discussed in Chapter 9, hundreds of biotypes of *Lolium rigidum* from Australia have evolved cross and multiple resistance to a range of herbicides, including ACCase inhibitors, ALS inhibitors, carbamates, chloroacetamides, and dinitroanilines.[97,98] In R biotype SLR 31, resistance to ACCase and ALS inhibitors is due to non-target site mechanisms. In field experiments, R biotype SLR 31 had similar phenology, biomass production, and fecundity to an S biotype, whether grown in isolation or in competition.[99] Seedling growth rate studies using unselected field accessions of *L. rigidum* indicated that gene(s) conferring resistance were not confined to portions of the population with lower mean growth rates.[99] The few experiments with R *L. rigidum* indicate that the mechanisms conferring multiple resistance in this species do not carry a substantial fitness penalty. This is consistent with the observed high frequency of R biotypes in untreated field populations.[100]

SUMMARY AND CONCLUSIONS

Research on the many cases of evolved herbicide resistance in weeds provides evidence that, in some cases, resistance is accompanied by a pleiotropic cost to fitness at the whole plant level. This principle is most clearly shown for weeds resistant to triazine herbicides in which the mechanism of resistance is a target site mutation which impairs the normal function at that site. Nevertheless, in the field environment, selection occurs for traits unrelated to resistance which may attenuate detrimental effects of the resistance mutation.

In other cases, for example weeds resistant to ALS or PS I inhibitors, it appears that any pleiotropic effects associated with resistance are of little practical significance in the field. Studies with biotypes resistant to ACCase inhibitors demonstrate the often overriding influence of agricultural practices on fitness of weeds. In the case of multiple resistance in *L. rigidum* from Australia, fitness differences between R and S biotypes are not apparent. Even if differences in fitness occurred between particular biotype pairs, it is likely that they would be masked by the substantial intraspecific variation in this species.

The data reviewed here demonstrate that herbicide resistance does not necessarily incur a cost to fitness at the whole plant level. Since the cases where R biotypes are similar to S biotypes include both target site and non-target site mechanisms of resistance, the relationship of resistance mechanisms to fitness effects is not yet clear. Thus, the impact of herbicide resistance on fitness must be evaluated on a case-by-case basis for weed biotypes in the field environment where they occur.

This review has revealed significant gaps in information on growth and productivity of herbicide-resistant weeds. For some recently discovered cases, such as multiple resistance in *L. rigidum*, little data has been published on ecological aspects of R and S biotypes. The myriad of mechanisms and modes of inheritance of herbicide resistance in weeds prevent generalization about effects of resistance on fitness in the field until more data are available. Such information is essential for development and utilization of effective resistance management strategies.[101,102] It is hoped that this review will inspire further research in the area of ecology and population dynamics of herbicide resistance in weeds.

REFERENCES

1. Silvertown, J. *Introduction to Plant Population Ecology*, 2nd ed. (Essex, England: Longman Scientific & Technical, 1987), Chapters 1, 7.

2. Haldane, J. B. S. "More Precise Expressions for the Cost of Natural Selection," *J. Genet.* 57:351-360 (1960).
3. Uyenoyama, M. K. "Pleiotropy and the Evolution of Genetic Systems Conferring Resistance to Pesticides," in *Pesticide Resistance: Strategies and Tactics for Management,* (Washington, D.C.: National Academy Press, 1986), pp. 207-221.
4. Gressel, J., and L. A. Segel. "The Paucity of Genetic Adaptive Resistance of Plants to Herbicides: Possible Biological Reasons and Implications," *J. Theor. Biol.* 75:349-371 (1978).
5. Holt, J. S. "History of Identification of Herbicide-Resistant Weeds," *Weed Technol.* 6:615-620 (1992).
6. Holt, J. S. "Fitness and Ecological Adaptability of Herbicide-Resistant Biotypes," in *Managing Resistance to Agrochemicals: From Fundamental Research to Practical Strategies,* ACS Symposium Series 421, M. B. Green, H. M. LeBaron, and W. K. Moberg, Eds. (Washington, D.C.: American Chemical Society, 1990), pp. 419-429.
7. Warwick, S. I. "Herbicide Resistance in Weedy Plants: Physiology and Population Biology," *Annu. Rev. Ecol. Syst.* 22:95-114 (1991).
8. Gressel, J., and L. A. Segel. "Herbicide Rotations and Mixtures: Effective Strategies to Delay Resistance," in *Managing Resistance to Agrochemicals: From Fundamental Research to Practical Strategies,* ACS Symposium Series 421, M. B. Green, H. M. LeBaron, and W. K. Moberg, Eds. (Washington, D.C.: American Chemical Society, 1990), pp. 430-458.
9. Maxwell, B. D., M. L. Roush, and S. R. Radosevich. " Predicting the Evolution and Dynamics of Herbicide Resistance in Weed Populations," *Weed Technol.* 4:2-13 (1990).
10. Putwain, P. D., and A. M. Mortimer. "The Resistance of Weeds to Herbicides: Rational Approaches for Containment of a Growing Problem," in *Proceedings of the Brighton Crop Protection Conference – Weeds* (Farnham, U.K.: British Crop Protection Council, 1989), pp. 285-294.
11. Gressel, J., and L. A. Segel. "Modelling the Effectiveness of Herbicide Rotations and Mixtures as Strategies to Delay or Preclude Resistance," *Weed Technol.* 4:186-198 (1990).
12. Maxwell, B. D., M. L. Roush, and S. R. Radosevich. "Prevention and Management of Herbicide Resistant Weeds," in *Proceedings of the 9th Australian Weed Control Conference* (Adelaide, Australia, 1990), pp. 260-267.
13. Bowyer, J. R., P. Camilleri, and W. F. J. Vermaas. "Photosystem II and Its Interaction with Herbicides," in *Herbicides. Topics in Photosynthesis,* N. R. Baker, and M. P. Percival, Eds. (Amsterdam, Holland: Elsevier, 1991), pp. 27-86.
14. Holt, J. S., S. B. Powles, and J. A. M. Holtum. "Mechanisms and Agronomic Aspects of Herbicide Resistance," *Annu. Rev. Plant Physiol. Plant Mol. Biol.* 44:203-229 (1993).
15. Hirschberg, J., A. Bleeker, D. J. Kyle, L. McIntosh, and C. J. Arntzen. "The Molecular Basis of Triazine-Resistance in Higher Plant Chloroplasts," *Z. Naturforsch.* 39c:412-419 (1984).
16. Souza Machado, V., J. D. Bandeen, G. R. Stephenson, and P. Lavigne. "Uniparental Inheritance of Chloroplast Atrazine Tolerance in *Brassica campestris,*" *Can. J. Plant Sci.* 58:977-981 (1978).
17. Pfister, K., K. E. Steinback, G. Gardner, and C. J. Arntzen. "Photoaffinity Labeling of an Herbicide Receptor Protein in Chloroplast Membranes," *Proc. Natl. Acad. Sci. U.S.A.* 78:981-985 (1981).

18. Bowes, J., A. R. Crofts, and C. J. Arntzen. "Redox Reactions on the Reducing Side of Photosystem II in Chloroplasts with Altered Herbicide Binding Properties," *Arch. Biochim. Biophys.* 200:303-308 (1980).
19. Holt, J. S., A. J. Stemler, and S. R. Radosevich. "Differential Light Responses of Photosynthesis by Triazine-Resistant and Triazine-Susceptible *Senecio vulgaris* Biotypes," *Plant Physiol.* 67:744-748 (1981).
20. Jursinic, P. A., and R. W. Pearcy. "Determination of the Rate Limiting Step for Photosynthesis in a Nearly Isonuclear Rapeseed (*Brassica napus* L.) Biotype Resistant to Atrazine," *Plant Physiol.* 88:1195-1200 (1988).
21. Ort, D., W. H. Ahrens, B. Martin, and E. W. Stoller. "Comparison of Photosynthetic Performance in Triazine-Resistant and Susceptible Biotypes of *Amaranthus hybridus*," *Plant Physiol.* 72:925-930 (1983).
22. van Oorschot, J. L. P., and P. H. van Leeuwen. "Comparison of the Photosynthetic Capacity between Intact Leaves of Triazine-Resistant and -Susceptible Biotypes of Six Weed Species," *Z. Naturforsch.* 39c:440-442 (1984).
23. Hobbs, S. L. A. "Comparison of Photosynthesis in Normal and Triazine-Resistant *Brassica*," *Can. J. Plant Sci.* 67:457-466 (1987).
24. Holt, J. S., S. R. Radosevich, and A. J. Stemler. "Differential Efficiency of Photosynthetic Oxygen Evolution in Flashing Light in Triazine-Resistant and Triazine-Susceptible Biotypes of *Senecio vulgaris* L.," *Biochim. Biophys. Acta* 722:245-255 (1983).
25. Ireland, C. R., A. Telfer, P. S. Covello, N. R. Baker, and J. Barber. "Studies on the Limitations to Photosynthesis in Leaves of the Atrazine-Resistant Mutant of *Senecio vulgaris* L.," *Planta* 173:459-467 (1988).
26. Burke, J. J., R. F. Wilson, and J. R. Swafford. "Characterization of Chloroplasts Isolated from Triazine-Susceptible and Triazine-Resistant Biotypes of *Brassica campestris* L.," *Plant Physiol.* 70:24-29 (1982).
27. De Prado, R., C. Dominguez, I. Rodriguez, and M. Tena. "Photosynthetic Activity and Chloroplast Structural Characteristics in Triazine-Resistant Biotypes of Three Weed Species," *Physiol. Plant.* 84:477-485 (1992).
28. Holt, J. S., and D. P. Goffner. "Altered Leaf Structure and Function in Triazine-Resistant Common Groundsel (*Senecio vulgaris*)," *Plant Physiol.* 79:699-705 (1985).
29. Lemoine, Y., J.-P. Dubacq, G. Zabulon, and J.-M. Ducruet. "Organization of the Photosynthetic Apparatus from Triazine-Resistant and -Susceptible Biotypes of Several Plant Species," *Can. J. Bot.* 64:2999-3007 (1986).
30. Elliott, J. R., and D. R. Peirson. "Growth Analysis of Atrazine-Resistant and Atrazine-Sensitive Biotypes of *Chenopodium album*," *Ann. Bot.* 51:727-739 (1983).
31. Holt, J. S. "Reduced Growth, Competitiveness, and Photosynthetic Efficiency of Triazine-Resistant *Senecio vulgaris* from California," *J. Appl. Ecol.* 25:307-318 (1988).
32. Holt, J. S., and S. R. Radosevich. "Differential Growth of Two Common Groundsel (*Senecio vulgaris*) Biotypes," *Weed Sci.* 31:112-120 (1983).
33. Warwick, S. I. "Differential Growth between and within Triazine-Resistant and Triazine-Susceptible Biotypes of *Senecio vulgaris* L.," *Weed Res.* 20:299-303 (1980).
34. Zanin, G., and M. Lucchin. "Comparative Growth and Population Dynamics of Triazine-Resistant and Susceptible Biotypes of *Solanum nigrum* L. in Relation to Maize Cultivation," *J. Genet. Breed.* 44:207-216 (1990).

35. Weaver, S. E., and S. I. Warwick. "Competitive Relationships between Atrazine Resistant and Susceptible Populations of *Amaranthus retroflexus* and *A. powellii* from Southern Ontario," *New Phytol.* 92:131-139 (1982).
36. Conard, S. G., and S. R. Radosevich. "Ecological Fitness of *Senecio vulgaris* and *Amaranthus retroflexus* Biotypes Susceptible or Resistant to Atrazine," *J. Appl. Ecol.* 16:171-177 (1979).
37. Warwick, S. I., and L. Black. "The Relative Competitiveness of Atrazine Susceptible and Resistant Populations of *Chenopodium album* and *C. strictum*," *Can. J. Bot.* 59:689-693 (1981).
38. Jansen, M. A. K., J. H. Hobe, J. C. Wesselius, and J. J. S. van Rensen. "Comparison of Photosynthetic Activity and Growth Performance in Triazine-Resistant and Susceptible Biotypes of *Chenopodium album*," *Physiol. Veg.* 24:475-484 (1986).
39. Schonfeld, M., T. Yaacoby, O. Michael, and B. Rubin. "Triazine Resistance without Reduced Vigor in *Phalaris paradoxa*," *Plant Physiol.* 83:329-333 (1987)
40. Weaver, S. E., S. I. Warwick, and B. K. Thomson. "Comparative Growth and Atrazine Response of Resistant and Susceptible Populations of *Amaranthus* from Southern Ontario," *J. Appl. Ecol.* 19:611-620 (1982).
41. Warwick, S. I., and P. B. Marriage. "Geographical Variation in Populations of *Chenopodium album* Resistant and Susceptible to Atrazine. I. Between- and Within-Population Variation in Growth and Response to Atrazine," *Can. J. Bot.* 60:483-493 (1982).
42. Warwick, S. I., and P. B. Marriage. "Geographical Variation in Populations of *Chenopodium album* Resistant and Susceptible to Atrazine. II. Photoperiod and Reciprocal Transplant Studies," *Can. J. Bot.* 60: 494-504 (1982).
43. McCloskey, W. B., and J. S. Holt. "Triazine Resistance in *Senecio vulgaris* Parental and Nearly Isonuclear Backcrossed Biotypes is Correlated with Reduced Productivity," *Plant Physiol.* 92:954-962 (1990).
44. Beversdorf, W. D., D. J. Hume, and M. J. Donnelly-Vanderloo. "Agronomic Performance of Triazine-Resistant and Susceptible Reciprocal Spring Canola Hybrids," *Crop Sci.* 28: 932-934 (1988).
45. Darmency, H., and J. Pernes. "Agronomic Performance of a Triazine Resistant Foxtail Millet (*Setaria italica* (L.) Beauv.)," *Weed Res.* 29:147-150 (1989).
46. Gressel, J., and G. Ben-Sinai. "Low Intraspecific Competitive Fitness in a Triazine Resistant, Nearly Nuclear-Isogenic Line of *Brassica napus*," *Plant Sci.* 38:29-32 (1985).
47. Stowe, A. E., and J. S. Holt "Comparison of Triazine-Resistant and -Susceptible Biotypes of *Senecio vulgaris* and Their F1 Hybrids," *Plant Physiol.* 87:183-189 (1988).
48. van Oorschot, J. L. P., and P. H. van Leeuwen. "Photosynthetic Capacity of Intact Leaves of Resistant and Susceptible Cultivars of *Brassica napus* L. to Atrazine," *Weed Res.* 29:29-32 (1989).
49. Darmency, H., and J. Gasquez. "Differential Temperature-Dependence of the Hill Activity of Isolated Chloroplasts from Triazine Resistant and Susceptible Biotypes of *Polygonum lapathifolium* L.," *Plant Sci. Lett.* 24:39-44 (1982).
50. Ducruet, J.-M., and Y. Lemoine. "Increased Heat Sensitivity of the Photosynthetic Apparatus in Triazine-Resistant Biotypes from Different Plant Species," *Plant Cell Physiol.* 26:419-429 (1985).
51. Fuks, B., F. van Eycken, and R. Lannoye. "Tolerance of Triazine-Resistant and Susceptible Biotypes of Three Weeds to Heat Stress: a Fluorescence Study," *Weed Res.* 32:9-17 (1992).

52. Havaux, M. "Comparison of Atrazine-Resistant and -Susceptible Biotypes of *Senecio vulgaris* L.: Effects of High and Low Temperatures on the *in vivo* Photosynthetic Electron Transfer in Intact Leaves," *J. Exp. Bot.* 40:849-854 (1989).
53. Polos, E., G. Laskay, Z. Szigeti, Sz. Pataki, and E. Lehoczki. "Photosynthetic Properties and Cross-Resistance to Some Urea Herbicides of Triazine-Resistant *Conyza canadensis* Cronq (L.)," *Z. Naturforsch.* 42c:783-793 (1987).
54. Ducruet, J.-M., and D. R. Ort. "Enhanced Susceptibility of Photosynthesis to High Leaf Temperature in Triazine-Resistant *Solanum nigrum* L. Evidence for Photosystem II D1 Protein Site of Action," *Plant Sci.* 56:39-48 (1988).
55. Jacobs, B. F., J. H. Duesing, J. Antonovics, and D. T. Patterson. "Growth Performance of Triazine-Resistant and -Susceptible Biotypes of *Solanum nigrum* over a Range of Temperatures," *Can. J. Bot.* 66:847-850 (1988).
56. McCloskey, W. B., and J. S. Holt. "Effect of Growth Temperature on Biomass Production of Nearly Isonuclear Triazine-Resistant and -Susceptible Common Groundsel (*Senecio vulgaris* L.)," *Plant Cell Environ.* 14:699-705 (1991).
57. Vencill, W. K., C. L. Foy, and D. M. Orcutt. "Effects of Temperature on Triazine-Resistant Weed Biotypes," *Environ. Exp. Bot.* 27:473-480 (1987).
58. Hart, J. J., and Stemler, A. "Similar Photosynthetic Performance in Low Light-Grown Isonuclear Triazine-Resistant and -Susceptible *Brassica napus* L.," *Plant Physiol.* 94: 1295-1300 (1990).
59. Hart, J. J., S. R. Radosevich, and A. Stemler. "Influence of Light Intensity on Growth of Triazine-Resistant Rapeseed (*Brassica napus*)," *Weed Res.* 32:349-356 (1992).
60. Hart, J. J., and A. Stemler. "High Light-Induced Reduction and Low Light-Enhanced Recovery of Photon Yield in Triazine-Resistant *Brassica napus* L.," *Plant Physiol.* 94: 1301-1307 (1990).
61. Gasquez, J., H. Darmency, and J. P. Compoint. "Comparaison de la Germination et de la Croissance de Biotypes Sensibles et Resistants aux Triazines chez Quartre Especes de Mauvaises Herbes," *Weed Res.* 21:219-225 (1981).
62. Mapplebeck, L. R., V. S. Machado, and B. Grodzinski. "Seed Germination and Seedling Growth Characteristics of Atrazine-Susceptible and Resistant Biotypes of *Brassica campestris*," *Can. J. Plant Sci.* 62:733-739 (1982).
63. Weaver, S. E., and A. G. Thomas. "Germination Responses to Temperature of Atrazine-Resistant and -Susceptible Biotypes of Two Pigweed (*Amaranthus*) Species," *Weed Sci.* 34:865-870 (1986).
64. Watson, D., A. M. Mortimer, and P. D. Putwain. "The Seed Bank Dynamics of Triazine Resistant and Susceptible Biotypes of *Senecio vulgaris* – Implications for Control Strategies," in *Proceedings of the British Crop Protection Conference – Weeds* (Farnham, U.K.: British Crop Protection Council, 1987), pp. 917-924.
65. Fuerst, E. P., and K. C. Vaughn. "Mechanisms of Paraquat Resistance," *Weed Technol.* 4:150-156 (1990).
66. Powles, S. B., E. S. Tucker, and T. W. Morgan. "Eradication of Paraquat-Resistant *Hordeum glaucum* Steud. by Prevention of Seed Production for 3 Years," *Weed Res.* 32:207-211 (1992).
67. Morgan, T. W., E. S. Tucker, and S. B. Powles, unpublished results (1993).
68. Tucker, E. S., and S. B. Powles, unpublished results (1993).
69. Purba, E., C. Preston, and S. B. Powles, unpublished results (1993).

70. Itoh, K., and S. Matsunaka. "Parapatric Differentiation of Paraquat Resistant Biotypes in Some Compositae Species," in *Biological Approaches and Evolutionary Trends in Plants*, S. Kawano, Ed. (San Diego: Academic Press, 1990), pp. 33-49.
71. Szigeti, Z., and E. Lehoczki. "P700 Measurements on Different Paraquat-Resistant Horseweed Biotypes," *Plant Physiol. Biochem.* 30:115-118 (1992).
72. Vaughn, K. C., and S. O. Duke. "Ultrastructural Alterations to Chloroplasts in Triazine-Resistant Weed Biotypes," *Physiol. Plant.* 62:510-520 (1984).
73. Preston, C. and S. B. Powles, unpublished results, (1993).
74. Ghersa, C. M., M. A. Ghersa, M. L. Roush, and S. R. Radosevich. "Fitness Studies of Italian Ryegrass Resistant to Diclofop-Methyl: 3. Selection Pressures for Resistance and Germination Time," in *Proceedings of the Western Society of Weed Science* (Newark, CA: WSWS, 1991), p. 42.
75. Roush, M. L., C. M. Ghersa, M. A. Ghersa, and S. R. Radosevich. "Fitness Studies of Italian Ryegrass Resistant to Diclofop-Methyl: 1. Seedling Demography and Plant Growth," in *Proceedings of the Western Society of Weed Science* (Newark, CA: WSWS, 1991), p. 41.
76. Ghersa, C. M., M. A. Ghersa, M. L. Roush, and S. R. Radosevich. "Fitness Studies of Italian Ryegrass Resistant to Diclofop-Methyl: 2. Pollen Phenology and Gene Flow," in *Proceedings of the Western Society of Weed Science* (Newark, CA: WSWS, 1991), p. 42.
77. Roush, M. L., unpublished results, (1993).
78. Marshall, G., R. C. Kirkwood, and G. E. Leach. "Studies on the Biology and Control of Graminicide-Resistant and Susceptible Biotypes of Goosegrass (*Eleusine indica* L. Gaertn.)," in *Proceedings of the Weed Science Society of America* (Champaign, IL: WSSA, 1993), pp. 62.
79. Roush, M. L., C. M. Ghersa, M. A. Ghersa, T. Brewer, and S. R. Radosevich, unpublished results (1993).
80. Hall, L. M., and M. Devine. "Cross-resistance of a Chlorsulfuron-resistant Biotype of *Stellaria media* to a Triazolopyrimidine Herbicide," *Plant Physiol.* 93:962-966 (1990).
81. Saari, L., J. C. Cotterman, W. F. Smith, and M. M. Primiani. "Sulfonylurea Herbicide Resistance in Common Chickweed, Perennial Ryegrass, and Russian Thistle," *Pestic. Biochem. Physiol.* 42:110-118 (1992).
82. Swanson, E. B., M. J. Herrgessell, M. Arnoldo, D. W. Sippell, and R. S. C. Wong. "Microscope Mutagenesis and Selection: Canola Plants with Field Tolerance to the Imidazolinones," *Theor. Appl. Genet.* 78:525-530 (1989).
83. Newhouse, K., T. Wang, and P. Anderson. "Imidazolinone-Tolerant Crops," in *The Imidazolinone Herbicides*, D. L. Shaner and S. L. O'Connor, Eds. (Ann Arbor, MI: CRC Press, 1991), pp. 139-150.
84. Anderson, P. C., and M. Georgeson. "Herbicide-Tolerant Mutants of Corn," *Genome* 31:994-999 (1989).
85. Alcocer-Ruthling, M., D. C. Thill, and B. Shafii. "Seed Biology of Sulfonylurea-Resistant and -Susceptible Biotypes of Prickly Lettuce (*Lactuca seriola*)," *Weed Technol.* 6:858-864 (1992).
86. Mallory-Smith, C. A., D. C. Thill, M. Alcocer-Ruthling, and C. R. Thompson. "Growth Comparisons of Sulfonylurea Resistant and Susceptible Biotypes," in *Proceedings of the First International Weed Control Congress*, Vol. 2 (Melbourne, Australia: Weed Science Society of Victoria, 1992), pp. 301-303.

87. Dyer, W. E., P. W. Chee, and P. K. Fay. "Low Temperature Seed Germination Characteristics of Sulfonylurea Herbicide-Resistant *Kochia scoparia* L. Accessions," in *Proceedings of the Western Society of Weed Science* (Newark, CA: WSWS, 1992), pp. 117-118.
88. Dyer, W. E., P. W. Chee, and P. K. Fay. "Rapid Germination of Sulfonylurea-Resistant *Kochia scoparia* L. Accessions is Associated with Elevated Seed Levels of Branched Chain Amino Acids," *Weed Sci.* 41:18-22 (1993).
89. Alcocer-Ruthling, M., D. C. Thill, and B. Shafii. "Differential Competitiveness of Sulfonylurea Resistant and Susceptible Prickly Lettuce (*Lactuca serriola*)," *Weed Technol.* 6:303-309 (1992).
90. Thompson, C. R., and D. C. Thill. "Sulfonylurea Herbicide-Resistant and -Susceptible Kochia (*Kochia scoparia* L. Schrad.) Growth Rate and Seed Production," in *Proceedings of the Weed Science Society of America* (Champaign, IL: WSSA, 1992), p. 44.
91. Christoffoleti, P. J., and P. Westra. "Fitness and Ecological Adaptability of Chlorsulfuron Resistant and Susceptible *Kochia* Biotypes," in *Proceedings of the Western Society of Weed Science* (Newark, CA: WSWS, 1991), pp. 81.
92. Christoffoleti, P. J., and P. Westra. "Competition and Coexistence of Sulfonylurea Resistant and Susceptible Kochia (*Kochia scoparia*) Biotypes in Unstable Environments," in *Proceedings of the Weed Science Society of America* (Champaign, IL: WSSA, 1992), p. 17.
93. Thompson, C. R. and D. C. Thill, unpublished results (1993).
94. Vaughn, K. C., and M. A. Vaughan. "Dinitroaniline Resistance in *Eleusine indica* May Be Due to Hyperstabilized Microtubules," in *Herbicide Resistance in Weeds and Crops*, J. C. Caseley, G. W. Cussans, and R. K. Atkin, Eds. (Oxford, England: Butterworths-Heinemann, Ltd., 1991), pp. 177-186.
95. Murphy, T. R., B. J. Gossett, and J. E. Toler. "Growth and Development of Dinitroaniline-Susceptible and -Resistant Goosegrass (*Eleusine indica*) Biotypes Under Noncompetitive Conditions," *Weed Sci.* 34:704-710 (1986).
96. Valverde, B. E., S. R. Radosevich, and A. P. Appleby. "Growth and Competitive Ability of Dinitroaniline-Herbicide Resistant and Susceptible Goosegrass (*Eleusine indica*) Biotypes," in *Proceedings of the Western Society of Weed Science* (Newark, CA: WSWS, 1988), 41:81.
97. Heap, I. M., and R. Knight. "Variations in Herbicide Cross-Resistance Among Populations of Annual Ryegrass (*Lolium rigidum*) Resistant to Diclofop-Methyl," *Aust. J. Agric. Res.* 41:121-128 (1990).
98. Holtum, J. A. M., and S. B. Powles. "Annual Ryegrass: An Abundance of Resistance, A Plethora of Mechanisms," in *Brighton Crop Protection Conf. – Weeds* (Farnham, U.K.: British Crop Protection Council, 1991), pp. 1071-1078.
99. Matthews, J. M., and S. B. Powles, unpublished results, (1993).
100. Matthews, J. M., and S. B. Powles. "Integrated Weed Management for the Control of Herbicide-Resistant Annual Ryegrass, *Lolium rigidum* (Gaud.)," *Pestic. Sci.* 34:365-377 (1992).
101. Maxwell, B. D. "Weed Thresholds: The Space Component and Considerations for Herbicide Resistance," *Weed Technol.* 6:205-212 (1992).
102. Roush, M. L., S. R. Radosevich, and B. D. Maxwell. "Future Outlook for Herbicide-Resistance Research," *Weed Technol.* 4: 208-214 (1990).

CHAPTER 12

Management of Herbicide Resistant Weed Populations

J. M. Matthews

INTRODUCTION

Cropping Systems and Herbicide Resistance

There are many benefits associated with the use of herbicides including the control of weeds in industrial situations, in perennial crops and control of noxious weeds. The impact of herbicides, however, has been greatest in short-season, arable crop production systems. Herbicides generally give more consistent weed control compared to tillage in many environments and consequently, the use of both nonselective herbicides for weed control prior to crop establishment and selective herbicides for in-crop weed control is increasing.[1] In Australia, North America, and many other countries, herbicide technology has facilitated earlier planting of short-season crops and improved weed control in many cropping systems. Selective herbicides are those herbicides generally used for the suppression of growth of certain plant species within another species (usually a crop), whereas nonselective herbicides kill or inhibit the growth of all actively growing plant material.

With the development of chemical crop protection and the increasing availability of effective selective herbicides, monocultures of crops have become common, which has led to repeated applications of the same or similar herbicides to monoculture crops.[2] In conservation or zero tillage crop establishment systems, cultivation for weed control has largely been replaced by the use of selective and nonselective herbicides. Thus, two potent factors for the development of herbicide resistance are present in these cropping systems, the frequent use of a limited range of effective herbicides and reliance upon these herbicides to the exclusion of other forms of weed control. Under crop regimes where these conditions prevail, resistance will occur if there is heritable variability in response to herbicide application in weed populations and selective mortality from the herbicides.

As outlined in previous chapters, cases of resistance have developed to nearly all of the selective herbicides used in cereal and small grain production. When weed populations become sufficiently enriched with resistant biotypes such that they cannot be controlled by the usual rate of herbicide and the weed burden causes, or threatens, loss of crop production, then changes in weed control techniques must be implemented. This chapter will focus on the difficulties of herbicide use on multiple and cross resistant weeds and review some weed control techniques that are appropriate to herbicide-resistant weed populations.

Fitness and Control of Resistant Weeds

The relative fitness of R and S biotypes has been studied for some of the resistant species. Although many weed species resistant to the triazine herbicides are less "fit" than their susceptible counterparts, this is not generally the case with weed biotypes resistant to the selective herbicides of cereal crops (see Chapter 11). Unless there is a substantial fitness penalty, resistant biotypes will not be eliminated from a population by selective mortality in the absence of herbicide application within an economically viable period of time. Also, it will not be possible to return resistant populations to susceptibility unless there are substantial and exploitable differences between S and R biotypes. The absence of such differences suggests that management of resistant populations will become a permanent aspect of crop production on infested sites.

Under these conditions, successful field management of herbicide-resistant weed populations depends upon the integration of alternative weed control methods. Available control strategies include the scope for effective herbicide use, exploitation of unique aspects of the biology of the weed species, and manipulating the cropping system to maximize both chemical and nonchemical weed control. With changing economic pressures and land use ethics, control of herbicide-resistant weeds may not be as simple as returning to the traditional methods, such as frequent cultivations, employed prior to the introduction of selective herbicides.

Distribution of Resistant Weed Biotypes of Small Grain Crops

Weed species from temperate small grain cropping areas for which resistant populations have documented are listed in Table 1. The numbers of species that

Table 1. Distribution of herbicide-resistant weeds reported in small grain crops

Weed species	Approximate number of documented sites			Resistance status[a]	Ref.
	Australia	N. America	Europe		
Alopecurus myosuroides			51	M, NTCR	3,4
Arctotheca calendula	1			S	5
Avena fatua	9			S	6
Avena fatua		>50		S, TCR, NTCR	7, 8, 9
Avena sterilis	10			TCR	10
Hordeum glaucum	5			S	11
Hordeum leporinum	4			S	12
Kochia scoparia		>75		TCR	13
Lactuca serriola		10		TCR	14
Lolium rigidum	>2,000			M, TCR, NTCR	15, 16
Lolium multiflorum		100		TCR	17
Setaria viridis		>45		S, M, TCR	18, 19
Sinapsis arvensis	5			TCR	20
Sonchus oleraceus	4			TCR	21
Sisymbrium orientale	5			TCR	21
Vulpia bromoides	1			S	22

[a]S, single resistance; M, multiple resistance; TCR, target site cross resistance; NTCR, non-target site cross resistance.

have resistant populations are listed, as are the number of sites or reports for each species. Triazine resistance in coarse grain crops is not included.

Of the species listed in Table 1 that have developed herbicide-resistant biotypes, *Lolium rigidum* (L.) Gaud. is probably the most important due to the number of sites with resistance, the rate of onset of resistance, and the variability of cross and multiple resistance encountered. Other economically important weed species with herbicide-resistant biotypes are *Avena fatua* L., *Avena sterilis* L. ssp. *ludoviciana, Lolium multiflorum,* Lam., *Setaria viridis* (L.) Beauv., and *Kochia scoparia* (L.) Schrad. Although both *Alopecurus myosuroides* Huds. and *Sinapis arvensis* L. have complex resistance patterns they are of limited distribution. Herbicide-resistant *Arctotheca calendula* (L.) Levyns, *Hordeum glaucum* Steud., *Hordeum leporinum* Link, *Lactuca serriola* L., *Sonchus oleraceus* L., and *Vulpia bromoides* (L.) S. F. Gray. are of limited distribution and have little economic or agronomic importance at this time.

HERBICIDE USE ON RESISTANT POPULATIONS

When faced with an initial failure of herbicide performance in the field due to herbicide resistance, the alternative herbicide options are usually the first management consideration. This is understandable in successful crop production systems which rely upon selective herbicides for post-planting weed control. The success and range of alternative herbicides depends upon the resistance spectrum or the extent of multiple or cross resistance of the resistant population.

Resistance to One Herbicide Group

Resistant biotypes listed in Table 1 which are resistant only to a single class of herbicide should not be difficult to control. Successful management of these resistant biotypes can be implemented by using an alternative herbicide to which there is no resistance. However, total reliance on herbicides for weed control must be avoided if selection for further resistance is to be minimized. The propensity of these biotypes for development of sequential resistance as a result of the persistent use of any alternative herbicides must be considered.

Resistance to More than One Herbicide Group

The majority of herbicide-resistant weed populations listed in Table 1 exhibit resistance to more than one herbicide group. Many resistant weed biotypes exhibit resistance to some or all of the herbicides from the chemical group(s) or herbicidal modes of action that were used in the selection process. For example, the *L. serriola*[14] and *K. scoparia*[13] populations from the northwest of the U.S. were selected with chlorsulfuron and have target site cross resistance to many acetolactate synthase (ALS) inhibitors (see Chapter 4). Resistant *Avena* spp. from

Australia and North America with target site cross resistance to cyclohexanedione (CHD) herbicides following selection with an aryloxyphenoxypropionate (APP) herbicide have been reported.[9,10] Many of the resistant *L. rigidum* populations in Australia display target site cross resistance as well as non-target site cross resistance simultaneously (see Chapter 9).

With non-target site cross resistant biotypes, resistance due to the use of one herbicide or class of herbicide has caused resistance to other unrelated chemical groups and modes of herbicide action. Many resistant *L. rigidum* populations,[15,16] *A. myosuroides* populations,[3] and some biotypes of the *Avena* spp.[7,8] display this type of resistance. Cross resistance and multiple resistance are the most difficult types of resistance from a field management viewpoint, although all resistant weed biotypes with "complex resistance," or resistance to more than one herbicide, are potentially difficult to manage.

Herbicide Use on Populations With Complex Resistance Patterns

The practical significance of both cross resistance and multiple resistance in weed populations is that many potential herbicides may be ineffective when applied to resistant populations, thereby severely limiting herbicide choice. This is the case with herbicide-resistant *L. rigidum*, *A. myosuroides,* and one biotype of *A. fatua* from Canada. Examples are discussed in Chapter 9 of this volume. The spectrum and the level of resistance of *L. rigidum* populations are extremely variable which increases the difficulty of choosing effective, alternative herbicides. Effective choices of herbicides for use on cross resistant and multiple resistant weeds are only possible with knowledge of the complete extent of resistance in a population.

If alternative herbicides are used on a resistant population and do not achieve a high level of weed mortality because a proportion of the population is resistant to the alternative herbicide, the surviving plants will contribute seed with a more complex resistance spectrum to the seed bank, thus confounding the prospects for control in the future. With allogamous species of high fecundity, such as *L. rigidum*, the contribution of even a small proportion of multiple or cross resistant survivors is important. When the intrinsic rate of increase of a species is added to the spread of resistance via pollen flow, the result is a rapid shift of the population to a more complex resistance spectrum, especially if resistance traits are genetically dominant. There should be an urgent approach to controlling herbicide-resistant populations of allogamous species as the rate of spread of resistance genes via pollen distribution from survivors of herbicide applications to any receptive plants is likely to be high (see Chapter 1). As discussed by Powles and Matthews,[24] if the species is genetically variable for resistance traits, the aggregation of multiple resistance mechanisms into the survivors will rapidly complicate the prospects for future herbicide use.

The random presence of genes conferring resistance and the response within a population to differing selection pressures are presumed to account for the variability of the herbicide resistance spectrum. The almost concurrent development of cross and multiple resistance in many hundreds of separate *L. rigidum* populations preclude migration or gene flow between widely separated populations as an important factor in widespread resistance development. However, for successful management of resistant *L. rigidum* populations, intra-population gene flow should be minimized. The same prospects for development of complex resistance apply to all allogamous resistant weed populations.

In contrast, only a few of the resistant *Avena* populations so far described display complex resistance. As discussed in Chapter 5, a few North American populations exhibit non-target site cross resistance; some *Avena* populations which have developed resistance to the herbicide triallate have been reported to display a low level of cross resistance to difenzoquat.[7,8,23] Also, Thai et al.[25] reported variability of response to triallate in populations of *A. fatua* exposed to triallate. One of these populations had a low level of cross resistance to diclofop-methyl. More recently, resistant *Avena* populations from both North America and Australia selected with ACCase inhibiting herbicides have been described which display target site cross resistance to CHD herbicides.[9,10] Devine et al.[26] reported a non-target site resistance mechanism in an APP- and CHD-resistant *A. fatua* biotype (see Chapter 5), and a similar observation was made with biotypes of multiple and cross resistant *L. rigidum*.[27]

The genus *Avena* clearly has the capacity to display both non-target site cross resistance to some chemical groups as well as target site cross resistance following exposure to some herbicides. These early observations of complex resistance in *Avena* spp. should serve as a warning and indicate that there is a high possibility of further resistance with continued herbicide use in these species. Populations of resistant *Avena* spp. which are subject to continuing herbicide selection pressures will almost certainly develop more complex patterns of resistance and the onset of such resistance should be monitored frequently. However, due to lower fecundity and preferential autogamy in the *Avena* spp., complex resistance will take longer to develop than for species such as *L. rigidum*.

Herbicide Application Rates and Management of Resistance

An important aspect of successful herbicide usage on weed populations is the application rate of the alternative herbicides and the use of herbicide mixtures. The effect of dose on the rate of initial selection for resistance has been discussed in Chapter 1. Similar conditions may be expected to apply to the selection of sequential resistance. Maxwell and Mortimer (Chapter 1) suggest that lower herbicide doses allow the accumulation of resistance alleles in the surviving population. At low herbicide rates there can be lower weed mortality, meaning that individuals expressing relatively weak resistance mechanisms will survive and contribute to the resistance gene pool. Additionally, at low herbicide rates, both heterozygous and homozygous resistant individuals are likely to survive.

However, when herbicides are applied at higher rates, heterozygous resistant individuals and individuals possessing only weak resistance mechanisms may be killed. Hence, higher herbicide doses result in higher weed mortality and less diversity of resistant genes in the surviving population. Herbicide application rate, therefore, affects both the size and the resistance gene spectrum of a population undergoing selection. For example, selection of a *L. rigidum* population for 3 consecutive years with a high rate of a CHD herbicide resulted in a resistant ACCase mutant with no other mechanisms evident in the population.[28] Conversely, hundreds of *L. rigidum* populations were selected with relatively low rates of diclofop methyl and a range of different resistance mechanisms have been selected both within and between populations (Chapter 9). Increased herbicide doses are not likely to be a viable strategy to overcome established resistance as model simulation shows the use of maximum dose rates only limit the increase of resistance for a short period.[29]

As previously mentioned, the mating behavior of a species is important in resistance development. With allogamous species, resistance alleles may spread rapidly throughout the population by cross pollination, whereas with autogamous weed species the surviving plants will rarely acquire resistance mechanisms from other survivors because there is limited gene flow between plants. Thus, survivors from an allogamous species can disperse resistance genes to any receptive individual within effective pollination range, whereas the autogamous species disperse genes mainly through seed increase.

Herbicide Mixtures and Management of Resistance

The use of mixtures of herbicides with different modes of action for control of herbicide-resistant weed populations has not often been addressed on a theoretical basis. It is recognized that the likelihood of failure with the use of mixtures depends on the presence in individuals of genes conferring resistance to all mixture components.[30] The likelihood of target site resistance developing in a population to a mixture of herbicides is the mathematical product of the frequency of any genes conferring resistance to the mixture components. Mixtures can, therefore, be an effective tool in resistance management of autogamous species, because individual plants bearing multiple mutations will be extremely rare and gene flow between surviving plants is low. However, mixtures will also select for any mechanism(s) that could confer resistance to all the components, including non-target site mechanisms. This has already been documented with a *L. rigidum* population selected for 10 years with a herbicide mixture which resulted in a resistant biotype[31] expressing metabolism-based non-target site cross resistance mechanism, rather than target site alterations (Chapter 9). In general, when mixtures are used, the components should span different target sites yet have approximately the same active life in the field to minimize differential selection by a residual component.

From the above discussion, it is clear that when alternative herbicides are applied to an already resistant population, continuing herbicide effectiveness depends on: (1) the mating behavior of the species; (2) the herbicide-induced mortality, of which dose rate is the major factor; and (3) the frequency of genes conferring resistance mechanisms to the alternative herbicides. The latter information is usually not known; however, the likelihood of the presence of resistance mechanisms in a population can be inferred by past experience with a herbicide or a particular weed species. For example, since resistance to auxin analog herbicides occurs at a low frequency (see Chapter 6), the probability of developing resistance to these herbicides in populations of *K. scoparia* resistant to ALS inhibiting herbicides is low. However, by comparison, the risk of encountering resistance to ALS inhibiting herbicides in populations of *L. rigidum* resistant to diclofop-methyl is extremely high (Chapter 9).

Prolonging the useful life of herbicides is a sound objective of resistance management and the development of complex resistance patterns in weed populations due to inappropriate choice of alternative selective herbicides should be avoided. For successful field control of herbicide-resistant populations it is important to identify which modes of herbicide action are rendered either totally or partially ineffective and therefore which chemical groups cannot be used. Herbicide resistance testing to establish the resistance spectrum is important as an aid for management of field populations. Users are often not familiar with the grouping of herbicides into chemical modes of action and container labels and product information may not show the mode of action, chemical group, or resistance risk. This increases the difficulty of appropriate herbicide selection.

MANAGEMENT OF RESISTANT POPULATIONS

Seed Bank Life

With fewer effective herbicide options available because of resistance, methods for control of resistant populations must focus on longer-term weed management techniques. In many intensive cropping systems the simplest management option may involve techniques to reduce weed populations such that expected weed numbers are below the economic threshold. This can be achieved by introducing periods of land use in the crop rotation where resistant weeds are prevented from producing seed. The optimum land use during such periods will be governed by local factors, including economic considerations, but the duration and effectiveness of such a change in rotations is dependent upon the seed bank or propagule life of the resistant weed species. Knowledge of the seed bank dynamics is crucial as the life of the seed or propagules in the soil will dictate the length of time that control measures need to be maintained to reduce the reservoir of resistant seed in the seed bank to a low level.

L. rigidum has a seed bank life of 3 to 4 years with an annual rate of decline of about 80%.[32] Fresh seed is strongly summer dormant with germination and emergence of the majority of seeds occurring in the year after seed fall. As is typical of many annual grasses, the germination of *L. rigidum* in Australia mostly occurs in late autumn following sufficient rainfall, with a small proportion germinating throughout the growing season.[32] Cultivation has a limited effect on reducing primary dormancy in *L. rigidum*.[32] Similarly, *L. multiflorum* has a seed bank life of about 3 years[33] and *S. viridis* has a nominal 3- to 5-year seed bank life with variable germination rates depending on temperature and soil moisture.[34] *K. scoparia* has a short seed bank life, about 1 to 2 years under normal field conditions,[35] as does *L. serriola*.[36]

Different germination rates between R and S biotypes have been observed in two *K. scoparia* biotypes resistant to ALS inhibitors which germinated faster than a susceptible accession at cool temperatures.[37] If this phenomenon is consistent across all resistant *K. scoparia* biotypes, there may be opportunities for selective control of the R biotypes. A similar phenomenon was noted for *L. serriola* resistant to ALS inhibitors.[37]

The persistence of *Avena* spp. due to short- and long-term dormancy and the environmental conditions promoting germination and seedling emergence have been widely discussed.[38] Wilson and Peters[39] suggest a 50% annual decline for fresh seeds and up to a 90% annual decline for older seeds under European conditions. In northern Australia, Martin and Felton reported a decline of 80% in the first year.[40] Similar rates of seed bank decline were noted in North American studies,[41] although genetic variability for dormancy between differing environments and management regimes was apparent.[42] However, most studies indicate the viable population in the seed bank can be substantially reduced by preventing seed set for 3 to 5 years.[41,42] Occasional soil disturbance with shallow cultivation in winter or early spring when temperatures are optimum for the germination of the *Avena* spp. has been shown to stimulate emergence.[43] Inevitably, there will be a small percentage of viable seeds remaining and persistence for at least 9 years has been reported.[44]

Successful Exploitation of Seed Bank Exhaustion

As the annual rate of seed bank decline of herbicide-resistant annual grass weeds ranges between 50 to 80%, prevention of seed production is an effective method to reduce the number of viable seeds in the soil.[45] In Australia, a common method of reducing seed production of herbicide-resistant *L. rigidum* on crop land involves rotations with planted pasture used for direct grazing with sheep or cattle. Control of *L. rigidum* and other weeds by grazing with livestock is effective, although variable. The application of low rates of paraquat or glyphosate at or immediately following anthesis is a cost-effective technique to prevent the development of fertile seed in many weed species.[46,47] The combination of livestock grazing and the application of low rates of a nonselective herbicide can greatly reduce viable seed production resulting in a rapid decline of the seed bank population if sustained for a 2- to 3-year period.

Except for Australia, direct grazing by livestock on arable land is not widely practiced; however, the prevention of seed production by early harvest of forage crops for hay or silage can be equally successful in limiting seed return to the seed bank. Wilson and Phipps[48] reported that 3 years of spring barley removed as silage was effective in eliminating emergence of *Avena* spp. The most suitable weed species for this management approach are those species which do not shed seed early, although with adequate forage density and early cutting many annual weeds could be controlled in this way. Soil disturbance during the planting of forage species may also stimulate germination of some weed species. Due to the light requirement for germination, undisturbed pasture is not considered effective for reducing *Avena* spp. seed burden in the soil.[44] The use of chemical or ploughed fallow can be an effective method if regrowth is controlled as seed return to the soil is totally prevented.

Burning dry undisturbed pastures or crop residues after harvest has been shown to be an effective method of destroying weed seed. Reductions of up to 99% of *L. rigidum* seeds still retained in the floret have been documented depending on the temperature and the duration of the fire.[49] The density of recently shed *Avena* spp. seed can be reduced by stubble burning, although the results of trials have been variable.[50] Legislative restrictions to burning and concerns about soil degradation have limited the use of stubble burning in many areas.

When land is taken out of crop production for a period of time to reduce the seed bank, strategies must be considered to limit the population increase from the remaining few seeds when cropping is resumed. Most of the weed species that have developed resistance have high fecundity and unchecked growth and seed production by a few remaining individuals can result in a rapid return to high weed densities. Knowledge of the herbicide resistance spectrum of remaining seeds and choice of crop species to maximize the effectiveness of any alternative herbicides used are essential to prevent a rapid increase in weed density. Such a program is a logical integration of nonselective control with any useful selective herbicides. In addition to alternative herbicides, the planting of vigorous and competitive crop swards will reduce the growth rate of weeds and therefore limit seed increase.[51]

Limiting Dispersal of Resistant Weed Seed

Weed seed dispersal mechanisms are varied and can influence the rate of spread of resistance. Seeds of some species can be rapidly disseminated and can cause serious and unexpected control failures. *K. scoparia* is an outstanding example in which the wind-driven dispersal of mature plants can disseminate seeds over a considerable distance and disperse resistant seeds into uninfested areas (Chapter 4). Other species, such as *A. myosuroides* or *S. viridis*, disperse seed over extremely limited areas. There are no reported differences between dispersal of seed from either resistant or susceptible biotypes.

The spread of resistant weed seeds via contaminated equipment is always a risk and strict hygiene standards should be enforced to minimize this eventuality. Harvesting and grain handling equipment are the usual sources of introduced weed seeds, but seeds can be dispersed by adhering to spraying equipment and other machinery. The introduction of resistant weed seeds at planting time from contaminated seed reserves retained from the previous harvest is also a high risk. As the number of sites and the proportion of resistance individuals increase, the possibility that mature weeds present at harvest are resistant also increases. Regulations covering crop seed for sale may not totally exclude foreign seeds and there are no requirements that the herbicide resistance status of contaminating seeds must be declared.

All of the major weed species displaying resistant biotypes are annual species and depend upon frequent replenishment of the seed bank for continuing occupation of a site. Seeds of some of these species are not shed at maturity but are retained on the standing plant. The retention of seed on the plant until harvest means that the weed is cut with the crop and the seed processed through the harvesting equipment. In most modern harvesting machinery the weed seeds are immediately dispersed back onto the field during the harvesting process and continue to reinfest the site.[52] Collection and removal from the field of all weed seeds processed through the harvester during the harvesting process can dramatically reduce reinfestation.

Removal of weed seeds during the harvesting operation is gaining acceptance in Australia for management of herbicide-resistant *L. rigidum*. Mature *L. rigidum* does not shed seed at maturity and under normal conditions mature seed remains in the floret (unless dislodged by severe weather conditions). About 70 to 80% of the *L. rigidum* seed present at harvest time can be processed through a harvester and retained instead of being dispersed from the harvest machinery with the crop residues.[53,54] There are two devices that are proving commercially successful depending upon the amount of crop residues generated: (1) a trailer drawn behind the harvester and equipped to receive the crop residues containing the weed seeds; (2) harvester modification to place an additional separating sieve within the harvester which removes any seeds or grains from the winnowing air stream and retains them in a container on the harvester. The collected weed seeds can be disposed of without reinfesting the field.

Collection of weed seed will also prevent crop grains from being returned to the field where they can be a "weed" in the following crop. This may reduce the need to apply herbicides to remove volunteer crop plants in the subsequent season. The application of herbicides to remove volunteer crop plants can account for a substantial proportion of herbicide application costs and also maintains herbicide selection pressure on non-target weed species. Removal of weed seed during harvest operations may reduce the seed dispersal of some herbicide-resistant species in addition to *L. rigidum*. For example, seed of species such as *S. arvensis* and *S. orientale* are retained in the silique until dislodged by impact and can be captured in the harvesting process. Seeds of other intractable weed species still remaining at harvest time may be suitable for removal during the harvesting process.

This method of reducing weed seed reinfestation may not be successful for the resistant *Avena* spp. Seed of *A. fatua* and *A. sterilis* are shed at maturity so many of the earlier maturing seeds fall to the ground before harvest.[55] Similarly, seeds of *A. myosuroides*[56] and *S. viridis*[15] usually fall prior to crop harvest. Windrowing crops prior to weed seed fall may increase the amount of *Avena* seed retained during the harvest operation, but as the usual fate of seeds of *Avena* spp. at harvest is to be carried with the cereal grains, the cost of recleaning *Avena*-contaminated grain must be evaluated against the potential benefits.

The continuing importance of *Avena* spp. as weeds of crops worldwide indicates the problems inherent with current control methods. The propensity for an extended period of emergence contributes to the lack of success of early in-crop control by herbicides. Medd[57] suggests *Avena* control must change from a short-term to a planned, longer-term approach and must focus on limiting seed production. The use of selective herbicides for late season control of wild oats in cereal crops has been shown to reduce viable seed output.[57] Also, the use of nonselective herbicides applied to maturing weeds in legume crops may become an important strategy for limiting seed production of late maturing weed species, especially in these crop species.

Influence of the Cropping System

The cropping system is important in the management of herbicide resistance. With the current emphasis on continuous cereal cropping, minimal cultivation and the frequent use of similar herbicides, it is no surprise that resistance has emerged in the common weeds of cereal crops. The adoption of minimum tillage and retention of crop residues has increased farmer reliance upon selective herbicides for in-crop weed control. With declining efficacy of selective herbicides due to resistance this reliance will have to change. The increasing presence of *A. myosuroides* in reduced tillage systems in the U.K. and the subsequent development of resistance due to reliance on selective herbicide use, is an example. The management of resistant *A. myosuroides* may require a return to cultivation and reduction in the use of the ACCase inhibitor herbicides while the populations are still limited in distribution and in size (see Clarke and Moss[58] for a review of control measures).

The range of crop species that can be grown will influence both the onset and the control of herbicide-resistant weeds. Some crop species or varieties are more competitive than others and will limit the growth and reduce the fecundity of weed species.[51,59,60] The introduction of alternative crops into a crop rotation may give the opportunity to change herbicides, to alter the herbicide application rate, or change other weed control techniques. Such tactics may be effective in reducing the population density of herbicide-resistant weeds.

A shift during the course of a rotation to late season crops, where possible, will reduce selective herbicide use in winter crops and provide an opportunity for control by nonselective means. This strategy has proved successful in the control of *A. sterilis* in the northern wheat belt of Australia where growing crops in different seasons than the principal weeds can give opportunities for weed control and thereby reduce weed densities.[40]

The possibility of using mixed crops should also be explored. Because of more complete occupation of the available niches, mixtures of crop species can display improved competitiveness towards weeds. Legumes grown in combination with barley or canola can suppress more weeds than when either crop is grown alone.[61] Innovations such as these may improve nonherbicidal methods of weed suppression.

Identifying Resistance In Field Populations

Prompt identification of the resistance status of surviving weeds before the seed bank becomes enriched with resistant seed is an important aspect of resistance management. If herbicide efficacy is declining and if the weed species and herbicides involved have been implicated in resistance at other sites then it is possible that resistance may be the cause. It should be noted that resistant weeds, as for weeds in general, are not usually spread evenly across the whole field in the early stages of resistance development. Patches of resistant weeds around a site of origin of resistance can be expected and if resistance is detected early, the size of the infested area can be restricted. This is especially so with autogamous weed species. Prevention of weed seed set in the "resistant patches" by nonselective herbicides or by cutting is preferable to allowing the whole field to become infested with a large population of herbicide-resistant weed seeds. Under most field conditions the potential seed set of the weed population on maturing plants exceeds the actual resistant seed population in the soil seed bank many times over.

Identification of the extent of resistance and the spectrum of resistance can be achieved with the use of herbicide test plots in the field or from laboratory-based testing services. Field plots can confirm or exclude resistance to some herbicides in the same season the problem is first noticed. Collection of mature seed allows herbicide resistance testing to assess the effectiveness of herbicides applied to seedlings grown from populations suspected of resistance. Laboratory tests can be performed to identify the resistance spectrum and to identify effective substitute herbicides. The accuracy of the test results depends on good sampling techniques to ensure that the sample being submitted for testing is representative of the area involved. Test results need to be interpreted with a knowledge of the emergence pattern of the weed, the likely proportion of R and S seeds in the soil, and the likely efficacy of chosen herbicides in the field.

INTEGRATED WEED MANAGEMENT (IWM)

The mounting incidence of herbicide-resistant weed biotypes and the increasingly complex resistance spectrum of many biotypes indicate the inevitability of herbicide-resistant weeds becoming a limiting factor to crop production in many areas. When these conditions develop, farm managers and weed control professionals will have to adopt a stategic or long-term approach to weed control, incorporating the use of any appropriate methods of weed control with new developments in methods of population containment. These techniques can be combined with the continued use of any effective herbicides in an IWM program.

Techniques to reduce the weed seed bank which can be incorporated into existing crop rotations are

1. Pasture or forage production
2. Fallow periods using nonselective herbicides or cultivation
3. Periods of green manure or plough down

The following will reduce weed populations in crops:

1. Varying planting dates
2. The use of cultivation or nonselective herbicides prior to crop establishment
3. Establishing competitive crops

The reduction of weed seed infestation in some species may be achieved by:

1. Removal of weed seed during the harvesting process
2. Burning weed residues containing seed,
3. Application of selective or nonselective herbicides to late developing weeds in crop

The integration of these methods, where suitable, into crop rotations, coupled with the use of effective alternative herbicides, will aid in reducing the herbicide-resistant weed burden. The particular resistant weed species, local conditions, and economic considerations will dictate the extent of adoption of integrated weed management techniques. Resistant *L. rigidum,* in which the rapidity of the onset of resistance and the wide spectrum of resistance has rendered most selective herbicides ineffective, has forced many farmers to adopt an integated approach to weed management.

HUMAN FACTORS AND RESISTANCE MANAGEMENT

Successful control of herbicide-resistant weed populations in the field can be complicated and requires both management skills and an understanding of the biology of the weeds. The ability to correctly identify the onset and extent of a problem, and a willingness to adopt the inevitable changes in crop management that accompany herbicide resistance, can be challenging, especially when the existing herbicide regime was part of a successful crop production system.

Farm managers may have varied objectives when faced with herbicide-resistant weed populations but usually it is to maintain production of existing crops. However, successful control of herbicide-resistant weeds may require changes in management emphasis or objectives, with the introduction of different crops or production systems. As herbicide resistance is a biological response to reliance upon herbicides, changing the farming system is a logical remedial step but can require complex management skills to implement in the field.

SUMMARY

Herbicide resistance in weeds of arable crop production systems are being reported in an increasing number of species and locations. Replacing tillage with selective herbicides for short-term weed control and the intensification of crop rotations with continual herbicide applications are responsible for this phenomenon. Reduced tillage systems of crop production have many benefits but where herbicide resistance has developed these benefits have to be balanced with nonselective weed control measures for control of the herbicide-resistant weeds.

When herbicide-resistant weeds occur at sufficient densities to limit crop production, changes to farm management practices are essential. The necessary changes depend upon the resistance status of the specific weed population. If herbicide use is continued, the likelihood of the development of further resistance must be taken into account.

The selection of resistant biotypes by repeated herbicide applications and the difficulties of controlling resistant biotypes in current farming systems are compelling reasons to introduce more diverse weed control techniques (IWM). The planned use of nonselective herbicides, delayed planting or out of season planting, judicious cultivation, and more complexity in crop rotations may be long-term solutions to the problem of herbicide resistance. The further integration of nonselective weed control methods targetted at specific exploitable characteristics of the principal weeds is a logical adaption to local needs.

Perhaps a positive outcome from the problems of controlling herbicide-resistant weeds is the realization that no single weed control technique is likely to provide continued control when used exclusively. Weed control on arable land should be a long-term strategic process involving a range of management techniques.

REFERENCES

1. Shear, G. "Introduction and History of Limited Tillage," in *Weed Control in Limited Tillage Systems*, A. F. Wiese, Ed. (Champaign, IL: Weed Science Society of America, 1985), pp. 1-14.

2. Wicks, G. A. " Weed Control in Conservation Tillage Systems — Small Grains," in *Weed Control in Limited Tillage Systems*, A. F. Wiese, Ed. (Champaign, IL: Weed Science Society of America, 1985), pp. 77- 92.
3. Moss, S. R. "Herbicide Cross Resistance in Slender Foxtail (*Alopecurus myosuroides*)," *Weed Sci.* 38:492-496 (1990).
4. Hall, L. M., S. R. Moss, and S. B. Powles. "Towards an Understanding of Resistance to APP Herbicides in *Alopecurus myosuroides*," Weed Management — Towards Tomorrow. 10th Council of Australian Weeds Science Societies and 14th Asian Pacific Weeds Science Society, Brisbane, Queensland (1993) p. 4.
5. Powles, S. B., E. S. Tucker, and T. R. Morgan. "A Capeweed (*Arctotheca calendula*) Biotype in Australia Resistant to Bipyridyl Herbicides," *Weed Sci.* 37:60-62 (1989).
6. Boutsalis, P. B., J. A. M. Holtum, and S. P. Powles. "Diclofop-methyl Resistance in a Biotype of Wild Oats (*Avena fatua* L.)," in *Proceedings 9th Australian Weeds Conference*, J. W. Heap Ed. (Adelaide, S.A., Crop Science Society of South Australia, 1990) p. 216.
7. Morrison, I. N., Heap, I. M., and Murray, B. 1992. "Herbicide Resistance in Wild Oat — The Canadian Experience," *Proceedings of the 4th International Oat Conference*, Vol 2. *Wild Oats in World Agriculture*, A. R. Barr, and R. W. Medd, Eds. (Adelaide, S. A. Robee Bureau Services, 1992) pp 36 - 41.
8. Malchow, W. E., P. K. Fay, B. D. Maxwell, and W. E. Dyer,"Wild Oat (*Avena fatua* L.) Resistance to Triallate in Montana," *Abstracts, Vol. 33, Proceedings of the 1993 Meeting of the Weed Science Society of America*, (Champaign, IL: WSSA, 1993), p. 18.
9. Seefeldt, S. S., D. R. Gealy, and B. D. Brewster. "Diclofop-methyl resistant Wild Oat (*Avena fatua* L.) in the Willamette Valley of Oregon: Distribution, Case Histories and Cross Resistance to Other Herbicides," *Abstracts, Vol. 33, Proceedings of the 1993 Meeting of the Weed Science Society of America*, (Champaign, IL: WSSA, 1993), p. 19.
10. Mansooji, A. M., J. A. M. Holtum, P. Boutsalis, J. M. Matthews, and S. B. Powles. "Resistance to Aryloxyphenoxypropionate Herbicides in Two Wild Oats Species (*Avena fatua* and *Avena sterilis* ssp. *ludoviciana*)," *Weed Sci.*, 40:599-605 (1992).
11. Powles, S. B. "Appearance of a Biotype of the Weed *Hordeum glaucum* Steud., Resistant to the Herbicide Paraquat," *Weed Res.* 26:167-172 (1986).
12. Tucker, E. S ., and S. B. Powles. "A Biotype of Hare Barley (*Hordeum leporinum*) Resistant to Paraquat and Diquat," *Weed Sci.* 39:159-162 (1991).
13. Primiani, M. M., J. C. Cotterman, and L. L. Saari. "Resistance of Kochia (*Kochia scoparia*) to Sulfonylurea and Imidazolinone Herbicides," *Weed Technol.* 4:169-172 (1990).
14. Mallory-Smith, C. A., D. C. Thill, and M. J. Dial. "Identification of Sulfonylurea Herbicide Resistant Prickly Lettuce (*Lactuca serriola*)," *Weed Technol.* 4:163-168 (1990).
15. Heap, I., and R. J. Knight. "The Occurrence of Herbicide Cross Resistance in a Population of Annual Ryegrass (*Lolium rigidum*) Resistant to Diclofop-methyl," *Aust. J. Agric. Res.* 37:239-247 (1986).
16. Powles S. B., and P. D. Howat. "Herbicide Resistant Weeds in Australia," *Weed Technol.* 4:178-185 (1990).
17. Stanger, C. E., and A. P. Appleby. "Italian Ryegrass (*Lolium multiflorum*) accessions Tolerant To Diclofop," *Weed Sci.* 37:350 (1989).

18. Morrison, I. N., B. G. Todd, and K. M. Nawolsky. "Confirmation of Trifluralin Resistant Green Foxtail (*Setaria viridis*) in Manitoba," *Weed Technol.* 3:544-551 (1989).
19. Heap, I. M., and I. N. Morrison. "Resistance to Aryloxyphenoxypropionate and Cyclohexanedione Herbicides in Green Foxtail (*Setaria viridis* (L.) Beauv.)," *Abstracts Vol. 33, Proceedings of the 1993 Meeting of the Weed Science Society of America*, (Champaign, IL: WSSA, 1993), p. 62.
20. Heap, I. M. and I. N. Morrison. "Resistance to Auxin Type Herbicides in Wild Mustard (*Sinapsis arvensis* L.)," *Abstracts, Vol. 32, Proceedings of the 1993 Meeting of the Weed Science Society of America*, (Champaign, IL: WSSA, 1992), p. 55.
21. Boutsalis, P. B. and S. B. Powles. University of Adelaide, South Australia. Unpublished data.
22. Purba, E., C. Preston, and S. B. Powles. "Paraquat Resistance In a Biotype Of *Vulpia bromoides* (L.) S. F. Gray," *Weed Res.* 33:409-413 (1993).
23. O'Donovan, J. T., M. P. Sharma, J. C. Newman, and H. Feddema, "Triallate Resistance Wild Oats (*Avena fatua*) are Cross resistant to Difenzoquat," *Abstracts, Vol. 32, Proceedings of the 1992 Meeting of the Weed Science Society of America*, (Champaign, IL: WSSA, 1992), p. 89.
24. Powles, S. B., and J. M. Matthews. "Multiple Herbicide Resistance in Annual Ryegrass *Lolium rigidum*: A Driving Force for the Adoption of Integrated Weed Management," in *Resistance 91, Achievements and Developments in Combating Pesticide Resistance*, I. Denholm, A. L. Devonshire, and D. W. Hollomon, Eds. (London, Elsevier Science Publishers, 1991) pp. 75-88.
25. Thai, K. M., S. Jana, and J. M. Naylor. "Variability for response to Herbicides in Wild Oat (*Avena fatua*) populations," *Weed Sci.* 33:829-835 (1985).
26. Devine, M. D., J. C. Hall, M. L. Romano, M. A. S. Marles, L. W. Thomson, and R. S. Shimabukuro. "Diclofop and Fenoxaprop Resistance in Wild Oats Is Associated with an Altered Affect on the Plasma Membrane Electrogenic Potential," *Pestic. Biochem. Physiol.*, 45:167-177 (1993).
27. Häusler, R. E., J. A. M. Holtum, and S. B. Powles."Cross Resistance to Herbicides in Annual Ryegrass (*Lolium rigidum*). IV. Correlation between Membrane Efeects and Resistance to Graminicides," *Plant Physiol.* 97:1035-1043 (1991).
28. Tardiff, F. J., J. A. M. Holtum, and S. B. Powles. "Occurrence of a Herbicide Resistant Acetyl-coenzyme A Carboxylase Mutant in Annual Ryegrass (*Lolium rigidum*) Selected by Sethoxydim," *Planta* 190:176-181 (1993).
29. Roush, R., J. M. Matthews, and S. B. Powles, Unpublished.
30. Gressel, J. "Why Get Resistance? It can be Delayed or Prevented." in *Herbicide Resistance in Weeds and Crops*, J. C. Caseley, G. W. Cussans, and R. K. Atkin, Eds. (Oxford, Butterworths-Heinemann, 1991) pp. 1-27.
31. Burnet, M. W. M., O. B. Hildebrand, J. A. M. Holtum, and S. B. Powles. "Amitrole, Triazine, Substituted Urea and Metribuzin Resistance in a Biotype of Rigid Ryegrass (*Lolium rigidum*)," *Weed Sci.* 39:3170-323 (1991).
32. Heap, I. M., "Herbicide Resistance in *Lolium rigidum*." Ph. D. Thesis, University of Adelaide, South Australia (1988).
33. Appleby, A. Oregon State University, Oregon. Personal communication.
34. Douglas, B. J., A, G. Thomas, I. N. Morrison, and M. G. Maw. "The Biology of Canadian Weeds. 70. *Setaria viridis* (L.) Beauv," *Can. J. Plant Sci.* 65:669-690 (1985).

35. Zorner, P. S., R. L. Zimdahl, and E. E. Schweizer,. "Effect of Depth and Duration of Seed Burial on *Kochia scoparia*," *Weed Sci.* 32: 602-607 (1984).
36. Alcocer-Ruthling, M., D. C. Thill, and B. Shafi. "Seed Biology of Sulfonylurea Resistant and Susceptible Biotypes of Prickly Lettuce (*Lactuca serriola*)," *Weed Sci.* 6:858-864 (1992).
37. Dyer, W. E., P. E. Chee, and P. K. Fay. "Rapid Germination of Sulfonylurea Resistant *Kochia scoparia* (L.) Accessions is Associated with Elevated Seed Levels of Branched Chain Amino Acids," *Weed Sci.* 41:18-22 (1993).
38. Chancellor, R. J. "Growth and Development of Wild Oat Plants," in *Wild Oats in World Agriculture*, D. Price-Jones, Ed. (London: Ministry of Agriculture, 1976) pp. 65-66.
39. Wilson, B. J., and N. C. B. Peters. "Biological and Agronomic Reasons for the Continuing Importance of Wild Oats in the United Kingdom." *Proceedings of the 4th International Oat Conference, Vol 2. Wild Oats in World Agriculture*, A. R. Barr, and R. W. Medd, Eds. (Adelaide, S. A., Robee Bureau Services, 1992) pp. 9-14.
40. Martin, R. J., and W. L. Felton. "Effect of Crop Rotation, Tillage Practice and Herbicide Use on the Population Dynamics of Wild Oats," *Proceedings 9th Australian Weeds Conference*, J. W. Heap, Ed., (Adelaide, S. A., Crop Science Society of South Australia, 1990) pp 20-23.
41. Zorner, P. S., R. L. Zimdahl, and E. E. Schweizer. "Sources of Viable Seed Loss in Buried Dormant and Non-dormant Populations of Wild Oat (*Avena fatua* L.) Seed in Colorado," *Weed Res.* 24:143-150 (1984).
42. Miller, S. D., J. D. Nalewaja, and C. E. G. Mulder. "Morphological and Physiological Variation in Wild Oat," *Agron. J.* 74:771-775 (1982).
43. Thurston, J. M. "The Effect of Depth of Burying and Frequency of Cultivation on Survival and Germination of Seeds of Wild Oats (*Avena fatua* L. and *Avena ludoviciana* Dur.)," *Weed Res.* 1:19-31 (1961).
44. Chancellor, R. J. "Seed Behaviour," in *Wild Oats in World Agriculture*, D. Price-Jones, Ed. (London: Agricultural Research Council, 1976) pp. 65-78.
45. Powles, S. B., E. S. Tucker, and T. R. Morgan. "Eradication of Paraquat Resistant *Hordeum glaucum* Steud. by Prevention of Seed Production for 3 Years," *Weed Res.* 32:207-211 (1992).
46. Leys, A. R., B. R. Cullis, and B. Plater. "Effect of Spraytopping Applications of Paraquat and Glyphosate on the Nutritive Value and Regeneration of *Vulpia bromoides* (L.) S. F. Gray," *Aust. J. Agric. Res.* 42:1405-1415 (1991).
47. Jones, S. M., W. M. Blowes, P. England, and P. K. Fraser. "Pasture Topping Using Roundup® Herbicide," *Aust. Weeds* 3:150-151 (1984).
48. Wilson, B. J., and P. A. Phipps, "A Long Term Experiment on Tillage, Rotation and Herbicide Use for the Control of *Avena fatua* in Cereals." *Proceedings of the British Crop Protection Conference — Weeds*. (Farnham, U.K.: The British Crop Protection Council, 1985) pp. 693-700.
49. Reeves, T. G., and I. S. Smith. "Pasture Management and Cultural Methods for the Control of Annual Ryegrass (*Lolium rigidum*) in Wheat," *Aust. J. Exp. Agric. Anim. Husb.* 15:527-530 (1975).
50. Wilson, B. J., and G. W. Cussans. "A Study of the Population Dynamics of *Avena fatua* L. as Influenced by Straw Burning, Seed Shedding and Cultivations," *Weed Res.* 15:249-258 (1975).

51. Cussans, G. W., and B. J. Wilson. "Cultural Control," in *Wild Oats in World Agriculture*, D. Price-Jones, Ed. (London: Agricultural Research Council, 1976) pp. 127-142.
52. Petzold, K. "Combine Harvesting and Weeds," *J. Agric. Eng. Res.* 1:178-181 (1956).
53. Matthews, J. M., and S. B. Powles. University of Adelaide, South Australia. Unpublished.
54. Gill, G. Department of Agriculture, Western Australia. Personal communication.
55. Thurston, J. M. "Wild Oats as Successful Weeds," in *Biology and Ecology of Weeds*, W. Holzner and N. Numata, Eds. (The Hague, Dr. W. Junk Publishers, 1982), pp. 191-199.
56. Moss, S. R. "The Production and Shedding of *Alopecurus myosuroides* Huds. Seed in Winter Cereal Crops," *Weed Res.* 23:45-51 (1983).
57. Medd, R. W. "New Developments in the Control of Wild Oats: Australian Advances," *Proceedings of the 4th International Oat Conference, Vol 2. Wild Oats in World Agriculture*, A. R. Barr, and R. W. Medd, Eds. (Adelaide, S.A., Robee Bureau Services, 1992) pp. 27-34.
58. Clark, J. H., and S. R. Moss. "The Distribution and Control of Herbicide Resistant Blackgrass (*Alopecurus myosuroides*) in Eastern and Central England," *Proceedings of the British Crop Protection Conference — Weeds*. (Farnham, U.K.: The British Crop Protection Council, 1985), pp. 301-308.
59. Evans, R. M., D. C. Thill, L. Tapia, B. Shafi, and J. M. Lish. "Wild Oat (*Avena fatua*) and Spring Barley (*Hordeum vulgare*) Density Affect Spring Barley Grain Yield," *Weed Technol.* 5:33-39 (1991).
60. Medd, R. W., B. A. Auld, D. R. Kemp, and R. D. Morrison. "The Influence of Wheat Density and Spatial Arrangement on Annual Ryegrass (*Lolium rigidum* Gaud.), Competition," *Aust. J. Agric. Res.* 36:361-371 (1985).
61. Mohler, C. L., and M. Liebman, "Weed Productivity and Composition in Sole Crops and Intercrops of Barley and Field Pea," *J. Appl. Ecol.* 24:685-699 (1987).

INDEX

A

ABP, see Auxin-binding proteins
Abscisic acid, 160, 189
Absorption
   acetolactate synthase herbicide resistance, 108–111
   acetyl coenzyme A carboxylase herbicides, 157
   auxin analog herbicides, 179, 193, 194, 201
   bipyridyl herbicides, 67
ABT, see 1-Aminobenzotriazole
Abutilon theophrasti, 28, 41–42, 216, 275, 276
Acc1 gene, 161, see also Acetyl coenzyme A carboxylase herbicides
ACCase, see Acetyl coenzyme A carboxylase herbicides
Acetate uptake, 147, 149–151
Acetohydroxyacid synthase, see Acetolactate synthase herbicides
Acetolactate synthase gene, 87, 105–106
Acetolactate synthase (ALS) herbicides
   biology and genetics, 117–124
   characteristics of resistant weeds, 306–307
   chemistry and use in agriculture, 84–86
   crop resistance, 116–117
   cross resistance, 95
   evolution and distribution, 89–93
   growth and competition in resistant weeds, 307–308
   Mendelian inheritance of resistance, 280–281, 283
   mode of actions, 86–87
   modeling and resistance, 93–95
   mutation rate of resistance in *Kochia scoparia*, 6
   non-target site cross resistance
      Lolium rigidum, 248
      metabolism, 111–116
      uptake and translocation, 108–111
   overview, 83–84
   resistance evaluation, 87–89
   seeds in resistant weeds, 307–309
   summary, 124–125
   target site cross resistance, 244, 246–247
   target site resistance
      binding domains, 107–108
      laboratory mutants, 101–102
      mutations resulting in resistance, 102–105
      overexpression and double mutants, 105–106
      weeds, 96–101
Acetyl coenzyme A carboxylase (ACCase) herbicides
   agricultural use, 142–143
   chemistry, 141–142
   conclusions of study, 162
   evolution and development of resistance, 154–156
   genetics and biology of resistant plants, 161
   Mendelian inheritance of resistance, 281–282
   mode of action, 144–152
   non-target site cross resistance
      Alopecurus myoduroides, 255
      Lolium rigidum, 247–248, 249, 252
   phenology and growth characteristics of resistant weeds, 305–306
   resistance mechanisms in weeds, 156–161
   seed germination characteristics of resistant weeds, 305
   selectivity mechanisms in resistant crops, 152–154
   target site cross resistance, 244, 246
Acid growth theory, 149
Acid trapping, 175
Acidification, 147, 149, 151
Adenosine triphosphatases (ATPases), 173
Aerobacter aerogenes, 230
Age, plant, 38, 179, 287
Agrobacterium spp., 13, 176, 192
Agroecology, 299–301
Alachlor, 220, see also Dinitroaniline herbicides
Alanine mutations, 102, 231
Alcaligenes eutrophus, 192
Alfalfa, see *Medicago sativa*
Algae, 37, 286
Aliphatic hydroxylation, 111, 113
Alkalinization, 149, 151
Allogamous species, 244, see also *Alopecurus myosuroides*; *Lolium rigidum*
Alopecurus myocuroides
   acetolactate synthase herbicide resistance, 114, 115
   dinitroaniline herbicide resistance, 218
   distribution in small grain crops, 319, 320
   multiple resistance, 256

natural selection and herbicide resistance, 6
non-target site cross resistance, 254–256
overview of resistance mechanisms, 253–254
photosystem II herbicide resistance, 28, 44–45, 271, 275, 276
resistance spreading, 16
target site cross resistance, 254
ALS, see Acetolactate synthase
ALS I_{50} ratios, 96–101, see also Acetolactate synthase herbicides
Amaranthus bouchonii, 266, 268
Amaranthus hybridus, 265
Amaranthus palmeri, 218, 220
Amaranthus powellii, 302
Amaranthus retroflexus, 109, 286, 302, 303
Amaranthus spp., 66, 180, 302
Amidases, 180
Amide formation, 180–181
Amino acids, aromatic, 230, see also Glyphosate
Amino acids, branched chain, 86
1-Aminobenzotriazole (ABT), 34, 43, 49, 196–197
Aminomethylphosphonate (AMPA), 233, 235, see also Glyphosate
Amiprophosmethyl, 220, 244
Amitrole, 248
Anilides, 28
Anisomycin, 178
Anthoxanthumodoratum sp., 11
Anthroquinone, 62
APP, see Aryloxyphenocypropanoates
Arabidopsis thaliana
    acetolactate synthase herbicides
        chlorsulfuron resistance in seed, 94
        cross resistance, 98
        gene mutations and resistance, 101–104, 117
    auxin resistance and root geotropism, 189
    glyphosate resistance, 235
    Mendelian inheritance of herbicide resistance, 283
    target site cross resistance, 245
Arctotheca calendula
    distribution in small grain crops, 319, 320
    Mendelian inheritance of herbicide resistance, 276, 280
    paraquat resistance, 66, 69, 70, 73, 74
    seedbank and photosystem I herbicide resistance, 304
aroA gene, 231, 232, 235
Arthrobacter sp., 233
Aryl hydroxylation

acetolactate synthase herbicide resistance in crops, 111–113, 248
auxin analog herbicides, 182
diclofop-methyl, 153
photosystem II herbicides, 33, 45
Aryl oxidation, 157–158
Aryloxyphenocypropanoates (APP), see also Acetyl coenzyme A carboxylase herbicides
    chemistry, 141–142
    mode of action, 151, 152
    resistance
        Alopecurus myosuroides, 45, 254
        Lolium rigidum, 43
    target site cross resistance, 244, 246, 247
Ascorbate-glutathione cycle, 63
Ascorbate peroxidase, 71
Assortative mating, 11
ATPases, see Adenosine triphosphatases
Atrazine, see also Photosystem II herbicides
    activity, 32
    inheritance of resistance
        maternal, 267–268
        Mendelian, 276, 277
        quantitative, 273
    plastoquinone interaction, 31–32
    point mutations and resistance, 36–37
    resistance
        Chenopodium album, 41
        Abutilon theophrasti, 41–42
        Lolium rigidum, 43
        velvetleaf, 115
        transfer to crops, 45
        Zea mays, 33, 34
*aux*1 genes, 189
Auxin analog herbicides
    case studies of resistance mechanisms, 189–202
    chemical structure, 171
    discussion, 202–204
    modes of action, 171–178
    resistance, 184–189
    usage, selectivity, metabolism, 178–184
Auxin-binding proteins (ABP), 175, 201
Auxin compounds, 144, 149, 151, 160, see also Acetyl coenzyme A carboxylase herbicides
Auxin overdose, 174
Avena barbata, 6
Avena fatua
    acetyl coenzyme A carboxylase herbicide resistance, 148, 153–156, 158, 161
    distribution in small grain crops, 319, 320
    herbicide resistance frequency, 6

INDEX 339

imazamethabenz-methyl sensitivity, 113
membrane response recovery, 249–250
quantitative inheritance to thiocarbamate herbicides, 270–272, 274
Avena sativa, 161, 148, 277
Avena spp., 325, 328
Avena sterilis
    acetyl coenzyme A carboxylase herbicide resistance, 155, 158, 161
    distribution in small grain crops, 319, 320
    Mendelian inheritance of herbicide resistance, 276, 281
*axr*1/2 genes, 189

B

Bacteria, 88, 231, 233–234, see also Glyphosate
bar gene, 116
Barban
    Mendelian inheritance of resistance, 277, 282
    non-target cross resistance, 256
    quantitative inheritance, 270, 272
    resistance frequency in weeds, 6
Barley, see *Hordeum vulgare*
Beet, see *Beta vulgare*
Benazolin, 172
Bensulfuron, 273, 277, 278
Bentazon, 36, 278
Benzyl viologen, 62
Beta vulgare, 182, 183
Binding domains, 107–108
Biomass production
    acetolactate synthase herbicide resistance, 121, 307
    acetyl coenzyme A carboxylase herbicide resistance, 161, 306
    dinitroaniline herbicide resistance, 222
    photosystem II herbicide resistance, 37, 302, 303
Biotin carboxylation, 147
Bipyridyl herbicides, 62–65, see also Individual entries
Birdsfoot trefoil, see *Lotus corniculatus*
Biscarbamates, 28, 31
Brassica campestris, 45, 265, 266, 270, 303
Brassica juncea, 109
Brassica napus
    acetolactate synthase herbicide resistance, 104, 116, 117
    atrazine resistance, 45
    fitness, 120, 287, 306
    *psb*A gene mutation, 38–40

genetic transfer and resistance evolution rate, 13
inheritance of resistance, 270, 283
triazine resistance, 264–266, 303
Brassica spp., 302
Bromoxynil, 6, 46, 47, 116
Bromus inermis, 182
Brown mustard, see *Brassica juncea*
Burning (field), 326, see also Seed bank
Butachlor, 277
bxn gene, 116

C

Cadaverine, 69
Calcium, 176, 223
Canola, see *Brassica napus*
Capsicum annuum, 277
Carbamate herbicides, 45, 220, 256, see also Photosystem II herbicides
Carbetamide, 256
Carbon, 38, 39, 230
Carbon dioxide fixation, 32, 37, 67–69, 302, see also Paraquat; Photosystem II herbicides
Carbonyl cyanide-m-chlorophenyl hydrazone (CCCP), 148–149
Carboxylesterase, 153
Cardaria arvensis, 190
Cardaria chalepensis, 185, 190
Carduus nutans, 6, 184, 190
Carrot, see *Daucus carota*
Catalase, 71
Catalytic competency, 121, 236
Catharanthus roseus, 231
Cauliflower mosaic virus, 106, 235
CCCP, see Carbonyl cyanide-m-chlorophenyl hydrazone
Cell walls, 69, 174
Ceratoptensis richardii, 66, 74
Cereals, 142, see also Individual entries
Cerulenin, 145, 151
Chara spp., 148
CHD, see Cyclohexanediones
Chemical fallow, 326, see also Seed bank
Chemical hoeing, 230, see also Glyphosate
Chenopodium album
    atrazine resistance, 41
    auxin analog herbicide resistance, 185
    mutation rate and resistance inheritance, 286
    photosystem II herbicide resistance and fitness, 302
    seed germination and herbicide resistance, 303

thifensulfuron-methyl sensitivity, 109
triazine resistance, 181, 265, 266, 268, 269
triclopyr sensitivity, 181
Chenopodium polyspermum, 266, 268
Chicory, see *Cichorium intybus*
Chlamydomonas reinhardtii
  acetolactate synthase gene mutations, 101, 117
  dinitroaniline herbicide resistance, 218
  paraquat resistance, 66, 67, 71, 73
  photosystem II herbicide resistance, 37
Chlamydomonas spp., 225–226
Chloramben, see also Auxin analog herbicides
  chemical structure, 172
  Cucumis sativus resistant biotypes, 187, 201
  detoxification in resistant species, 182, 183
  quantitative inheritance, 273
Chlorimuron-ethyl, 111
Chloroacetanilides, 45
p-Chloromercuribenzenesulfonic acid (PCMBS), 148
2,4-Chlorophenoxyacetic acid, 193, 195
Chlorophyll *a*/*b* ratio, 36, 61, 269, 302
Chlorophyll P680, 29
Chloroplast
  plastome mutator role, 286
  photosystem II herbicide resistance, 36, 264–269, 302
Chlorosis, 144, 152, 157, 174
Chlorpropham (CIPC), 220, 221
Chlorsulfuron, see also Acetolactate synthase herbicides
  crop management, 83
  cross insensitivity and cross resistance patterns, 98–101, 103
  non-target site cross resistance
    Alopecurus myoduroides, 255
    Lolium rigidum, 109–110, 248, 252
    weeds, 114
  residual activity, 94
  target site resistance in *Kochia scoparia*, 87, 89
Chlortoluron, see also Photosystem II herbicides
  detoxification by wheat, 33, 34
  inheritance of resistance
    Mendelian, 276–278, 283
    quantitative, 270, 271
  non-target site cross resistance, 254
  resistance
    Alopecurus myosuroides, 44, 218
    Lolium rigidum, 43

Cichorium intybus, 116, 231
Cinosulfuron, 105
CIPC, see Chlorpropham
Cirsium arvense, 186
Clethodim, 143
Climate, 287
Cline, 11
Clomazone, 273
Clover, see *Trifolium*
Colchicine, 216–217, see also Dinitroaniline herbicides
Commelina diffusa, 185
Compartmentation, see Sequestration
Competition
  acetolactate synthase herbicide resistance, 121, 307–308
  dinitroaniline herbicide resistance, 309
  photosystem I herbicide resistance, 304
  photosystem II herbicide resistance, 302–303
Conifers, 112
Conjugation, 157, see also Acetyl coenzyme A carboxylase herbicides
Convolvulus arvensis
  auxin analog herbicide resistance, 186
  glyphosate resistance, 232–233, 270, 272, 274
Conyza bonariensis
  bipyridyl herbicide absorption, 67
  paraquat resistance, 65, 66
    carbon dioxide fixation, 69
    inheritance, 74
    metabolism, 70–72
    sequestration, 69, 70
    translocation of herbicides and resistance, 67
Conyza canadensis, 304
Conyza spp., 276, 279
Corn, 218
Corydalis sempervirens, 231, 232
Cotton, see *Gossypium hirsutum*
Crassocephalum spp., 66
Crop rotation, 117, 324, 325
Crop selectivity, 111–113, 116–117, see also Acetolactate synthase herbicides
Cropping systems, 317–318, 328–329, see also Weeds, management
Crops, see also Individual entries
  acetolactate synthase herbicide resistance, 84, 85, 89, 116–117
  acetyl coenzyme A carboxylase herbicide resistance, 142–143
  detoxification and herbicide resistance, 28
  dinitroaniline herbicide use, 215

glyphosate resistance, 235–237
inheritance of resistance
   Mendelian, 277, 282–284
   quantitative, 272–274
photosystem II herbicide resistance, 45–46
small grain, distribution of resistant weeds, 318–320
Cross insensitivity
   acetolactate synthase herbicides, 96–101, 103–105
   resistance evaluation, 87
Cross resistance, see also Individual entries
   acetolactate synthase herbicides
      binding domains, 107–108
      non-target site and metabolism, 115–116
      patterns, 87, 95, 98–101, 103, 104
      proline mutations, 103–105
   acetyl coenzyme A carboxylase herbicides, 155, 156
   auxin analog herbicides, 184, 187
   bipyridyl herbicides, 72
   dinitroaniline herbicides, 219
   glycine mutations, 36
   non-target site mechanisms, 254–256
   mechanisms
      Alopecurus myosuroides, 253–256
      overview, 243–245
      Lolium rigidum, 246–252
   protocol for target sites, 88–89
   photosystem II herbicides, 42, 43, 45, 48–49
csr1 gene, 104
Cucumber, see *Cucumis sativus*
Cucumis sativus, 179, 188, 201, 270
Cucurbita moschata, 277
Currants, see *Ribes* spp.
Cyanoacrylates, 28
Cyanobacteria, 37, 286, see also *psb*A gene
Cyclohexanediones (CHD), see also Acetyl coenzyme A carboxylase herbicides
   chemistry, 142, 143
   target site cross resistance, 45, 244, 246, 247, 254
Cycloheximide, 178
Cytochrome b6/f complex, 29, 304
Cytochrome P450 (Cyt P450)
   acetolactate synthase herbicides, 111–113, 116, 248
   acetyl coenzyme A carboxylase herbicides, 153
   auxin analog herbicides, 182
   photosystem II herbicides, 33, 34, 43, 45, 49, 64

Cytoskeleton proteins, 224, see also Dinitroaniline herbicides

D

2,4-D, see also Auxin analog herbicides
   acetyl coenzyme A carboxylase herbicide antagonism, 144, 149, 160
   agricultural use, 178
   chemical structure, 171–172
   Mendelian inheritance of resistance, 277, 278
   mRNA regulation, 177
   O-demethylation, 181
   replacement hormone role, 202
   resistance in
      crops, 184, 187, 191–193
      weeds, 186, 189–190
   selectivity, 178, 182
   sequestration, 183
2,4-DB, 172, see also Auxin analog herbicides
D1 protein, 28, 32, 36, 37, 301, see also Photosystem II herbicides
Dactylis glomerata, 182
Datura innoxia, 101
Datura stramonium, 181
Daucus carota, 184, 216, 231, 232
N-Dealkylation, 33, 43, 248–249
O-Dealkylation, 33, 111, 181
DCPA, 219, 220, see also Dinitroaniline herbicides
Decarboxylation, 181–182, 201
Deesterification, 111, 113, 179
N-Demethylation, 45
Diaminodurene, 62
Dicamba, see also Auxin analog herbicides
   acetyl coenzyme A carboxylase herbicide antagonism, 144, 149
   chemical structure, 172
   excretion from root of soybean, 179
   metabolism in wheat and *Poa annua*, 182
   resistance in *Sinapis arvensis*, 201
   resistance in weeds, 186
Dichloroanisole, 18
Dichlorophenolindophenol, 62
2,4-Dichlorophenoxypropionic acid, 182
3,6-Dichloropicolinic acid, 172
Diclofop, see also Acetyl coenzyme A carboxylase herbicides
   inheritance of resistance, 273, 276, 277, 281
   nucleotide uptake in *Zea mays*, 149
   resistance to R(+) vs S(-) enantiomers, 150, 159

target site sensitivity, 158, 159
transmembrane proton gradient perturbation, 148
Diclofop-methyl, see also Acetyl coenzyme A carboxylase herbicides
auxinic compound antagonism, 157
lipid biosynthesis inhibition, 146
metabolism in plants, 153
non-target site cross resistance in *Alopecurus myoduroides*, 255
resistance in *Lolium rigidum*, 114, 157
non-target site cross resistance, 248, 249, 252
target site cross resistance, 96, 246, 247, 254
triallate and cross resistance, 156
weed control in crops, 143, 154
Diclorprop, 184
Difenzoquat, 271, 277, 283
Differential metabolism theory, 203, see also Auxin analog herbicides; Metabolism
Digitaria sanguinalis, 155
2,4-Dihydroxy-7-methoxy-1,4(2H)-benzoxazin-3–4H)-one (DIMBOA), 33, 34, see also Photosystem II herbicides
Dinitroaniline herbicides
characteristics of resistant weeds, 309
conclusions and prospects, 226
cross resistance, 45
development of resistance, 217–222
genetic studies, 224–226
mechanisms of resistance, 223–224
mode of action, 216–217
overview, 215
selectivity, 216
target site cross resistance, 244
Diquat, 62, 64–66, 72, 304, see also Photosystem I herbicides
Dithiopyr, 220, see also Dinitroaniline herbicides
Diurnal cycle, 38, 39, see also Photosystem II herbicides
Diuron, 32, 37, 43, see also Photosystem II herbicides
Dormancy, 2, 304, 305, 325, see also Seed bank
Double mutants, 105–106, 117
Drought, 72, 160

E

Echinochloa crus galli, 266, 267
Ecological fitness, see Fitness

Electrolyte leakage, 152
Electron transfer
glycine mutation role, 36, 38–40
photosystem II, 28–29
Electron transport, see also Individual entries
photosystem I, 61, 62
photosystem II, 28, 31, 32, 301, 302
Eleusine indica
acetolactate synthase herbicide resistance, 89
acetyl coenzyme A carboxylase herbicide resistance, 155, 161, 306
dinitroaniline herbicide resistance, 217, 218, 222, 224–225, 309
Mendelian inheritance of herbicide resistance, 282
target site cross resistance, 220
Elongation, stem, 174
Elymus repens, 154
Elytrigia repens, 143
Em, see Membrane potential
Emetine, 178
Endoplasmic reticulum, 175
5-Enolpyruvylshikimate-3-phosphate synthase (EPSPS), 230–233, 235, 236
Epilobium ciliatum, 65, 66
Epinasty, 174, 194, 197–198
Equisetum arvense, 234, 236
Erechitites hieracifolia, 185
Erigeron canadensis
herbicide translocation and resistance relation, 67, 73
Mendelian inheritance of herbicide resistance, 276, 279
paraquat resistance, 65, 66, 68–69, 71, 74
Erigeron philadelphicus
herbicide translocation and resistance relation, 67, 73
Mendelian inheritance of resistance, 278
paraquat resistance, 65, 66, 74
Escherichia coli, 101, 102, 104, 231, 235
Esterase, 179
Ethalfluralin, 256
Ethylene, 173, 174, 194, 197–198
Euglena gracilis, 37, 231
Evolution
acetolactate synthase herbicide resistance, 89, 91–94
acetyl coenzyme A carboxylase herbicides, 154–156
population models, 300
requirements for resistance, 3
resistance rate, 8–14
Excretion, herbicide, 201

F

Farming, 6, 156, see also Crops; Weeds, management
Fatty acid biosynthesis, 144–147
Fatty acid synthetase (FAS), 151
Fd, see Ferrodoxin
Fecundity, 17–20, 308, see also Fitness; Seed
Fenaxoprop, 276
Fenchlorazole-ethyl, 154, see also Acetyl coenzyme A carboxylase herbicides; Safener
Fenoxaprop, 281–282, see also Acetyl coenzyme A carboxylase herbicides
Fenoxaprop-ethyl, 143, 154, 157, 254, 255, see also Acetyl coenzyme A carboxylase herbicides
Fenton reaction, 64
Ferredoxin (Fd), 61, see also Photosystem I herbicides
Fescue, 232
Field plots, 329, see also Weeds, management
Fitness
  acetolactate synthase herbicide resistance, 119, 120, 306
  acetyl coenzyme A carboxylase herbicide resistance, 161, 305, 306
  agroecology of herbicide resistance, 299–301
  auxin analog herbicide resistance, 184, 185
  definition, 2
  dinitroaniline herbicide resistance, 222
  photosystem I herbicide resistance, 73–74, 304–305
  photosystem II herbicide resistance, 37–40, 47–48, 269, 287, 302
  selectivity for herbicide resistance, 14–15, 17–20
  shifts and resistance evolution, 3
  weed control, 318
Flavobacterium sp., 234
Flax, see *Linum usitatissimum*
Flowering, 305, 306
Fluazifop, see also Acetyl coenzyme A carboxylase herbicides
  Mendelian inheritance of resistance, 276, 281–282
  nucleotide uptake in *Zea mays,* 149
  target site cross resistance in *Lolium rigidum,* 246, 247
  wheat susceptibility, 153
Fluazifop-butyl, 143, 254, 255, see also Acetyl coenzyme A carboxylase herbicides
Flumetsulam, 112
Fluorescence, 68, 69, 254
Flurmeturon, 220, see also Dinitroaniline herbicides
Fluroxypry, 172
Founder effect, 225, 286
Fungi, 13–14

G

Galium aparine, 181
Gene activation hypothesis, 176
Gene flow
  acetolactate synthase herbicide resistance, 89, 93, 122–124
  frequency of resistance in natural populations, 6
  herbicide resistance evolution rate, 11–14
  population models, 300
Genes, see also Individual entries
  acetolactate synthase herbicide resistance evolution, 94
  auxin role in expression, 176–178
  photosystem II resistance inheritance, 40–41
  selection for herbicide resistance, 4, 5, 9, 13–14
Genetics (of resistance)
  chloroplast and triazine resistance, 264–269
  Mendelian inheritance
    crops, 277, 282–284
    overview, 274–275
    prospects, 284–285
    weeds, 275, 276, 278–282
  origin and dynamics of resistance genes, 285–288
  overview, 263–264
  quantitative inheritance, 269–274
Genotypic fitness, 14, see also Fitness
Germination, see also Seeds
  acetyl coenzyme A carboxylase herbicide resistance, 305
  photosystem I herbicide resistance, 304
  photosystem II herbicide resistance, 42, 303–304, 307
  rates and resistance management, 325
Glucose conjugation, 33, 112, 113
Glutathione, 42, 49
Glutathione reductase, 71, 73, 279
Glutathione *S*-transferase (GST), 33, 34, 115, 275

Glycine mutations, 36, 233
Glycine max
    acetolactate synthase herbicide resistance, 109–113, 116, 117
    acetyl coenzyme A carboxylase herbicide resistance, 153, 154
    auxin analog herbicide resistance, 179, 191, 197
    glyphosate metabolism, 232, 234, 235
    inheritance of resistance, 270, 277, 283
Glycine spp., 192, 233
Glycoside conjugation, 183, 191
Glyphosate
    agricultural use, 229–230
    application and seed bank exhaustion, 325
    chemistry, 229
    engineering resistance, 235–236
    future developments, 236–237
    metabolism, 233–235
    mode of action, 230–231
    quantitative inheritance, 270, 272
    resistance, 231–233
    toxicity, 46
Glyphosine, 229
Goosegrass, see *Eleusine indica*
Gossypium hirsutum
    2,4-D resistance, 192
    cross insensitivity to acetolactate synthase herbicides, 96, 98
    dinitroaniline herbicide use, 216, 217
    bromoxynil resistance, 46
    glyphosate metabolism, 234
Grana stacking, 36, 302
Growth, see also Individual entries
    acetolactate synthase herbicides, 86, 120–121, 307–308
    acetyl coenzyme A carboxylase herbicides, 305–306
    auxin analogs, 173
    dinitroaniline herbicides, 218, 309
    measure of herbicide activity, 269
    photosystem I herbicides, 73–74, 304–305
    photosystem II herbicides, 302–303
GST, see Glutathione *S*-transferase

H

Hairy root, 176
Haloxyfop, see also Acetyl coenzyme A carboxylase herbicides
    Mendelian inheritance of resistance, 276, 278
    target site cross resistance in *Lolium rigidum*, 246, 247

transmembrane proton gradient perturbation, 148
wheat susceptibility, 153
Zea mays resistance, 146, 149, 156, 158
Hardy-Weinberg equilbrium, 7
Harvesting, 326, 327, see also Weeds, management
H^+-ATPase, 147, 148, 149, 151
Herbicides, 320-324
Hill reaction, 28, 254
Homogamous species, 244
Homoglutathione conjugation, 111
Hordeum glaucum
    distribution in small grain crops, 319, 320
    herbicide translocation and resistance relation, 67
    Mendelian inheritance of herbicide resistance, 276, 279
    paraquat resistance, 65, 66, 70, 73–74
    seed bank and photosystem I herbicide resistance, 304
Hordeum jubatum, 275, 276
Hordeum leporinum
    distribution in small grain crops, 319, 320
    herbicide translocation and resistance relation, 67
    Mendelian inheritance of herbicide resistance, 276, 279
    natural selection and herbicide resistance, 6
    paraquat resistance, 66, 73–74
Hordeum spp., 68, 69
Hordeum vulgare, 182, 183, 277, 282
HRA, see *Nicotiana tabacum*
Hydrogen ion pumping, 198–200
Hydrogen peroxide, 63, 64
Hydroxyl radicals, 64, see also Radicals
Hydroxylation, 112, 113, 190, see also Individual entries
Hydroxymethyl mecoprop, 196, 198
Hygiene standards, 326–327, see also Weeds, management
Hyperpolarization, 147, 149
Hypersensitivity, 101–102

I

IAA, see Indol-3-yl acetic acid
*iba*1 gene, 188, 189
Imadazolinones, 95, 111, see also Acetolactate synthase herbicides
Imazamethabenz, 255, see also Acetolactate synthase herbicides
Imazamethabenz-methyl, 113, see also Acetolactate synthase herbicides

Imazapyr, see also Acetolactate synthase herbicides
  cross insensitivity and cross resistance patterns, 98, 100, 103
  Mendelian inheritance of resistance, 278
  mutation frequency and resistance, 94
  proline mutations and resistance, 104
Imazaquin, 94, 104, 112, see also Acetolactate synthase herbicides
Imazethapyr, 112, see also Acetolactate synthase herbicides
Imidazolinones, see also Acetolactate synthase herbicides
  acetolactate synthase gene mutations for resistance, 101, 102
  acetolactate synthase inhibition, 83–86, 306
  cross insensitivity and cross resistance, 96, 98–101, 105
  cross resistance in *Alopecurus myosuroides*, 45
  inheritance of resistance in microbes and crops, 117–118
  resistance in *Lolium rigidum*, 43, 246
  target site cross resistance, 244, 245
Imomoea purpurea, 109
Indol-3-yl acetic acid (IAA), 171, 172, 187–189
Integrated weed management (IWM), 329–330, see also Weeds, management
Introgression, 12
Ioxophorus unisetus, 89
Ioxynil, 186
Ipomea batatas, 270
Irrigation, waterways, 64, see also Bipyridyl herbicides
Isogenic lines, 268–269, see also Triazine
Isoleucine, 307
Isopropalin, 256
Isoproturon, 44, 218, see also Photosystem II herbicides
Isoxaben, 277
Isozyme polymorphisms, 286, 287
IWM, see Integrated weed management

J

Johnsongrass, see *Sorghym halepense*

K

Klebsiella ozaenae, 46
Klebsiella pneumoniae, 231
Kochia scoparia
  acetolactate synthase herbicide resistance
  cross insensistivity and cross resistance patterns, 95, 98–101
  evaluation, 87
  evolution, 89–93
  catalytic competency, 121
  fitness, 306–309
  gene flow, 122–123
  gene mutations and resistance, 102–105
  inheritance, 118
  non-target site, 110
  seeds, 120, 121, 122
  target site, 96
  auxin analog herbicide resistance, 185
  distribution in small grain crops, 319, 320
  Mendelian inheritance of herbicide resistance, 280–281
  natural selection and herbicide resistance, 6
  seed bank life, 325

L

Lactuca sativa, 119, 120
Lactuca serriola
  acetolactate synthase herbicide resistance cross resistance, 95
  evolution and distribution, 89–92
  fitness, 306–309
  gene flow, 123
  gene mutations, 102
  growth, 120–121
  inheritance, 118–119
  seeds, 121, 122
  distribution in small grain crops, 319, 320
  Mendelian inheritance of herbicide resistance, 276, 280
Lemna spp., 230
Leucine, 149, 307
Liatrisglindracea, 11
Light-harvesting proteins, 302
Linum usitatissimum, 116, 117, 264, 270
Linuron, 44, see also Photosystem II herbicides
Lipids
  alteration in triazine-resistant biotypes, 36, 38
  biosynthesis inhibition and acetyl coenzyme A carboxylase herbicides, 145, 151
  seed content and dinitroaniline herbicides, 216, 223
Livestock grazing, 325, see also Weeds, management

Lolium multiflorum
  acetyl coenzyme A carboxylase herbicide resistance, 155, 158, 161, 305–306
  distribution in small grain crops, 319, 320
  Mendelian inheritance of herbicide resistance, 276, 281
  natural selection and herbicide resistance, 6
  pollen dispersal and herbicide resistance evolution rate, 12, 13
  seed bank life, 325

Lolium perenne
  acetolactate synthase herbicide resistance, 95, 96, 108–109, 121
  gene flow and herbicide resistance evolution rate, 13
  paraquat metabolism, 66, 68, 70, 71, 73–74
  quantitative inheritance of herbicide resistance, 270

Lolium ridigum
  acetolactate synthase herbicide resistance, 90, 91, 95–96, 110, 114–115
  acetyl coenzyme A carboxylase herbicide resistance, 150, 154, 155, 157–159, 161
  dinitroaniline herbicide resistance, 218, 221, 222
  distribution in small grain crops, 319, 320
  frequency of resistance in natural populations, 5
  Mendelian inheritance of herbicide resistance, 276, 281
  natural selection and herbicide resistance, 6
  photosystem II herbicide resistance, 28, 43–44
  resistance mechanisms
    multiple, 252–253, 309–310
    non-target site cross resistance, 247–251
    overview, 245–246
    target site cross resistance, 246–247
  resistance spreading, 16
  seed bank life, 324–325
  seed removal and management, 327
  transmembrane proton gradient perturbation, 148

Lotus corniculatus, 106, 116, 187, 191, 233
Lycopersicon spp., 233
Lycosersicon esculentum, 277, 283

M

Maize, see *Zea mays*
Malathion, 248, 252
Malonyl coenzyme A, 145, 147, 151
Malonylation, 198, see also Auxin analog herbicides
Malonyltransferase, 198
Maternal inheritance, 40, 47, 265, 267, 268, 301
Mating behavior, 11, see also Reproduction
Matricaria perforata, 186
MCPA, see also Auxin analog herbicides
  acetyl coenzyme A carboxylase herbicide antagonism, 144
  agricultural use, 178
  NIH shift, 182
  quantitative inheritance, 273
  resistance in weeds and crops, 184–186, 193–195, 201
Mecoprop, see also Auxin analog herbicides
  chemical structure, 171, 172
  NIH shift, 182
  resistance in weeds and crops, 184, 186, 193–200
Medicago sativa, 64, 94
Mehler reaction, 67
Membrane-auxin interactions, 174–177
Membrane potential (Em)
  membrane function role, 147–151, 159, 160
  restoration and resistance relation, 249–251
Membrane response recovery, 249–253
Mendelian inheritance (resistance genes), 274–285
Meristem growth, 144, 149–151, see also Acetyl coenzyme A carboxylase herbicides
Messenger ribonucleic acid (mRNAs), 177, 178
Metabolism, see also Individual entries
  acetolactate synthase herbicide and resistance, 109, 114–115
  acetyl coenzyme A carboxylase herbicides in crops, 153
  auxin analog herbicides, 179, 190, 192, 196–197, 203
  photosystem II herbicide resistance, 41–45
Methylamine, 70
N-Methylation, 249
N-Methyl isonicotinic acid, 70
Methyl purple, 62
Metibuzin, 273
Metoxuron, 277
Metribuzin, 43, 277, 283, see also Photosystem II herbicides
Metsulfuron, 276, 280, see also Acetolactate synthase herbicides

Metsulfuron-methyl, 98, 100, see also Acetolactate synthase herbicides
Microorganisms, 13–14
Mitosis, 215, see also Dinitroaniline herbicides
Mitotic index, 218, see also Dinitroaniline herbicides
Mixed function oxidase, see Cytochrome P450
Monodehydroascorbate reductase, 71, see also Paraquat
Monopyridone, 70, see also Paraquat
Monoquat, 70, see also Photosystem I herbicides
Morfamquat, 72, see also Photosystem I herbicides
mRNAs, see Messenger ribonucleic acids
Multiple resistance, see also Individual entries
   acetolactate synthase herbicides, 116
   Alopecurus myosuroides, 256
   dinitroaniline herbicides, 218, 222
   Lolium rigidum, 252–253, 309–310
   mechanisms, overview, 243–245
Mungbean, 150
Mutagenesis
   laboratory, herbicide resistance, 278, 283
   site-directed
      acetolactate synthase gene, 102, 104
      glyphosate, 231, 236
Mutations, see also Individual entries
   acetolactate synthase gene, 94, 101–105
   acetyl coenzyme A carboxylase herbicides, 156
   frequency, resistance inheritance, 286
   genetic and herbicide resistance, 4–5
   photosystem II herbicide resistance, 36
   resistance evaluation role, 88
Mycoherbicides, 204

N

1-Naphthylacetic acid (NAA), 172, 177, 187, 202, see also Auxin analog herbicides
Natural selection, 3, 6
Necrosis, 152, 186
Nicosulfuron, 112, see also Acetolactate synthase herbicides
Nicotiana plumbaginifolia, 188
Nicotiana tabacum
   acetolactate synthase herbicide resistance
      cross insensitivity, 96, 98
      gene mutations, 101–104
      gene overexpression, 105–106
      inheritance, 117
      modeling, 94
   auxin analog herbicides, 187, 192–193, 195, 201–202
   glyphosate resistance, 231, 232, 235
   inherited triazine resistance, 265
   Mendelian inheritance of herbicide resistance, 283
   paraquat resistance, 66, 67, 71–73
   photosystem II herbicide resistance, 36–37, 46
NIH shift, 182
Nitralase, gene, 46, 47
Nitrate reductase, 186
Nitriles, 28
Nitrophenols, 28, 31
Nonheme iron, 29
Non-target site cross resistance, see also Cross resistance; Target site cross resistance
   chlorsulfuron, 114, 115
   definition, 245
   Lolium rigidum, 247–251
   metabolism relation, 115–116
Norway spruce, 234
Nucleic acids, 174, 176, 178

O

OAC Triton, see *Brassica napus*
Oat, see *Avena sativa*
Oilseed rape, see *Brassica napus*
Oryza sativa, 116, 270, 277
Oryzalin, 215, 218, 223
β-Oxidation, 182
Oxyfluorfen, 277
Oxygen consumption, 254
Oxygen evolution, 67–68
Oxygen radicals, 31, see also Radicals
Ozone, 72

P

P700, 61
Paclobutrazol, 189
Paraquat, see also Photosystem I herbicides
   application and seed bank exhaustion, 325
   characteristics of resistant weeds, 304–305
   inheritance of resistance, 74–75
      Mendelian inheritance of resistance, 276, 278, 279
      quantitative inheritance, 273

metabolism, 70
mode of action, 64
photosystem I electron carrier role, 62
resistance in weeds, 65–66
sequestration and resistance relation,
  69–70
Parthenium spp., 66
pat gene, 116
Paternal inheritance, 266–267, see also
  Maternal inheritance
PBO, see Piperonyl butoxide
PC, see Plastocyanin
PCMBS, see *p*-Chloromercuribenzenesulfonic
  acid
Pea, 146, 150
Peanuts, 64
Peldon, see *Alopecurus myosuroides*
Pendimethalin, 215, 218, 223, 255
PEP, see Phosphoenolpyruvate
Peroxidase, 71
Peroxygenase, 34
Petunia, 231, 232, 235
pH, 174
Phenazine methosulfonate, 62
Phenmedipham, 278
Phenology, 2, 305–306
Phenylcarbamates, 28
Phenylurea monouron, 28
Phenylureas, 43, 44, 115, see also
  Photosystem II herbicides
p-Phenylenediamines, 62
Pheophytin, 29
Phleum pratense, 182
Phloem, 142, 143, 181
Phosphinothricin, 116
Phosphoenolpyruvate (PEP), 230, 232
Phospholipids, 176
N-(Phosphonomethyl)glycine, see Glyphosate
Photoinhibition, 30–32, 40, 48, 303, see also
  Photosystem II herbicides
Photosynthesis
  auxin analog herbicides, 174, 194
  photosystem I herbicide resistance, 67–69,
    304
  photosystem II herbicide resistance, 38,
    39, 302–303
Photosystem I (PS I) herbicides, 61–75,
  304–305
Photosystem II (PS II) herbicides
  conclusions and future directions, 46–49
  Mendelian inheritance, 275, 278–280
  mode of action, 28–32
  non-target site cross resistance, 248–249
  overview, 27–28

photosynthesis, growth, competition in
  resistant weeds, 302–303
quantitative inheritance, 271
resistance, 35–46, 301–302
seed germination in resistant weeds,
  303–304
selectivity mechanisms in crop plants,
  33–35
Phylloquinone, 61
Picloram, 172, 187, 201–202, 278, see also
  Auxin analog herbicides
Piperonyl butoxide (PBO), 43
Plasmalemma, 175, 176
Plasmodiophora brassicae, 13
Plastocyanin (PC), 29, 61
Plastome mutator, 40–41, 286
Plastoquinones, 29, 31
Ploidy, 9, 11, 94
Ploughing, 326, see also Seed bank
pmf, see Proton motive force
Poa annua
  inherited triazine resistance, 266, 267
  metabolism of benzoic acid herbicides,
    182
  paraquat resistance, 65, 66
  polymorphic resistance populations, 286
  triazine resistance, 17
Point mutations, see Mutations
Pollen
  acetolactate synthase herbicide resistance,
    89, 122–123
  acetyl coenzyme A carboxylase herbicides,
    306
  dispersal and herbicide resistance
    evolution rate, 11–13
  triazine resistance inheritance, 265–266
Polygenic traits, see Quantitative variation
Polygonum lapathifolium, 303
Population models, 300–301
Population polymorphism, 286–287
Potasssium, 174
Potato, see *Solanum*
poxB gene, 107
Primisulfuron, 94, 105, 111–112, 118, 278,
  see also Acetolactate synthase
  herbicides
Productivity, 302
Proline mutations
  acetolactate synthase herbicide resistance,
    102–105, 280, 283
  glyphosate resistance, 231
Pronamide, 220
Propanil, 273
Propham, 220

Proteolysis, 31
Proton motive force (pmf), 147, 148, see also Membrane potential
Proton symport, auxin-associated, 175
Protonophores, 148
PS I, see Photosystem I
PS II, see Photosystem II
*psb*A gene
   atrazine resistance, 46, 286
   D1 protein encoding, 29
   photosystem II herbicide resistance, 36–38
   triazine resistance relation, 265, 267, 268, 301
Pseudomonas sp., 233, 234
Purple photosynthetic bacteria, 31
Putrescine, 69
Pyrazon, 41
Pyridate, 41, 47
Pyridazinones, 28, 31, see also Photosystem II herbicides
Pyrimidinylthiobenzoates, 84–86, see also Acetolactate synthase herbicides
Pyrimidinyloxybenzoates, 86, 98–101, see also Acetolactate synthase herbicides
Pyruvate oxidase, 107

Q

Q_B binding niche, 30–32, 36, see also Photosystem II herbicides
QB protein, see D1 protein
Qualitative variation, 4
Quantitative inheritance, 10, 269–274
Quantitative variation, 4, 43
Quantum yield, 37, 40, 46, 302, 303, see also Photosystem II herbicides
Quinclorac, 172
Quinones, 107
Quizalofop, 157, see also Acetyl coenzyme A carboxylase herbicides
Quizalofop-ethyl, 149, 254, see also Acetyl coenzyme A carboxylase herbicides

R

Radical formation, 40, 70–73
Ranunculus acris, 185, 194, 201
Receptors, auxin-binding proteins, 171, 173, 183
Recurrent selection, 3, 4, 233, see also Selection
Reproduction, 244, 323
Reproductive isolation, 11
Residual activity, 93, 94, 202

Resistance factor, 88, 98
Respiration, 174
Rhizobium japonicum, 230
Rhodopseudomonas viridis, 28
Ribes spp., 181
Rice, see *Oryza sativa*
Ring methyl hydroxylation, see Aryl hydroxylation
Roots
   auxin analog herbicides, 179, 186, 189
   chlorsulfuron half-life, 114
   dinitroaniline herbicides, 216
Rothamsted, see *Alopecurus myosuroides*
Rubber trees, 112

S

Saccharomyces cerevisiae, 101, 102, 117
Safener, 143, 154
Salmonella typhimurium, 104, 117, 231, 232, 235
Salsola iberica
   acetolactate synthase herbicide resistance
   catalytic competency, 121
   evolution and distribution, 89–91, 93
   gene flow, 122–123
   inheritance, 118
   target site resistance, 96
   Mendelian inheritance of herbicide resistance, 281
Saranine T, 62
Sarcosine, 233
Satura innoxia, 121
Sclerotinia sclerotiorum, 204
Screening, herbicide resistance, 6
Season, 74
Second messengers, 176, see also Messenger ribonucleic acid
Seed, see also Weeds, management
   dispersal and herbicide resistance evolution rate, 11, 13
   germination
      acetolactate synthase herbicide-resistant weeds and crops, 120, 307
      acetyl coenzyme A carboxylase herbicide-resistant weeds, 305
      photosystem I herbicide-resistant weeds, 304
      photosystem II herbicide-resistant weeds, 303–304
   resistance management, 325
   limiting dispersal and weed management, 326–328
   longevity, 122, 308–309

production, 14, 15, 17–20, 37, 121, 308
removal during harvesting and weed
    management, 327
Seed bank, see also Weeds, management
    dynamics
    acetolactate synthase-resistant weeds,
        308–309
    acetyl coenzyme A carboxylase
        herbicide-resistant weeds, 305
    photosystem II herbicide resistance,
        303, 304
    population models, 300
    exhaustion and weed management,
        325–326
    life and weed management, 324–325
Selection (herbicide resistance)
    basic concepts, 3–8
    challenges and lessons, 15–20
    conclusions, 20
    fitness, 14–15
    population genetics, 7–8
    rate of evolution, 8–14
    overview, 1–3
Selection duration, 3
Selection intensity, 3, 7
Selection pressure
    acetolactate synthase herbicide resistance,
        89, 93–96
    auxin analog herbicide resistance in crops,
        187
    dinitroaniline herbicides, 222
    glyphosate resistance, 272
    reduction to eliminate herbicide resistance,
        299, 300
    resistance evolution, 3
    urea-substituted herbicides, 275
Selectivity, 152–154, 179, 216
Senecia vulgaris
    carbon dioxide fixation, 302
    fitness and herbicide resistance relation, 15
    inherited triazine resistance, 264, 266
    natural selection and herbicide resistance, 6
    photosystem II herbicide resistance and
        fitness, 302, 303
    psbA gene mutation and ecological fitness,
        38
    quantitative inheritance to photosystem II
        herbicides, 270, 271, 274
    simazine resistance, 27
Sequestration
    acetyl coenzyme A carboxylase herbicides,
        160

auxin analog herbicides, 180–181, 183,
    191, 202
herbicides and resistance relation, 250–251
mecoprop, 197, 198, 200
paraquat, 69–70, 279
Serine mutations, 105, 285, 286
Setaria faberii, 155
Setaria glauca, 153
Setaria italica, 265
Setaria viridis
    acetyl coenzyme A carboxylase herbicide
        resistance, 155, 157–159
    chlorsulfuron sensitivity, 114
    dinitroaniline herbicide resistance,
        218–220, 222, 224, 225
    distribution in small grain crops, 319, 320
    inherited triazine resistance, 266
    Mendelian inheritance of herbicide
        resistance, 276, 282
    natural selection and herbicide resistance, 6
    seed bank life, 325
Sethoxydim
    cross resistance, 246, 247, 255
    membrane potential effect, 150
    Mendelian inheritance of resistance, 278
    multiple resistance, 252
    Setaria viridis resistant biotypes, 158, 159
    weed control in crops, 142–143
    Zea mays resistant cultures, 146, 156, 158
Shade chloroplasts, see Chloroplasts
Shikimic acid pathway, 230, 235, see also
    Glyphosate
Shoots, 114
Side chain degradation, 181–182, see also
    Aauxin analog herbicides
Side chain elongation, 182, see also Auxin
    analog herbicides
Siduron, 275, 276
Signal reception, 175, see also Auxin analog
    herbicides
Signal transduction, 176
Simazine
    Hill reaction inhibition, 28
    quantitative inheritance, 270, 273
    resistance
        Abutilon theophrasti, 41–42
        Lolium rigidum, 43, 248, 249
        Senecio vulgaris, 27
        Zea mays, 33
Sinapis alba, 264
Sinapis arvensis, 109, 186, 201, 319, 320
Sisymbrium orientale, 319, 320

SOD, see Superoxide dismutase
Solanum nigrum
  photosystem II herbicide resistance, 302, 303
  plastome mutator role, 286
  *psb*A gene mutation, 38
  seed germination, 303
  triazine resistance inheritance, 266, 267
Solanum spp., 36–37, 65, 66, see also Paraquat
Solanum tuberosum, 277
Sonchus arvensis, 185
Sonchus oleraceus, 319, 320
Sorghum bicolor, 154
Sorghum halepense, 155, 158, 218, 220
Soybean, see *Glycine max*
Sphenoclea zeylanica, 185
Spinach, 30, see also D1 protein
Spindle microtubules, 217, 220
Stellaria media
  acetolactate synthase herbicide resistance, 95, 96, 121
  auxin analog herbicide resistance, 181, 184, 193–200
Stomata, closure, 174, 194
Streptomyces griseolus, 116
Stress, 5, 32
Sucrose, 149
Sugarbeets, 116
Sulfometuron-methyl, 94, 98–101, see also Acetolactate synthase herbicides
Sulfonates, 62
Sulfonylureas, see also Acetolactate synthase herbicides
  acetolactate synthase gene mutations, 101–105
  acetolactate synthase inhibition, 83–86
  Alopecurus myosuroides resistant-biotypes, 45
  crop selectivity, 111
  cross resistance, 95
    and cross insensitivity in weeds, 98–101
    non-target site, 255
    target site, 244–246
  inheritance of resistance, 117–119
  metabolic half life in weeds, 115
  multiple resistance, 253
  resistance characteristics of weeds, 306
Sulfoxidation, 33
Sunflower, 150
Superoxide, generation, 63
Superoxide dismutase (SOD), 63, 70–73, 279

Synergists
  acetyl coenzyme A carboxylase herbicides, 154
  photosystem II herbicides, 34
  control of resistance weeds, 48–49, 203–204

T

2,4,5-T, 177, 184, 187
Target site
  acetolactate synthase herbicide resistance in weeds, 88–89, 96–101
  acetyl coenzyme A carboxylase herbicide resistance, 158–159
  photosystem II herbicides, 31–32, 35–41
Target site cross resistance, see also Cross resistance; *Lolium rigidum;* Non-target site cross resistance
  definition, 244–245
  dinitroaniline herbicides, 220–221
  mechanisms in Lolium rigidum, 246–247
  dicamba, 201
Taxaxacum offcinale, 185
Taxol, 224
Temperature, 38, 74, 120, 302–303
Terbutol, 220, 221, see also Dinitroaniline herbicides
*tfd*A gene, 192–194
Thiocarbamate herbicides, 271–272
Threonine mutations, 36–37
Thylakoid membrane, 29, 31
Tissue swellings, 174
Tobacco, see *Nicotiana tabacum*
α-Tocopherol acetate, 216
Tomato, 231, 232, 235, see also Glyphosate
Tralkoxydim, 143, 255, see also Acetyl coenzyme A carboxylase herbicides
Transcarboxylation, 147
Translocation
  acetolactate synthase herbicide resistance, 108–111
  acetyl coenzyme A carboxylase herbicides, 143, 157
  auxin, 175, 179, 181, 186, 193–195, 201
  bipyridyl herbicides, 67–69
  dinitroaniline herbicides, 223
Transmembrane proton gradient, 147–150, 159, see also Acetyl coenzyme A carboxylase herbicides
Transpiration, 194

Triacolopyrimidines, 96, 97, see also Acetolactate synthase herbicides; Cross insensitivity
Triallate, 156, see also Acetyl coenzyme A carboxylase herbicides
Triasulfuron, 98–101, 104, 108–109, see also Acetolactate synthase herbicides
Triazine, see also Photosystem II herbicides
  activity of, 32
  Chenopodium album resistant biotypes, 5
  coresistance with paraquat in weeds, 65
  fitness relation, 14, 300
  inheritance of resistance, 264–269, 282
  Lolium rigidum resistant biotypes, 43, 248
  photosystem II inhibition, 31, 301
  plastome mutator role, 40–41
  pleotropic effects, 36
  prevalence of resistance, 27
  resistance transfer to crops, 45–46
  target site cross resistance, 244
  temperature sensitivity, 38
Triazinones, 28, 244
Triazolopyrimidines
  acetolactate synthase inhibition, 84–86, 95
  crop selectivity, 111
  resistance characteristics of weeds, 306
  target site cross resistance, 244, 245
Triclopyr, 181, see also Auxin analog herbicides
Tridiphane, 49
Trifensulfuron-methyl, 106, see also Acetolactate synthase herbicides
Trifluralin, see also Dinitraniline herbicides
  lipid content in seeds, 216
  Mendelian inheritance of resistance, 276, 277, 282
  mitosis disruption, 215
  mode of action in crops and grasses, 218–219
  non-target cross resistance, 255
  translocation, 223
Trifolium repens, 185, 187
Triquat, 62
Trisulfuron, 111, 112, see also Acetolactate synthase herbicides
Triticum aestivum, 264, 270, 277
Triticum dicoccoides, 270, 271, 274
Tubulin, 217, 224, 282, 309, see also Dinitroaniline herbicides

U

Umbelliferae, 216
uni linkage group, 225

Uracil, 28
Uracil herbicides, 301, see also Photosystem II herbicides
Ureas, 31, 32, 301, 254–255, see also Photosystem II herbicides

V

Vacuoles, 70, 183, 197, 198, see also Sequestration
Valine, 307
Velvetleaf, 115
Viologens, 62
VLR 69, see *Lolium rigidum*
Vulpia bromoides, 65, 66, 319, 320

W

Weather, 179
Weeds
  characteristics of herbicide resistance, 114–115, 301–310
  definition, 1
  management
    cropping system influence, 328–329
    cropping systems, 317–318
    distribution in small grain crops, 318–320
    exploitation of seed bank exhaustion, 325–326
    fitness and control, 318
    herbicide use on resistant populations, 320–324
    human factors, 330–331
    identification of resistance in field populations, 29
    integrated weed management, 329–330
    limiting dispersal of seed, 326–328
    seed bank life, 324–325
  Mendelian inheritance of resistance genes, 275–276, 278–282
  quantitative inheritance of resistance, 271–272
  resistance mechanisms, 35–36
Weibull function, 12
Wheat
  acetolactate synthase herbicides, 108–109, 113, 116, 248
  acetyl coenzyme A carboxylase herbicides, 148, 153
  benzoic acid herbicide metabolism, 182
  glyphosate metabolism, 235
  inheritance of resistance, 271
  photosytem II herbicides, 33, 34

Wild mustard, see *Sinapis arvensis*
Wild oat, see *Avena fatus*
Wind, 12, see also Pollen
Winterbourn's reactions, 64
WLR 2, see *Lolium rigidum*

X

Xanthium strumarium, 95, 96, 105
Xylem, 67, see also Bipyridyl herbicides
X-ray crystallography, 28

Y

Yeast, see *Saccharomyces cerevisiae*

Z

Zea mays
   acetolactate synthase herbicides, 101, 109, 111–113, 116, 118
   acetyl coenzyme A carboxylase herbicides, 146, 156, 158, 161
   auxin-binding protein receptors, 175, 183
   glyphosate metabolism, 232, 234, 235
   inheritance of resistance, 264, 270, 277, 282, 283
   photosystem II herbicides, 33, 34
   sucrose uptake, 149
Zero tillage crop systems, 318